LINEAR
ACTIVE
CIRCUITS –
DESIGN
AND
ANALYSIS

LINEAR ACTIVE CIRCUITS— DESIGN AND ANALYSIS

William Rynone, Jr.

To Mom and Dad
I suspect that I had a choice . . .
And I picked the best!

Acknowledgements

My thanks to Dennis Ricci . . . an accomplished man in two professions.

His capable editorial abilities were demonstrated in the development of this text and his mastery of psychology became evident while he worked with a worrisome writer.

I would also like to thank Mrs. Alice Broda, Mrs. Bonnie Brophy, and Mrs. Catherine Sammon . . . truly professionals, as evidenced by their typing skills and their tenacity in successfully completing major cryptographic tasks.

Contents

Preface

Many electrical engineering or electronics engineering technology curricula require that a student complete a one- or two-semester sequence in a subject area that may be loosely termed "Active Devices." Depending upon the particular program, this course (or courses) usually contains a study of diodes and their uses in clipper and clamper circuits and in power supply circuits. The learning sequence is often followed by a discussion of transistors, and their use in amplifier, power supply regulator, and logic circuits. A discussion of dc biasing of bipolar transistors in addition to h parameter analysis of small signal amplifiers is also usually included. The customary discussion of field-effect transistors (FETs) may also be accompanied by a review of oscillators as well as operational amplifiers. An introduction to digital (binary) logic with Boolean algebra is sometimes included.

It is the author's belief that all of this material is important, but the material dealing with amplifiers may be sufficiently important to certain programs for it to be offered in a "stand-alone" course. There is certainly enough material available on an undergraduate level to devote an entire semester exclusively to the study of amplifiers.

This text deviates somewhat from the philosophy of covering only amplifiers in that it includes the study of oscillators as an extension of amplifier theory. There are also short chapters on operational amplifiers and tuned amplifiers.

The 15 chapters most likely comprise too much material to be covered in one semester, but the organization is such that the first few chapters may be omitted if it can be assumed that the elementary material has previously been covered, or if they are not relevant to the objectives of the course. For example, some topics may be covered in a second course dealing with specialized devices such as four-layer devices, display devices, control circuits, and specialized communications devices.

Students who are capable of assimilating material at a rapid pace may be able to conquer all of the material presented in this text within one semester.

The approach in writing this text has been to develop both the design and analysis formulas for the circuits. Following the development of each formula is an example that requires its use with values that are reasonably consistent with "real-life" parameters. One of the objectives of the problems for students is to present the techniques of design and analysis with equal emphasis because students will, upon graduation, be confronted with the need for both kinds of skills. Consequently, both exact and approximation formulas are developed, and corresponding examples are shown in order to emphasize the circumstances in which approximations can be used (and, of course, the percentage of error involved).

An understanding of the flow of majority carriers in n and p doped semiconductor materials such as silicon, germanium, or gallium arsenide, serves as an appropriate introduction to the "mechanics" of semiconductor diode and transistor operation. Building upon these concepts provides us with the necessary understanding of a diode's capability to produce unidirectional current flow, and this, in turn, is essential to the comprehension of bipolar transistor operation.

Chapter 1

Field-Effect and Bipolar Transistors

1.1 PHYSICS OF TRANSISTORS

A knowledge of basic semiconductor physics, in sufficient detail, should enable us to understand how a *p-n* junction allows for unidirectional current flow.

Let us picture an extension of diode construction where we "bolt together" three blocks of silicon as shown in Fig. 1.1.

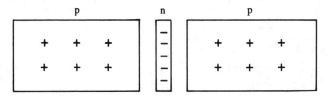

Fig. 1.1

If we construct our "slab" of silicon such that the middle *n* piece is very thin, then we have essentially a "swiss cheese sandwich." If we connect a bias battery to the left *p-n* pair, current will flow (with the bias battery polarity as shown in Fig. 1.2).

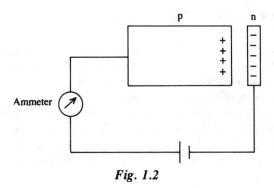

Fig. 1.2

If we connect a bias battery as shown in Fig. 1.3 to the right-hand *p-n* pair, no current will flow because we now have a back-biased junction.

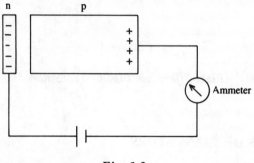

Fig. 1.3

If both circuits are together as shown in Fig. 1.4, a different set of electrical conditions occurs (as compared with individual circuit operation).

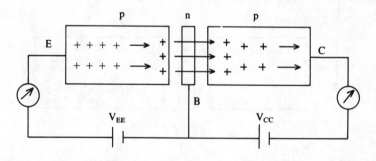

Fig. 1.4

1.2 BIPOLAR TRANSISTORS

We note in Fig. 1.4 that the left-hand *p-n* junction is still conducting electricity, but the right-hand *p* material also has majority carriers flowing. The majority carriers arriving at the left-hand *p-n* junction must flow out of terminal B, or proceed across the right-hand *p-n* junction. Because the *n* material is a very thin slice of high resistivity material, the majority carriers cross the right-hand junction and enter the *p* material. Once in the right-hand *p* material, they are attracted to the negative terminal of the V_{CC} power supply. Almost all of the majority carriers flowing from terminal E to the *n* material will continue the journey to terminal C. A small percentage of the majority carriers will be diverted out of the B lead. For a modern transistor it is not unrealistic to expect 99 percent or more of the majority carriers (I_E) that start their journey at the E (or emitter) terminal to arrive at the C (or collector) terminal (I_C). By Kirchoff's current law, we know that the remaining majority carriers (approximately one percent of those that started their journey at the emitter) will exit the B (or base) terminal (I_B). Thus we can write

$$I_E = I_C + I_B \qquad\qquad (1.1)$$

The voltage drop that exists between the emitter-to-base junction is that which would be anticipated for a forward-biased silicon diode; approximately 0.6 to 0.7 V. For germanium, 0.3 V is a commonly used figure. Let us compute the value of the dc resistance exhibited by the base-to-emitter junction. If we assume an emitter current of 6 mA, then the diode resistance is

$$R_{EB} = \frac{V_{EB}}{I_E} = \frac{0.6 \text{ V}}{6 \text{ mA}} = 0.1 \text{ k} = 100 \ \Omega$$

The collector-to-base junction, in contrast to the emitter-to-base junction, is a reversed-biased diode. Let us assume that the voltage across the junction is 12 V and compute the equivalent dc resistance. To make this computation, we will also assume that the collector current I_C is approximately equal to the emitter current, I_E.

$$R_{CB} = \frac{V_{CB}}{I_E} = \frac{12 \text{ V}}{6 \text{ mA}} = 2 \text{ k} = 2000 \ \Omega$$

We note here the difference in resistances—a low value input resistance has been *translated* into a high value output *resistance*. Thus we have a *trans-resistor*, or simply a "transistor." If we vary the input current, the input voltage will vary by a very small amount. Referring to the $V_{forward}$ *versus* $I_{forward}$ curve plotted in Fig. 1.5, we see that the input voltage *change* may be less than one-tenth of a volt for a doubling of the input current.
However, the *change* in the output voltage from the base terminal to collector terminal is considerable. If we assume that the output resistance is a constant 2000 ohms (the value we computed earlier), then a change in input current ΔI_E will result in a change in output voltage ΔV_{CB} as follows:

$$\text{Assume: } \Delta I_C \doteq \Delta I_E \qquad = 2.2 \text{ mA}$$
$$\Delta V_{CB} = \Delta I_C \, R_{CB} = 2.2 \text{ mA (2 k)}$$
$$= 4.4 \text{ V}$$

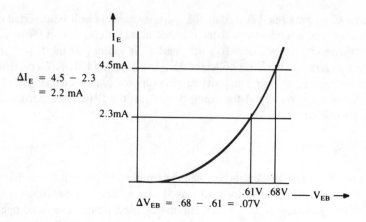

$\Delta I_E = 4.5 - 2.3$
$= 2.2 \text{ mA}$

$\Delta V_{EB} = .68 - .61 = .07V$

Fig. 1.5

The amount of increase in signal that has been gained, which is referred to as the voltage gain (or voltage amplification), is computed as a ratio

$$\frac{\Delta V_{out}}{\Delta V_{in}} = \frac{4.4 \text{ V}}{.07 \text{ V}} = 62.9 \qquad \text{(dimensionless)} \qquad (1.2)$$

The ratio is also designated by a "short-hand" notation, A_V, where A represents the word "amplification" and the subscript V represents "voltage."

This particular transistor connection, where the electrical signal to be amplified is connected between the emitter and base terminals, and the amplified signal is observed or measured between the base-to-collector terminals, is given a special name: the *common base connection* or the *common base configuration*.

Some observations about the common base connection are:

- The ac input resistance is a low value (typically 5 Ω to 50 Ω), while the ac output resistance is a high value (typically 100 kΩ to 2 MΩ)
- The value of the voltage amplification is generally a value greater than unity (possibly from 30 to 300 for a practical circuit)

In the above example, the voltage gain was computed:

$$A_V = 62.9$$

- The value of the current amplification is a number slightly less than, but almost equal to, unity:

$$A_I \doteq 1$$

A commonly defined ratio used when dealing with the common base configuration is

$$\alpha_{dc} \doteq \frac{I_C}{I_E}$$

or

$$\alpha_{ac} \doteq \frac{\Delta I_C}{\Delta I_E} \Bigg|_{V_{CB}\ =\ \text{constant}} \tag{1.3}$$

The values of α, both ac and dc, tend to be a number slightly less than unity.

If we were to construct the circuit shown in Fig. 1.6 and plot the three variables
indicated by the meter movements, we would expect the collector current I_C to
be approximately equal to the emitter current I_E, even though the V_{CC} power
supply voltage was varied over a wide range. If we were to use a high quality
ammeter to measure I_C, the voltage drop across the meter would be essentially
negligible, and the collector-to-base voltage V_{CB} would be almost identical to
the collector power supply voltage V_{CC}.

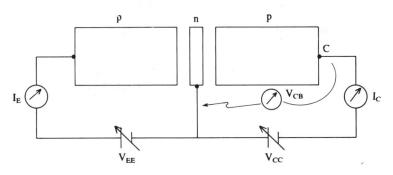

Fig. 1.6

If we plot I_C *versus* V_{CB}, while holding I_E constant, the curve of Fig. 1.7
would be obtained.

If we plot I_C *versus* V_{CB} for *various* values of I_E, a *set* of *output curves* (sometimes
referred to as an ''output family'') would be generated, as shown in Fig. 1.8.

Similarly, if we ''hold constant'' a different variable, the collector-to-base
voltage V_{CB}, but allow the emitter-to-base voltage V_{EB} to vary, while monitoring
the emitter current I_E (see Fig. 1.9), the set of curves in Fig. 1.10 would be
obtained.

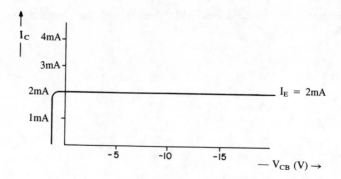

Fig. 1.7

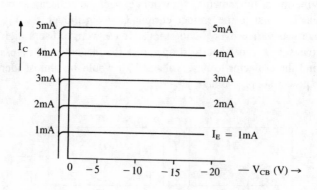

Fig. 1.8

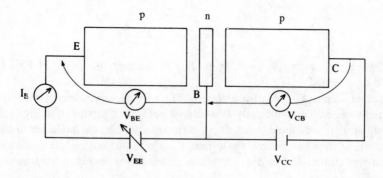

Fig. 1.9

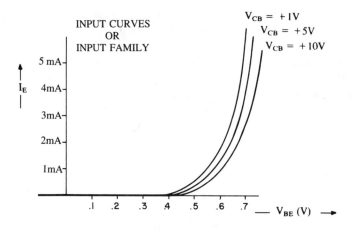

Fig. 1.10

After becoming familiar with the use of transistors in amplifier circuits, we could use the input and output curves for design or analysis purposes. The curves can also be of value in enabling a designer to estimate how well a transistor will perform in an amplifier circuit.

If we were to "stand a transistor up on end" and connect the bias batteries slightly differently from the common base connection, as in Fig. 1.11, we now have a circuit where the emitter is a common terminal to both bias batteries.

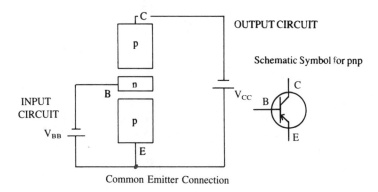

Fig. 1.11

As might be anticipated, this connection is referred to as the *common emitter connection.*

One major difference that can be noted about the common emitter connection

8

(see Fig. 1.11) is that the V_{CC} power supply is applied across two transistor *p-n* junctions. Because the base-to-emitter junction is a forward-biased junction with a low voltage drop compared with the collector-to-base junction voltage drop, the majority of the V_{CC} power supply voltage appears across the collector-to-base junction.

Referring to Fig. 1.12, if the base current I_B is maintained at a constant value, and the emitter-to-collector voltage is allowed to vary, there would be a corresponding change in the collector current I_C, as shown in Fig. 1.13.

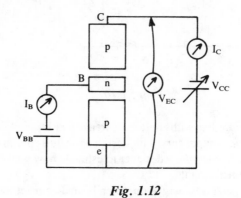

Fig. 1.12

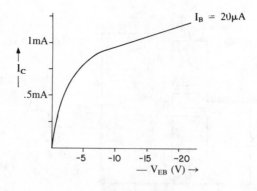

Fig. 1.13

If the base current I_B is allowed to "take on" different values, a set of output curves would be generated (see Fig. 1.14).

If we were to connect the circuit shown in Fig. 1.15, by varying the emitter-to-base voltage V_{EB}, different values of base current I_B would be observed for fixed values of emitter-to-collector voltage V_{EC}.

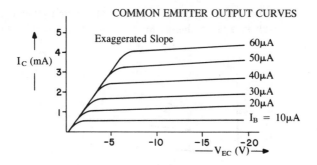

Fig. 1.14

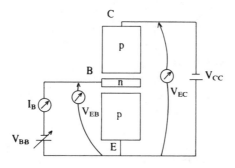

Fig 1.15

A complete set of input curves would be generated if V_{EC} were allowed to take on different values (Fig. 1.16).

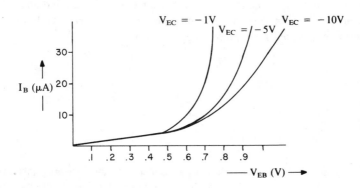

Fig. 1.16

If we examine the output curves, we note that a small change in the base current I_B will result in a large change in the collector current I_C (Fig. 1.17).

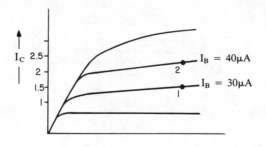

Fig. 1.17

For the curves shown in Fig. 1.17

$$\Delta I_C = 2 - 1.5 = 0.5 \text{ mA}$$

and

$$\Delta I_B = 40 \ \mu\text{A} - 30 \ \mu\text{A} = 10 \ \mu\text{A}$$

and if the collector current is the output current and the base current is the input current for an amplifier, then we can compute the current amplification

$$A_I = \frac{\Delta I_{out}}{\Delta I_{in}} = \frac{\Delta I_C}{\Delta I_B} = \frac{0.5 \text{ mA}}{10 \ \mu\text{A}} = \frac{500 \ \mu\text{A}}{10 \ \mu\text{A}} = 50 \qquad \text{(dimensionless)}$$

A commonly used parameter for transistor circuit design and analysis is the ratio of the collector current to the base current. The Greek symbol beta β has come to be accepted for representing this ratio.

$$\beta_{dc} \doteq \frac{I_C}{I_B} \tag{1.4}$$

$$\beta_{ac} \doteq \frac{\Delta I_C}{\Delta I_B}\bigg|_{V_{EC}} = \text{constant}$$

The collector-to-emitter terminal pair consists of the forward-biased base-to-emitter junction (low resistance) in series with the reversed-biased base-to-collector junction (high resistance). Because the sum of a low resistance value in series with a high resistance value is a high resistance according to Ohm's law, the voltage generated across the collector-to-emitter terminals is a large value:

$$\Delta V_{CE} = \Delta I_C R_{CE}$$

We can now estimate the overall voltage gain as

$$A_V = \frac{\Delta V_{out}}{\Delta V_{in}} = \frac{\Delta V_{CE}}{\Delta V_{BE}} \frac{\text{(large value)}}{\text{(small value)}} = \text{large value}$$

Voltage gain values from 20 to 300 are not unrealistic for the common emitter connection.

Summarizing, for a common emitter connection:

- The ac input resistance is moderate in value (typically 200 Ω to 5000 Ω), while the ac output resistance may be a value between 20k Ω to 200k Ω
- The voltage amplification may be a value from 30 to 300
- The current amplification may be a value from 30 to 300

There is one other method of connecting a transistor, referred to as the common collector configuration (or, as it is more often called, the emitter-follower configuration). For the emitter-follower connection, we again "stand the transistor on end," but connect the power supplies as shown in Fig. 1.18.

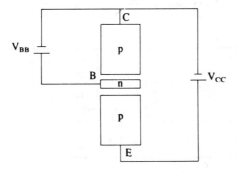

Fig. 1.18 Common Collector Configuration

When the common collector configuration is used as an amplifier, the ac input signal is applied between the base and collector, and the output signal is detected between the emitter and collector.

The characteristic curves of the common collector output are obtained by plotting the data for three variables I_B (input current), I_E (output current), and V_{EC}, with I_E replacing I_C as the dependent variable. Because I_C and I_E are almost equal values, the characteristic curves of the common collector output are very similar in appearance to the characteristic curves of the common emitter output (see Fig. 1.19).

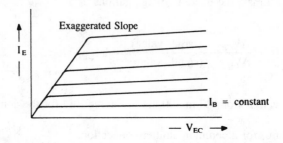

Fig. 1.19

Summarizing for the common collector:

- The ac input resistance is a fairly high value (200 Ω to 5000 Ω)
- The voltage gain is a value less than unity
- The current gain is moderate (30 to 90)

If we further examine the flow of charge in a transistor, the minority carrier flow must also be considered. Redrawing the transistor connected in the common base configuration, we have the schematic of Fig. 1.20.

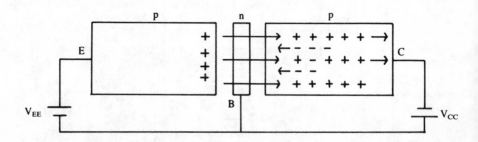

Fig. 1.20

Up to this point, only *pnp* transistors have been discussed. Early in the development of transistors, germanium was the most common semiconductor material used in the manufacturing of both diodes and transistors. Germanium is more suitable for the manufacturing of *pnp*-type transistors, but the high leakage currents and lower operating temperature range of these units soon paved the way for silicon to become the most commonly used material. Silicon transistors are superior to germanium units in terms of leakage current levels, operating voltage levels, and operating temperature extremes. Silicon is more suitable for manufacturing *npn* transistors. In the drawing shown in Fig. 1.21, the appropriate power supply polarities and symbols for both *pnp* and *npn* transistors are shown for common emitter connections.

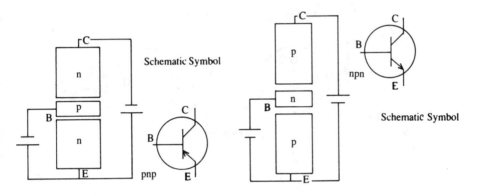

Fig. 1.21

We note that a *pnp* transistor requires a negative collector voltage, and *npn* requires a positive collector voltage. The requirement for a positive collector voltage for modern silicon transistors is fortunate for certain mobile receiver applications, such as in automobiles or aircraft, where it has become conventional to ground the negative battery terminal and designate the positive terminal as the "hot" terminal.

To remember whether the transistor symbol appearing on a schematic designates an *npn* or *pnp* transistor, various analogies have become popular. Students quite often use them as "remembering tools." Referring to the transistor symbol, we use the arrow as shown in referring to the "sayings": *npn—not* pouring i*n*; *pnp—pour* i*n* *p*ot.

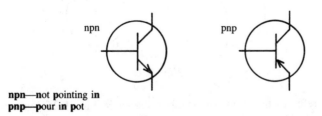

Fig. 1.22

14

We are constantly faced with the task of identifying the emitter, base, and collector leads of a transistor. Generally, for small transistors, the identification of the emitter, base, and collector leads may be accomplished by viewing the bottom of the transistor as shown in Fig. 1.23. If we line up two of the leads on a vertical center line, the base lead will be on either the left or right. Only the collector and emitter leads lie on the center line. If we are careful to orient the base on the left, then the lead orientation of the standard transistor symbol (with the emitter on the bottom) corresponds to the bottom transistor view.

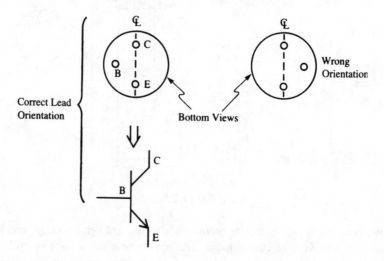

Fig. 1.23

On some small transistors the leads are all in line with different lead arrangements (Fig. 1.24). A simple ohmmeter check will help us to identify the unit as a *pnp* or *npn* transistor, and gain a somewhat subjective idea as to whether the unit is operable or defective.

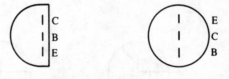

Fig. 1.24

If we refer to the transistor symbol shown in Fig. 1.25, we see how the transistor may be thought of as two diodes connected back-to-back. By use of an ohmmeter connected to the base lead and set on the R × 1 scale, a low resistance reading will be obtained between B and E, and B and C for the red lead connected to the base lead. (American-made ohmmeters have " + " terminal of the internal ohmmeter battery connected to " + " lead of the ohmmeter case. Japanese meters are usually opposite, but check with a sample diode.) For a *pnp* transistor, the results are opposite, as indicated in Fig. 1.26.

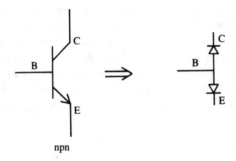

npn

Fig. 1.25

pnp

Fig. 1.26

As the result of thermal agitation, a number of electrons become available in the right-hand *p* silicon material. These charges are referred to as minority carriers, since the holes injected from the emitter that eventually transition into the right-hand *p* silicon are predominate in quantity (hence, the term "majority carriers"). The current flow out of the collector thus consists of two distinct quantities: *majority carriers* and *minority carriers*. The minority carrier flow is called the "leakage current." Referring to Fig. 1.27, the leakage flow path is around the "right-hand loop." However, each electron that is generated, which flows around the right-hand loop in a counter-clockwise direction, may be thought of as positive charge flowing in a clockwise direction around the loop. Thus, we can see where the equation for collector current results in two additive quantities

$$I_C = \alpha I_E \qquad + \qquad I_{CBO}$$

Majority carriers   Minority carriers (leakage current)   (1.5)

The subscripts for the leakage current term, I_{CBO}, are representative of CB = *collector to base*, O = with emitter *open*.

The collector-to-base junction is essentially a reversed-biased diode, and because the leakage current in a reversed-biased diode increases with temperature and applied voltage for a given transistor type, manufacturers will specify the conditions under which the leakage current measurement was taken. If a transistor is connected in a common emitter connection but the base is disconnected, a small amount of collector current will still flow. This small amount of current is quite often negligible compared with circuit currents, particularly if the transistors are manufactured from silicon. For transistors manufactured from germanium, however, the "cut-off" current may not be negligible.

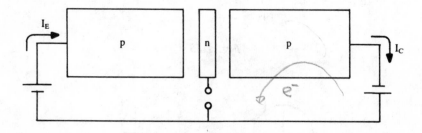

Fig. 1.27

To compute the value of current when $I_B = 0$, we use equations (1.5) and (1.1)

$$I_C = \alpha I_E + I_{CBO}$$
$$I_E = I_C + I_B$$
$$I_C = \alpha(I_C + I_B) + I_{CBO}$$
$$I_C = \alpha I_C + \alpha I_B + I_{CBO}$$
$$I_C (1-\alpha) = \alpha I_B + I_{CBO}$$
$$I_C = \frac{\alpha}{1-\alpha} I_B + \frac{I_{CBO}}{1-\alpha} \qquad (1.6)$$

The term $\alpha/(1-\alpha)$ is assigned the letter β. In equation (1.4), we stated that

$$\beta_{dc} \doteq \frac{I_C}{I_B}$$

This equality (1.4) neglected the last term in (1.6). Therefore, we can write

$$I_C = \beta I_B + \frac{I_{CBO}}{1-\alpha} \qquad (1.7)$$

If I_B is set equal to zero and we assume a value of .99 for α_{dc}, then

$$I_C \mid_{I_B\,=\,0} = \frac{I_{CBO}}{1-.99} = 100\, I_{CBO}$$

We can see that the leakage current which would normally flow out of (or into) the base terminal appears as much larger leakage current in the collector-to-emitter circuit. The value of I_C when the base terminal is open is given a special designation:

$$\left.\frac{I_{CBO}}{1-\alpha}\right|_{I_B\,=\,0} = I_{CEO} \qquad (1.8)$$

As might be anticipated, I stands for dc current, and the subscripts CE may be interpreted to mean "from collector to emitter" and O implies that the base terminal is open-circuited.

Referring to the output curves, we note that each curve represents a particular constant base current value. The lowest base current curve is plotted for $I_B = 0$. We note that at $I_B = 0$, some collector current is flowing and the value of collector current increases slightly with increasing emitter-to-collector voltage. The small amount of collector current flowing when $I_B = 0$ is the I_{CEO} value determined in equation (1.8).

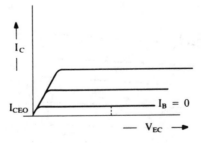

Fig. 1.28 Common Emitter Output Curves

18

The leakage current I_{CBO} computed earlier appears in the common-base curves. Because it is 100 times less than I_{CEO} (approximately), it is not noticeable and the $I_E = 0$ curve appears to lie on top of the $I_C = 0$ line.

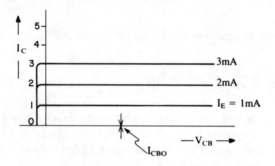

Fig. 1.29 Common Base Output Curves

Some additional transistor characteristics of importance to designers are the maximum ratings. A manufacturer will normally specify these values on a separate data sheet for each device that it manufactures. Quite often the specifications are registered with the Electronic Industries Association (EIA). The EIA records the data for a particular semiconductor device and assigns an EIA part number. Thereafter, any manufacturer may produce a similar unit and assign the same part number. Of course, any manufacturer who wishes to "copy" a device, must meet all of the registered EIA specifications for that device. Some maximum ratings of importance to designers are the maximum allowable power dissipation, maximum current, and maximum voltage. Maximum voltage may be specified under various conditions, for example, the voltage from collector-to-base with emitter open $V_{CBO\ max}$, or from collector-to-emitter with the base open $V_{CEO\ max}$. These maximum ratings may be depicted on a set of output curves as shown in Fig. 1.30.

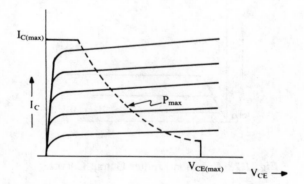

Fig. 1.30

Generally, the EIA will assign a 1N designation to diodes—the 1 indicates a single *p-n* junction. For example, a 1N914 is a small, silicon pulse diode, whose specifications are shown in Fig. 1.31. We note that $V_{R\ max}$ is 50V, $I_{dc\ (avg)} = 100$mA, $P_{D\ max} = 200$ mW. For transistors, the EIA assigns a 2N number. The 2 represents two *pn* junctions. A very commonly used small-signal silicon transistor is the 2N2222, whose specifications are shown in Figs. 1.32 to 1.35.

To determine whether we are exceeding the maximum power dissipation rating for a transistor, we must calculate all power dissipated in the device. If a transistor is operated in the common emitter configuration (the common emitter connection is the most frequently used configuration), the power dissipated due to the collector current is

$$P_{DC} = I_C V_{EC} \tag{1.9}$$

and the power dissipated due to base current is

$$P_{DB} = I_B V_{EB} \tag{1.10}$$

The total power is

$$P_{D\ total} = P_{DC} + P_{DB} \tag{1.11}$$

The collector-base junction power dissipation is considerably greater than the base-emitter junction power dissipation and, therefore, P_{DC} is usually used for $P_{D\ total}$.

To illustrate this, let us examine some typical values.

Example (1.1). Compute the total power dissipated in a 2N2222 transistor if the transistor is operated under the following condition:

$$I_C = 5\text{mA},\ I_B = .05\text{ mA},\ V_{EC} = 10\text{V},\ V_{EB} = 0.7\text{V}$$
$$P_{DC} = I_C \cdot V_{EC} = 5\text{mA}(10\text{V}) = 50\text{mW}$$
$$P_{DB} = I_B \cdot V_{EB} = .05\text{mA}(0.7\text{V}) = .035\text{mW}$$

The ratio is

$$\frac{P_{DC}}{P_{DB}} = \frac{50}{.035} = 1428$$

20

TYPES 1N914, 1N914A, 1N914B, 1N915, 1N916, 1N916A, 1N916B, 1N917
SILICON SWITCHING DIODES

BULLETIN NO. DL-S 7311954, MARCH 1973

FAST SWITCHING DIODES

- **Rugged Double-Plug Construction**

Electrical Equivalents

1N914 . . . 1N4148 . . . 1N4531
1N914A . . . 1N4446
1N914B . . . 1N4448
1N916 . . . 1N4149
1N916A . . . 1N4447
1N916B . . . 1N4449

mechanical data

Double-plug construction affords integral positive contacts by means of a thermal compression bond. Moisture-free stability is ensured through hermetic sealing. The coefficients of thermal expansion of the glass case and the dumet plugs are closely matched to allow extreme temperature excursions. Hot-solder-dipped leads are standard.

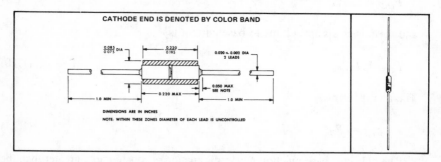

absolute maximum ratings at specified free-air temperature

	1N914 1N914A 1N914B	1N915	1N916 1N916A 1N916B	1N917	UNIT
Working Peak Reverse Voltage from −65°C to 150°C	75*	50*	75*	30*	V
Average Rectified Forward Current (See Note 1) at (or below) 25°C	75*	75*	75*	50*	mA
at 150°C	10*	10*	10*	10*	mA
Peak Surge Current, 1 Second at 25°C (See Note 2)	500*	500	500*	300	mA
Continuous Power Dissipation at (or below) 25°C (See Note 3)	250*	250	250*	250	mW
Operating Free-Air Temperature Range	−65 to 175				°C
Storage Temperature Range	−65 to 200*				°C
Lead Temperature 1/16 Inch from Case for 10 Seconds	300				°C

NOTES: 1. These values may be applied continuously under a single-phase 60-Hz half-sine-wave operation with resistive load.
2. These values apply for a one-second square-wave pulse with the devices at nonoperating thermal equilibrium immediately prior to the surge.
3. Derate linearly to 175°C free-air temperature at the rate of 1.67 mW/°C.

*JEDEC registered data

Fig. 1.31
Courtesy of Texas Instruments

YPES 1N914, 1N914A, 1N914B, 1N915, 1N916, 1N916A, 1N916B, 1N917
SILICON SWITCHING DIODES

1N914 SERIES AND 1N915

electrical characteristics at 25°C free-air temperature (unless otherwise noted)

PARAMETER		TEST CONDITIONS		1N914		1N914A		1N914B		1N915		UNIT
				MIN	MAX	MIN	MAX	MIN	MAX	MIN	MAX	
$V_{(BR)}$	Reverse Breakdown Voltage	$I_R = 100 \mu A$		100		100		100		65		V
I_R	Static Reverse Current	$V_R = 10 V$									25		nA
		$V_R = 20 V$			25		25		25			
		$V_R = 20 V$, $T_A = 100°C$							3		5	
		$V_R = 20 V$, $T_A = 150°C$			50		50		50			μA
		$V_R = 50 V$									5	
		$V_R = 75 V$			5		5		5			
V_F	Static Forward Voltage	$I_F = 5 mA$						0.62	0.72	0.6	0.73	V
		$I_F = 10 mA$			1							
		$I_F = 20 mA$	See Note 4				1					
		$I_F = 50 mA$									1	
		$I_F = 100 mA$							1			
C_T	Total Capacitance	$V_R = 0$, $f = 1 MHz$			4		4		4		4	pF

1N916 SERIES AND 1N917

electrical characteristics at 25°C free-air temperature (unless otherwise noted)

PARAMETER		TEST CONDITIONS		1N916		1N916A		1N916B		1N917		UNIT
				MIN	MAX	MIN	MAX	MIN	MAX	MIN	MAX	
$V_{(BR)}$	Reverse Breakdown Voltage	$I_R = 100 \mu A$		100		100		100		40		V
I_R	Static Reverse Current	$V_R = 10 V$									50		nA
		$V_R = 20 V$			25		25		25			
		$V_R = 20 V$, $T_A = 100°C$							3		25	
		$V_R = 20 V$, $T_A = 150°C$			50		50		50			μA
		$V_R = 75 V$			5		5		5			
V_F	Static Forward Voltage	$I_F = 0.25 mA$									0.64	V
		$I_F = 1.5 mA$									0.74	
		$I_F = 3.5 mA$									0.83	
		$I_F = 5 mA$							0.63	0.73			
		$I_F = 10 mA$			1						1	
		$I_F = 20 mA$	See Note 4				1					
		$I_F = 30 mA$							1			
C_T	Total Capacitance	$V_R = 0$, $f = 1 MHz$			2		2		2		2.5	pF

NOTE 4: These parameters must be measured using pulse techniques. $t_w = 300 \mu s$, duty cycle ≤ 2%.

DESIGNED FOR HIGH-SPEED, MEDIUM-POWER SWITCHING AND GENERAL PURPOSE AMPLIFIER APPLICATIONS

- h_{FE} ... Guaranteed from 100 μA to 500 mA
- High f_T at 20 V, 20 mA ... 300 MHz (2N2219A, 2N2222A)
  250 MHz (all others)
- 2N2218, 2N2221 for Complementary Use with 2N2904, 2N2906
- 2N2219, 2N2222 for Complementary Use with 2N2905, 2N2906

*mechanical data

Device types 2N2217, 2N2218, 2N2218A, 2N2219, and 2N2219A are in JEDEC TO-5 packages.
Device types 2N2220, 2N2221, 2N2221A, 2N2222, and 2N2222A are in JEDEC TO-18 packages.

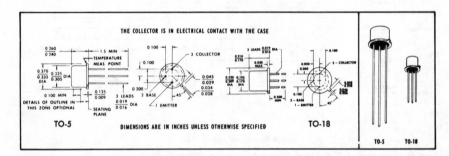

THE COLLECTOR IS IN ELECTRICAL CONTACT WITH THE CASE

TO-5   TO-18

DIMENSIONS ARE IN INCHES UNLESS OTHERWISE SPECIFIED

*absolute maximum ratings at 25°C free-air temperature (unless otherwise noted)

	2N2217 2N2218 2N2219	2N2218A 2N2219A	2N2220 2N2221 2N2222	2N2221A 2N2222A	UNIT
Collector-Base Voltage	60	75	60	75	V
Collector-Emitter Voltage (See Note 1)	30	40	30	40	V
Emitter-Base Voltage	5	6	5	6	V
Continuous Collector Current	0.8	0.8	0.8	0.8	A
Continuous Device Dissipation at (or below) 25°C Free-Air Temperature (See Notes 2 and 3)	0.8	0.8	0.5	0.5	W
Continuous Device Dissipation at (or below) 25°C Case Temperature (See Notes 4 and 5)	3	3	1.8	1.8	W
Operating Collector Junction Temperature Range	−65 to 175				°C
Storage Temperature Range	−65 to 200				°C
Lead Temperature 1/16 Inch from Case for 10 Seconds	230				°C

NOTES: 1. These values apply between 0 and 500 mA collector current when the base-emitter diode is open-circuited.
2. Derate 2N2217, 2N2218, 2N2218A, 2N2219, and 2N2219A linearly to 175°C free-air temperature at the rate of 5.33 mW/°C.
3. Derate 2N2220, 2N2221, 2N2221A, 2N2222, and 2N2222A linearly to 175°C free-air temperature at the rate of 3.33 mW/°C.
4. Derate 2N2217, 2N2218, 2N2218A, 2N2219, and 2N2219A linearly to 175°C case temperature at the rate of 20.0 mW/°C.
5. Derate 2N2220, 2N2221, 2N2221A, 2N2222, and 2N2222A linearly to 175°C case temperature at the rate of 12.0 mW/°C.

*JEDEC registered data. This data sheet contains all applicable registered data in effect at the time of publication.

USES CHIP N24

Fig. 1.32
Courtesy of Texas Instruments

TYPES 2N2217 THRU 2N2222, 2N2218A, 2N2219A, 2N2221A, 2N2222A
N-P-N SILICON TRANSISTORS

2N2217 THRU 2N2222

***electrical characteristics at 25°C free-air temperature (unless otherwise noted)**

PARAMETER		TEST CONDITIONS		2N2217 (TO-5) / 2N2220 (TO-18) MIN	MAX	2N2218 (TO-5) / 2N2221 (TO-18) MIN	MAX	2N2219 (TO-5) / 2N2222 (TO-18) MIN	MAX	UNIT
$V_{(BR)CBO}$	Collector-Base Breakdown Voltage	$I_C = 10\ \mu A$, $I_E = 0$		60		60		60		V
$V_{(BR)CEO}$	Collector-Emitter Breakdown Voltage	$I_C = 10\ mA$, $I_B = 0$,	See Note 6	30		30		30		V
$V_{(BR)EBO}$	Emitter-Base Breakdown Voltage	$I_E = 10\ \mu A$, $I_C = 0$		5		5		5		V
I_{CBO}	Collector Cutoff Current	$V_{CB} = 50\ V$, $I_E = 0$			10		10		10	nA
		$V_{CB} = 50\ V$, $I_E = 0$,	$T_A = 150°C$		10		10		10	µA
I_{EBO}	Emitter Cutoff Current	$V_{EB} = 3\ V$, $I_C = 0$			10		10		10	nA
h_{FE}	Static Forward Current Transfer Ratio	$V_{CE} = 10\ V$, $I_C = 100\ \mu A$				20		35		
		$V_{CE} = 10\ V$, $I_C = 1\ mA$		12		25		50		
		$V_{CE} = 10\ V$, $I_C = 10\ mA$	See Note 6	17		35		75		
		$V_{CE} = 10\ V$, $I_C = 150\ mA$	See Note 6	20	60	40	120	100	300	
		$V_{CE} = 10\ V$, $I_C = 500\ mA$	See Note 6			20		30		
		$V_{CE} = 1\ V$, $I_C = 150\ mA$	See Note 6	10		20		50		
V_{BE}	Base-Emitter Voltage	$I_B = 15\ mA$, $I_C = 150\ mA$	See Note 6		1.3		1.3		1.3	V
		$I_B = 50\ mA$, $I_C = 500\ mA$	See Note 6				2.6		2.6	V
$V_{CE(sat)}$	Collector-Emitter Saturation Voltage	$I_B = 15\ mA$, $I_C = 150\ mA$	See Note 6		0.4		0.4		0.4	V
		$I_B = 50\ mA$, $I_C = 500\ mA$	See Note 6				1.6		1.6	V
$\|h_{fe}\|$	Small-Signal Common-Emitter Forward Current Transfer Ratio	$V_{CE} = 20\ V$, $I_C = 20\ mA$,	$f = 100\ MHz$	2.5		2.5		2.5		
f_T	Transition Frequency	$V_{CE} = 20\ V$, $I_C = 20\ mA$,	See Note 7	250		250		250		MHz
C_{obo}	Common-Base Open-Circuit Output Capacitance	$V_{CB} = 10\ V$, $I_E = 0$,	$f = 1\ MHz$		8		8		8	pF
$h_{ie(real)}$	Real Part of Small-Signal Common-Emitter Input Impedance	$V_{CE} = 20\ V$, $I_C = 20\ mA$,	$f = 300\ MHz$		60		60		60	Ω

NOTES: 6. These parameters must be measured using pulse techniques. $t_w = 300\ \mu s$, duty cycle $\leq 2\%$.
7. To obtain f_T, the $|h_{fe}|$ response with frequency is extrapolated at the rate of −6 dB per octave from f = 100 MHz to the frequency at which $|h_{fe}| = 1$.

switching characteristics at 25°C free-air temperature

PARAMETER		TEST CONDITIONS†		TYP	UNIT
t_d	Delay Time	$V_{CC} = 30\ V$,	$I_C = 150\ mA$, $I_{B(1)} = 15\ mA$,	5	ns
t_r	Rise Time	$V_{BE(off)} = -0.5\ V$,	See Figure 1	15	ns
t_s	Storage Time	$V_{CC} = 30\ V$,	$I_C = 150\ mA$, $I_{B(1)} = 15\ mA$,	190	ns
t_f	Fall Time	$I_{B(2)} = -15\ mA$,	See Figure 2	23	ns

†Voltage and current values shown are nominal; exact values vary slightly with transistor parameters.

*JEDEC registered data

Fig. 1.33

Courtesy of Texas Instruments

TYPES 2N2217 THRU 2N2222, 2N2218A, 2N2219A, 2N2221A, 2N2222A
N-P-N SILICON TRANSISTORS

2N2218A, 2N2219A, 2N2221A, 2N2222A

***electrical characteristics at 25°C free-air temperature (unless otherwise noted)**

PARAMETER		TEST CONDITIONS	TO-5→ / TO-18→	2N2218A / 2N2221A MIN	2N2218A / 2N2221A MAX	2N2219A / 2N2222A MIN	2N2219A / 2N2222A MAX	UNIT
$V_{(BR)CBO}$	Collector-Base Breakdown Voltage	$I_C = 10\ \mu A$, $I_E = 0$		75		75		V
$V_{(BR)CEO}$	Collector-Emitter Breakdown Voltage	$I_C = 10$ mA, $I_B = 0$,	See Note 6	40		40		V
$V_{(BR)EBO}$	Emitter-Base Breakdown Voltage	$I_E = 10\ \mu A$, $I_C = 0$		6		6		V
I_{CBO}	Collector Cutoff Current	$V_{CB} = 60$ V, $I_E = 0$			10		10	nA
		$V_{CB} = 60$ V, $I_E = 0$,	$T_A = 150°C$		10		10	μA
I_{CEV}	Collector Cutoff Current	$V_{CE} = 60$ V, $V_{BE} = -3$ V			10		10	nA
I_{BEV}	Base Cutoff Current	$V_{CE} = 60$ V, $V_{BE} = -3$ V			−20		−20	nA
I_{EBO}	Emitter Cutoff Current	$V_{EB} = 3$ V, $I_C = 0$			10		10	nA
h_{FE}	Static Forward Current Transfer Ratio	$V_{CE} = 10$ V, $I_C = 100\ \mu A$		20		35		
		$V_{CE} = 10$ V, $I_C = 1$ mA		25		50		
		$V_{CE} = 10$ V, $I_C = 10$ mA	See Note 6	35		75		
		$V_{CE} = 10$ V, $I_C = 150$ mA	See Note 6	40	120	100	300	
		$V_{CE} = 10$ V, $I_C = 500$ mA	See Note 6	25		40		
		$V_{CE} = 1$ V, $I_C = 150$ mA	See Note 6	20		50		
		$V_{CE} = 10$ V, $I_C = 10$ mA, $T_A = -55°C$	See Note 6	15		35		
V_{BE}	Base-Emitter Voltage	$I_B = 15$ mA, $I_C = 150$ mA	See Note 6	0.6	1.2	0.6	1.2	V
		$I_B = 50$ mA, $I_C = 500$ mA	See Note 6		2		2	V
$V_{CE(sat)}$	Collector-Emitter Saturation Voltage	$I_B = 15$ mA, $I_C = 150$ mA	See Note 6		0.3		0.3	V
		$I_B = 50$ mA, $I_C = 500$ mA	See Note 6		1		1	V
h_{ie}	Small-Signal Common-Emitter Input Impedance	$V_{CE} = 10$ V, $I_C = 1$ mA	f = 1 kHz	1	3.5	2	8	kΩ
		$V_{CE} = 10$ V, $I_C = 10$ mA	f = 1 kHz	0.2	1	0.25	1.25	kΩ
h_{fe}	Small-Signal Forward Current Transfer Ratio	$V_{CE} = 10$ V, $I_C = 1$ mA	f = 1 kHz	30	150	50	300	
		$V_{CE} = 10$ V, $I_C = 10$ mA	f = 1 kHz	50	300	75	375	
h_{re}	Small-Signal Common-Emitter Reverse Voltage Transfer Ratio	$V_{CE} = 10$ V, $I_C = 1$ mA	f = 1 kHz		5×10^{-4}		8×10^{-4}	
		$V_{CE} = 10$ V, $I_C = 10$ mA	f = 1 kHz		2.5×10^{-4}		4×10^{-4}	
h_{oe}	Small-Signal Common-Emitter Output Admittance	$V_{CE} = 10$ V, $I_C = 1$ mA	f = 1 kHz	3	15	5	35	μmho
		$V_{CE} = 10$ V, $I_C = 10$ mA	f = 1 kHz	10	100	25	200	μmho
$\|h_{fe}\|$	Small-Signal Common-Emitter Forward Current Transfer Ratio	$V_{CE} = 20$ V, $I_C = 20$ mA,	f = 100 MHz	2.5		3		
f_T	Transition Frequency	$V_{CE} = 20$ V, $I_C = 20$ mA,	See Note 7	250		300		MHz
C_{obo}	Common-Base Open-Circuit Output Capacitance	$V_{CB} = 10$ V, $I_E = 0$,	f = 100 kHz		8		8	pF
C_{ibo}	Common-Base Open-Circuit Input Capacitance	$V_{EB} = 0.5$ V, $I_C = 0$,	f = 100 kHz		25		25	pF
$h_{ie(real)}$	Real Part of Small-Signal Common-Emitter Input Impedance	$V_{CE} = 20$ V, $I_C = 20$ mA,	f = 300 MHz		60		60	Ω
$r_b'C_c$	Collector-Base Time Constant	$V_{CE} = 20$ V, $I_C = 20$ mA,	f = 31.8 MHz		150		150	ps

NOTES: 6. These parameters must be measured using pulse techniques. $t_w = 300\ \mu s$, duty cycle ≤ 2%.

7. To obtain f_T, the $|h_{fe}|$ response with frequency is extrapolated at the rate of −6 dB per octave from f = 100 MHz to the frequency at which $|h_{fe}| = 1$.

*JEDEC registered data

Fig. 1.34

Courtesy of Texas Instruments

TYPES 2N2217 THRU 2N2222, 2N2218A, 2N2219A, 2N2221A, 2N2222A
N-P-N SILICON TRANSISTORS

*operating characteristics at 25°C free-air temperature

PARAMETER	TEST CONDITIONS	TO-5 → TO-18 →	2N2218A 2N2221A MAX	2N2219A 2N2222A MAX	UNIT
F Spot Noise Figure	V_{CE} = 10 V, I_C = 100 μA, R_G = 1 kΩ, f = 1 kHz			4	dB

*switching characteristics at 25°C free-air temperature

PARAMETER	TEST CONDITIONS†	TO-5 → TO-18 →	2N2218A 2N2221A MAX	2N2219A 2N2222A MAX	UNIT
t_d Delay Time	V_{CC} = 30 V, I_C = 150 mA, $I_{B(1)}$ = 15 mA,		10	10	ns
t_r Rise Time	$V_{BE(off)}$ = −0.5 V, See Figure 1		25	25	ns
τ_A Active Region Time Constant‡			2.5	2.5	ns
t_s Storage Time	V_{CC} = 30 V, I_C = 150 mA, $I_{B(1)}$ = 15 mA,		225	225	ns
t_f Fall Time	$I_{B(2)}$ = −15 mA, See Figure 2		60	60	ns

†Voltage and current values shown are nominal; exact values vary slightly with transistor parameters.

‡Under the given conditions τ_A is equal to $\dfrac{t_r}{10}$.

---

*PARAMETER MEASUREMENT INFORMATION

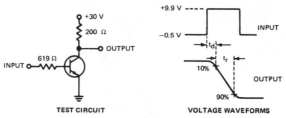

TEST CIRCUIT   VOLTAGE WAVEFORMS

FIGURE 1—DELAY AND RISE TIMES

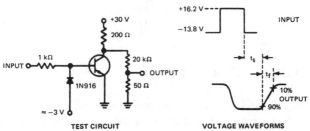

TEST CIRCUIT   VOLTAGE WAVEFORMS

FIGURE 2—STORAGE AND FALL TIMES

NOTES: a. The input waveforms have the following characteristics: For Figure 1, t_r ≤ 2 ns, t_w ≤ 200 ns, duty cycle ≤ 2%; for Figure 2,
t_f ≤ 5 ns, t_w ≈ 100 μs, duty cycle ≤ 17%.
        b. All waveforms are monitored on an oscilloscope with the following characteristics: t_r ≤ 5 ns, R_{in} ≥ 100 kΩ, C_{in} ≤ 12 pF.

*JEDEC registered data

Fig. 1.35
Courtesy of Texas Instruments

1.2.1 Transistor Cases

In order for a designer to use a commercially available transistor, he or she must be able to identify the terminals. Some pictures of commercially available bipolar transistors are shown in Fig. 1.36. On the left is an epoxy encapsulated, high frequency, low power transistor. The middle picture is of a medium power, plastic encapsulated, audio transistor. The right-hand picture is of a high power audio or switching transistor in a metal case (TO-3 case style as designated by the EIA).

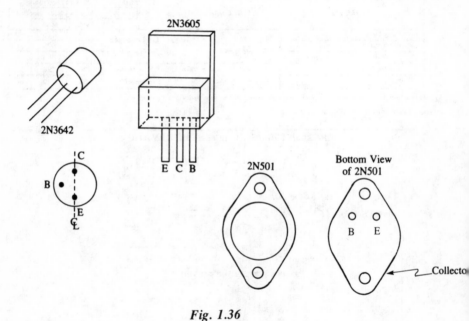

Fig. 1.36

1.3 FIELD-EFFECT TRANSISTORS

There are two basic types of transistors, the bipolar units that we have just discussed, which were commercially available for a considerable period (since approximately 1950), and, more recently field effect transistors (FET).

Field-effect transistors derive their name from their mode of operation. Let's examine an equivalent drawing for a junction field-effect transistor (see Fig. 1.37). If the bar were cut vertically along the axis, a cross section would appear as shown in Fig. 1.38. If a bias voltage were applied, current would flow as in Fig. 1.39, where the bottom of the transistor would be a source of electrons, and the electrons would be drained from the top of the transistor and proceed to the top (+) terminal of the battery.

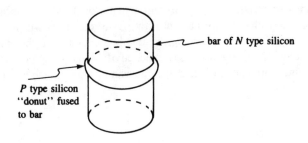

Fig. 1.37

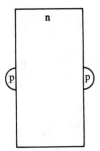

Fig. 1.38

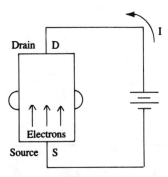

Fig. 1.39 Assumed Current Flow Direction for Conventional Current Flow

If an additional bias battery is connected as shown in Fig. 1.40, a reversed-biased diode exists between the *p* and *n* material. Thus, no current will flow between the *p* and *n* material. The presence of the negative terminal connected to the terminal marked "gate" will have the effect of removing the majority carriers in the *n* material, resulting in a region that is depleted of majority carriers. With the narrower channel for current flow, fewer electrons will flow from the source to the drain. In fact, if the V_{GG} source is great enough, the channel from source to gate is completely blocked or pinched off. This voltage value is referred to as the "pinch-off voltage," and it is measured from the source-to-gate terminals:

$$V_{SG}\bigg|_{I_D = 0} = V_P \tag{1.12}$$

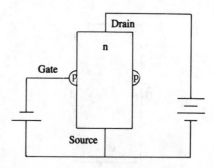

Fig. 1.40

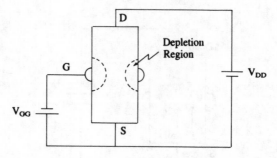

Fig. 1.41

Depending upon the transistor type, V_P may vary from -0.2 V to -3.0 V (typically). The pinch-off voltage is slightly dependent upon the source-to-gate voltage. As might be anticipated, if the source-to-gate voltage is a large value, some electrons may be drawn from the channel that was "blocked-off" for a lower source-to-drain voltage.

If we were to leave the gate terminal disconnected and gradually increase the source to drain voltage, the plotted data gathered would result in the curve given in Fig. 1.42.

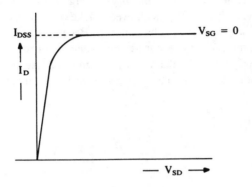

Fig. 1.42

We note that the curve in Fig. 1.42 is somewhat flat after an initially steep portion. This value of the drain current is given a special designation, I_{DSS}. This value is inexact because the curve slopes slightly upward with increasing V_{SD}. However, manufacturers will often specify I_{DSS} and the V_{SD} at which I_{DSS} was measured. The I represents current, where D is the symbol for drain, S represents common source, and the second S represents "saturation."

If a bias battery V_{GG} is connected in the gate circuit and the battery voltage is set at $-0.2\ V$, a different I_D versus V_{SD} curve will be produced, with I_D being less than when $V_{SG} = 0$. (See Fig. 1.43.)

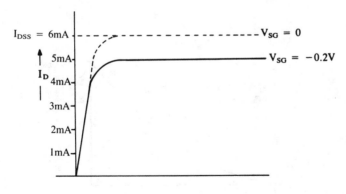

Fig. 1.43

If V_{SG} is increased in magnitude, the entire family of curves given in Fig. 1.44 will be produced. Because no current flows into the gate of a FET under normal operation, an input family of curves is not possible. However, a family of curves may be developed which relates the input to the output. If the source to drain voltage is held constant, the curve of Fig. 1.45 will be obtained if we monitor I_D and vary V_{SG}. If V_{SD} is allowed to take on various values, the family of curves given in Fig. 1.46 will be obtained. Because these curves nearly "lie on top of each other," a single transfer characteristic curve, which is recorded at some average or nominal value of V_{SD}, is published. An equation that relates the drain current I_D to the source to gate voltage V_{SG} is given below:

$$I_D \doteq I_{DSS} \left(1 - \frac{V_{SG}}{V_P} \right)^2 \tag{1.13}$$

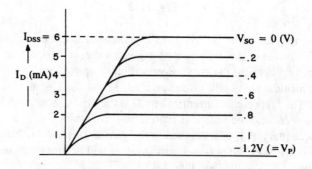

Fig. 1.44 Output Family for an N-Channel JFET

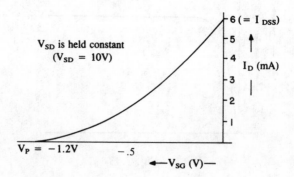

Fig. 1.45 Transfer Characteristic Curve

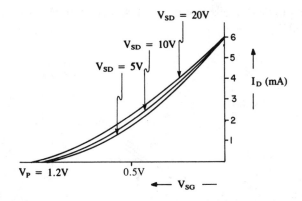

Fig. 1.46 Transfer Characteristic Curves

If the value of V_P and I_{DSS} are known for a particular transistor, V_{SG} may be allowed to take on different values, and I_D data can then be computed. Thus, an approximate transfer characteristic curve may be plotted. It should be noted that the actual transfer characteristic curve for a JFET follows a three-halves power relationship, rather than the square law relationship, as given in formula (1.13). The result of using equation (1.13), however, is sufficiently accurate for most design applications.

Let us develop an example of the use of the transfer characteristic curve.

Example (1.2). Determine the required V_{GG} gate bias battery to enable 2 mA of drain current to flow. Use the schematic shown in Fig. 1.47 and the curves of Fig. 1.46. We note that the source to gate voltage, V_{SG} is identically equal to the bias battery voltage V_{GG}. Therefore, if we can determine the value of V_{SG} that is necessary to enable 2 mA of current to flow, the value of the bias battery will have been determined.

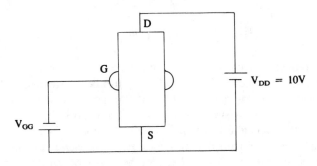

Fig. 1.47

Referring to the curve shown in Fig. 1.48 (based on Fig. 1.46) we note that the value of $V_{SG} = -0.5\ V$.

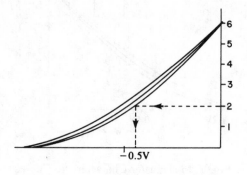

Fig. 1.48

To become familiar with the notation used in electrical schematics, let us redraw Fig. 1.47 with the equivalent electrical symbol for a junction field-effect transistor used in place of the "silicon bar." We also eliminate drawing the battery that represents the V_{DD} power supply by drawing an arrow pointing to V_{DD}, thereby implying that a connection is made to $+V_{DD}$.

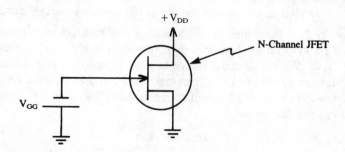

Fig. 1.49

With the main channel constructed of *n*-type silicon, the JFET symbol is as shown in Fig. 1.49. However, a JFET could also be constructed with the main channel made of *p* material. This type of transistor operates with a negative power supply (relative to ground) and its symbol is given in Fig. 1.50.

Another type of field effect transistor with an input resistance in the 10^{12} to 10^{16} ohms (as compared to 10^9 ohms for a JFET) is the *metal oxide semiconductor field-effect transistor* (MOSFET). Because the device operates in a manner similar to a JFET in that majority carriers are depleted from the main conducting channel, a more complete title is *N-channel, depletion mode MOSFET (DMOS-*

FET). Although this device is not made by junction alloy techniques, we can redraw a JFET for illustration purposes (Fig. 1.51).

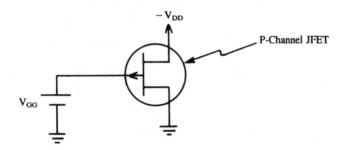

Fig. 1.50

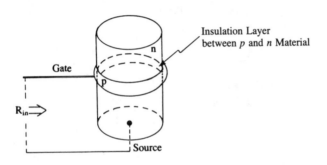

Fig. 1.51

If an insulation layer is sandwiched between the p and n material, two changes take place in the operation and electrical characteristic of a MOSFET. First, the input resistance between source and gate is increased from approximately 10^9 ohms for a JFET to approximately 10^{12} to 10^{16} ohms for a MOSFET, and, second, the gate may now be driven positive (for a MOSFET) without drawing gate current. Thus, a MOSFET can operate with both negative and positive voltages at the gate (relative to the source).

Thus, the output curves will appear as shown in Fig. 1.52 and the transfer characteristic is given in Fig. 1.53.

Data for the depletion mode MOSFET (DMOSFET) transfer characteristic curve also may be computed by using the equation,

$$I_D = I_{DSS} \left(1 - \frac{V_{SG}}{V_P} \right)^2 \tag{1.14}$$

34

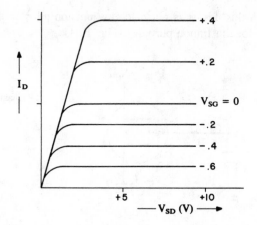

Fig. 1.52 Output Curves for N-Channel Depletion Mode MOSFET
(DMOSFET)

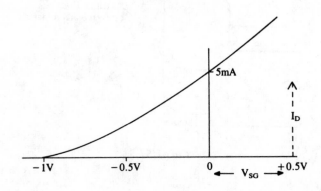

Fig. 1.53 Transfer Characteristic Curve for N-Channel Depletion Mode
MOSFET (DMOSFET)

However, we note that an equal sign is employed (instead of approximately equal), since the drain current I_D and the source to gate voltage V_{SG} follow a true square law relationship.

A depletion mode MOSFET, which is intended to operate with a positive drain power supply, is referred to as an N-channel MOSFET. The N-channel MOSFET and its P-channel complement are shown in Fig. 1.54.

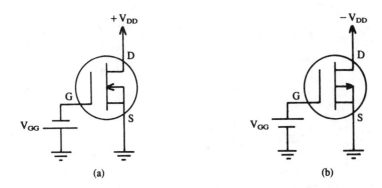

Fig. 1.54 (a) N-Channel and (b) P-Channel MOSFETs

An additional MOSFET type, which is deliberately fabricated such that no drain current will flow with zero source-to-gate voltage, is the enhancement mode MOSFET (EMOSFET), or insulated-gate field-effect transistor (IGFET). For an N-channel EMOSFET, the gate must be driven positive (relative to the source) in order for drain current to flow. Referring to the curves shown in Figs. 1.55 and 1.56, we note that the transfer characteristic curve appears to have been shifted from the left side of the vertical axis ($V_{SG} = 0$) to the right side, where V_{SG} is a positive value. The positive source-to-gate voltage value where drain current just begins to flow is given by the symbol V_T (voltage-threshhold or voltage turn-on). V_T for an EMOSFET corresponds to V_P for a depletion mode MOSFET. The drain current *versus* the source-to-gate voltage again follows a square law relationship:

$$I_D = k(V_{SG} - V_T)^2 \qquad (1.15)$$

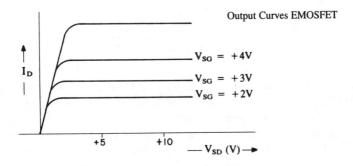

Fig. 1.55 Output Curves for EMOSFET (or IGFET)

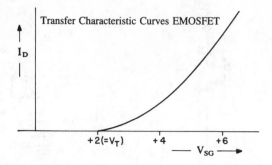

Fig. 1.56 Transfer Characteristic Curves for EMOSFET (or IGFET)

where k typically is a value of 0.3 mA/V. I_{DSS} does not appear in equation (1.15) because I_{DSS} was the drain current value for a DMOSFET when $V_{SG} = 0$ V. For an EMOSFET, when $V_{SG} = 0$, $I_D = 0$, thus I_{DSS} for EMOSFET is meaningless. The symbols for N- and P-channel EMOSFETs, in addition to some elementary bias arrangements, are shown in Figs. 1.57 and 1.58.

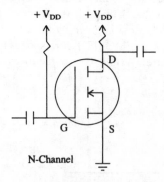

Fig. 1.57 N-Channel EMOSFET

A convenient method of remembering that the above symbols represent EMOSFETs is to recall that the spaces between the source and drain imply that drain current cannot flow with the gate open-circuited.

In some MOSFETS, the substrate is brought out of the transistor case as a separate connection. This allows additional design flexibility in that it may be desirable to have the substrate at a different electrical potential than the source. The symbol for a MOSFET with this construction is shown in Fig. 1.59.

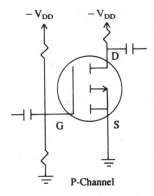

Fig. 1.58 P-Channel EMOSFET

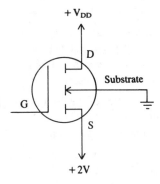

Fig. 1.59 N-Channel EMOSFET with Separate Substrate

PROBLEMS

1. A bipolar transistor is constructed such that the base-to-emitter junction is a _____ biased junction and the collector to base junction is ____ biased.

2. For an *npn* transistor connected in a common base configuration, the bias battery between the collector-to-base is connected with the positive terminal to the (base, collector)?

3. Approximately _____ % of the entire emitter current "arrives" at the collector terminal.

4. For a silicon transistor, the base-to-emitter voltage drop is often approximated at _____ volt.

5. The input resistance of a transistor connected in the common base configuration is a _____ value and the output resistance is a _____ value.

6. The voltage amplification that is demonstrated in a common base connection occurs because _____ .

7. The alpha (α) of a transistor is a ratio of _____ .

8. The value of the alpha (α) is approximately equal to _____ .

9. The common base output curves are approximately "flat." Thus, we can deduce that the collector current is relatively unaffected by changes in _____ if the emitter current is held constant.

10. The input curves for a common base connection appear to be similar to the current *versus* voltage curves for what two-terminal device?

11. The common emitter connection has a slightly _____ input resistance and slightly _____ output resistance than the common base connection. This fact may be deduced by drawing a tangent line to the input and output curves for each connection.

12. One of the major advantages of the common emitter connection over the common base connection is that both _____ and _____ are demonstrated by the common emitter connection.

13. The beta (β) of a transistor is defined as the ratio of _____ to the _____ .

14. The common collector configuration is more often referred to as _____ .

15. The common collector configuration will yield a current gain _____ than unity and a voltage gain _____ unity.

16. The symbol for a *pnp* transistor is _____ and for an *npn* transistor is _____ .

17. Label the terminals of the transistor leads shown below (emitter, etc.).

1. _____

2. _____

3. _____

18. To check whether the base-to-emitter junction of an *npn* transistor is in proper working order, if we connect the red lead of an American-made ohmmeter to the base and the black lead to the emitter, the meter needle will indicate a (low, high) resistance?

19. The two diode equivalent circuit for a PNP transistor is as shown below:

20. What do the *CBO* subscripts for the current I_{CBO} represent?

21. I_{CBO} can be represented as the offset of what variable in the output curves of a common _____ connection?

22. Which is greater leakage current, I_{CBO} or I_{CEO}?

23. By what amount is I_{CBO} or I_{CEO} greater than the other?

24. I_{CEO} may be noted on what set of characteristic curves?

25. What are three maximum transistor ratings which must be observed so that a transistor is not destroyed?

26. Which transistor junction dissipates the most power?

27. A junction field-effect transistor has how many *p-n* junctions?

28. For an N-channel JFET, the gate lead is connected to (*p, n*) material?

29. The source-to-gate voltage at which drain current ceases to flow is called

_____ .

30. For an N-channel JFET, the source-to-gate voltage is a (positive, negative) value?

31. The power supply connected to the gate is usually labeled "V __" (with what subscripts)?

32. The input resistance of a JFET is approximately _____ ohms.

33. The input resistance of a MOSFET is approximately _____ ohms.

34. In contrast to a bipolar transistor, FETs have no input curves. Why?

35. The curve relating the input characteristics to the output characteristics of a FET are called _____ curves.

36. For a JFET, an equation that relates the drain current to the source to gate voltage is

$$I_D \doteq I_{DSS} \left(1 - \frac{V_{SG}}{V_P} \right)^2$$

Why is an "approximately equals" sign employed?

37. I_{DSS} represents what variable?

38. V_P represents what variable?

39. Draw the symbols for an N-channel JFET and P-channel JFET and label them.

40. An N-channel JFET uses what type of drain power supply (negative, positive)?

41. What words do the initials "MOSFET" represent?

42. What is a major difference between a DMOSFET and JFET?

43. Draw the symbols for both N- and P-channel DMOSFETs.

44. What equation relates the drain current to the source-to-gate voltage of a DMOSFET?

45. An EMOSFET is different from a DMOSFET in that the gate power supply voltage polarity is _____ .

46. An equation that relates the drain current to the source to gate voltage for an EMOSFET is

$$I_D = k(V_{SG} - V_T)^2$$

where k is a constant that is a function of the transistor geometry and usually has a value of _____ .

47. For the equation shown in Problem 46, V_T represents what quantity?

Chapter 2

Heat Flow Analysis
in Electronic Components

2.1 DISCUSSION OF HEAT GENERATION

All electronic circuits that are actively performing a useful function are dissipating heat due to the electrical power being used to energize the devices. Many times we do not concern ourselves with the power dissipation, but our first indication that power dissipation must be reckoned with is when we accidentally "cook" a resistor while performing a dc electricity course laboratory experiment. We know that power is the rate of energy utilization, which is computed in electrical circuits by multiplying the voltage potential across an electronic component (or system) by the amount of current flowing into the component (or system).

High power levels dissipated in a device will tend to raise the temperature at the location where the power is being dissipated as heat. In semiconductor devices, a location such as the back-biased p-n junction of a diode or the collector-to-base junction of a transistor are very small in area. Consequently, the heat generated is concentrated and tends to increase the temperature locally. At a given temperature, the junction of dissimilar semiconductor materials (p and n) begins to change its properties, and eventually melting and permanent damage occur. This maximum temperature is usually specified by semiconductor manufacturers on their device data sheets ($T_{j\ max}$). For silicon devices, typical maximum junction temperatures are specified between 150°C to 200°C. For germanium devices, maximum junction temperatures between 75°C to 105°C are typical.

2.2 OHM'S LAW ANALOGY BETWEEN TEMPERATURE, POWER, THERMAL RESISTANCE, AND ELECTRICAL COUNTERPARTS

The physics of heat flow are somewhat analogous to those of electrical current flow. We know that the greater the potential difference that exists between two points, the greater the amount of electricity that will flow for a given electrical

41

42

resistance between the two points. Similarly, the greater the temperature difference that exists between two points, the greater the heat flow for a given thermal resistance. The concept of thermal resistance is similar to electrical resistance. From the practical experience of heating a copper pipe when performing a plumbing chore, we know that we are unable to hold the pipe for an extended period because the temperature increases as the heat rapidly flows along the pipe. Conversely, a pot holder used in the kitchen has a high thermal resistance and we are able to hold a hot pot handle for a considerable period before the heat flows through the cloth. Our experience in a dc electricity course familiarized us with the notion that electrical resistance is proportional to the total length of an object, the resistivity of the object, and inversely proportional to the cross sectional area of the object. The same factors affect thermal resistance.

For a very small transistor, the length of the path through which heat must flow is relatively short. This would tend to reduce the thermal resistance from the junction of the semiconductor to the outside air. (See Fig. 2.1). However, the cross sectional area of the heat flow path is quite small, and, therefore, the thermal resistance is quite large. In an attempt to reduce the junction temperature, a transistor or power diode is often bolted to a highly heat-conductive metal surface. The metal surface is referred to as a "heat sink," which is often made of aluminum. (See Fig. 2.2). Aluminum has a very low thermal resistivity and is low in cost. Copper has a more desirable thermal resistivity, but is significantly more expensive. To improve the heat flow out of the semiconductor device, we must do whatever is necessary to improve the heat dissipation away from the heat sink. From elementary physics, we know that heat is transmitted by three mechanisms: radiation, conduction, and convection (which is a form of conduction). Therefore, if we paint the heat sink with flat black paint, we will improve the heat flow away from it by radiation. If we enlarge the heat sink, blow air across its surface, or (in extreme cases) use water cooling, we will improve the heat flow by convection.

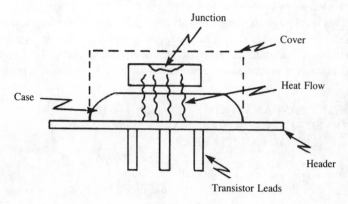

Fig. 2.1

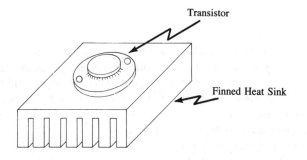

Fig. 2.2

The semiconductor device manufacturer can improve the power dissipation capabilities of transistors, diodes, or other devices by using high thermal conductivity (low thermal resistivity) materials, decreasing the length of the heat flow path (thin wafer and thin case), and by using a large cross-sectional area. A large cross-sectional area requires the use of a large semiconductor die and component case. High thermal conductivity case materials are typically aluminum, brass, or copper.

Let us diagram the total heat flow path for a transistor bolted to a heat sink that is suspended in space (ambient air). The resulting diagram is given in Fig. 2.3. We note that there are three thermal resistances, each denoted by the Greek letter, theta (θ), and a double subscript:

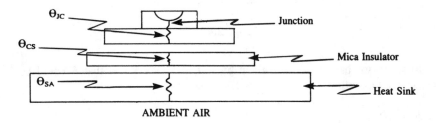

Fig. 2.3

θ_{JC} = The thermal resistance from junction to case;

θ_{CS} = The thermal resistance from the transistor case to the heat sink. θ_{CS} represents the thermal resistance of a mica or similar electrical insulator;

θ_{SA} = The thermal resistance from heat sink to the ambient air.

In the diagram (Fig 2.3) we also notice a mica insulator located between the transistor and heat sink. As much as we would like to omit this additional thermal resistance, it is sometimes necessary to include this electrical insulator because the transistor case and the heat sink may be at different electrical potentials. Normally, the transistor collector terminal is connected to the transistor case and

44

the collector may be at a potential of from 5 V to 100 V above ground. The heat sink is usually connected to some other mechanical device such as an equipment case. The equipment case may be connected to electric utility ground.

For the heat flow theory to be of use, we must quantify the analysis so that we can predict whether a transistor or other device will be able to operate safely under desired electrical and thermal conditions.

Reviewing from dc theory, we know that for the circuit shown in Fig. 2.4 the total current flowing through the resistor may be computed by dividing the voltage across it by the resistance value, or

$$I = \frac{10 \text{ V} - 3 \text{ V}}{7 \text{ }\Omega} = \frac{7 \text{ V}}{7 \text{ }\Omega} = 1 \text{ A}$$

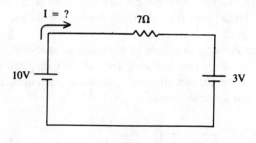

Fig. 2.4

Suppose we had the circuit given by Fig. 2.5. We know that E could be determined as follows:

$$E = IR + V_{gnd-A}$$
$$= 1(7) + 3$$

therefore,

$$E = 10 \text{ V}$$

Now suppose we had the circuit of Fig. 2.6. Again:

$$V_{gnd-J} = IR_{total} + V_{gnd-A} = 1 \text{ A} (2\Omega + 1\Omega + 4\Omega) + 3 \text{ V}$$
$$V_{gnd-J} = 10 \text{ V}$$

and the voltage V_{gnd-S} could be computed as follows:

$$V_{gnd\text{-}S} = IR_{SA} + V_{gnd\text{-}A}$$
$$= 1 \text{ A } (4\Omega) + 3 \text{ V}$$
$$V_{gnd\text{-}S} = 7 \text{ V}$$

Voltage $V_{gnd\text{-}C}$ may be determined:

$$V_{gnd\text{-}C} = I(R_{CS} + R_{SA}) + V_{gnd\text{-}A}$$
$$= 1 \text{ A } (1\Omega + 4\Omega) + 3 \text{ V}$$
$$V_{gnd\text{-}C} = 8 \text{ V}$$

In a thermal circuit, the temperature corresponds to the voltage in an electrical circuit, power flow to electrical current, and thermal resistance to electrical resistance. Thus,

$$\theta \rightarrow R$$
$$P \rightarrow I$$
$$T \rightarrow V$$

and Ohm's Law can be used for a thermal circuit.

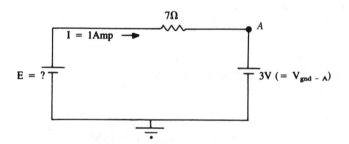

Fig. 2.5

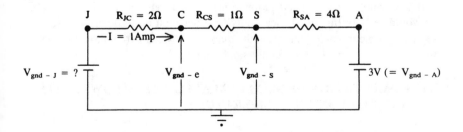

Fig. 2.6

Example (2.1). Determine the sink, case, and junction temperatures for the thermal circuit shown below if the ambient temperature is 25°C. In most instances, we would only be interested in the junction temperature.

A transistor is dissipating one watt of power. The manufacturer's data sheet specifies that the thermal resistance from junction to case is 75°C/watt. The manufacturer of a mica wafer specifies the wafer thermal resistance at 10°C/watt and the heat sink manufacturer specifies his product at 50°C/watt. See Fig. 2.7.

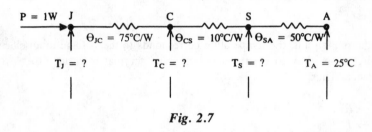

Fig. 2.7

According to Ohm's law and Kirchoff's voltage law, the sink temperature is

$$T_S = P\,\theta_{SA} + T_A$$
(Electrical Analogue: $V_{gnd-S} = IR_{SA} + V_{gnd-A}$) (2.1)
$$T_S = 1W\,(50°C/W) + 25°C = 75°C$$

$$T_C = P\,\theta_{CA} + T_A = P(\theta_{CS} + \theta_{SA}) + T_A$$
$$= 1W(10°C/W + 50°C/W) + 25°C$$ (2.2)
$$T_C = 85°C$$

$$T_J = P\,\theta_{JA} + T_A = P(\theta_{JC} + \theta_{CS} + \theta_{SA}) + T_A$$
$$= 1W(75°C/W + 10°C/W + 50°C/W) + 25°C$$ (2.3)
$$T_J = 160°C$$

If the manufacturer's specifications for a device stated that $T_{J\,max} = 200°C$, we would be operating the device safely with one watt of dissipation. A slightly different problem would be to pose the question, "What is the maximum power that the above thermal circuit could safely handle?"

2.3 COMPUTATION OF MAXIMUM ALLOWABLE POWER FLOW AND COMPONENT TEMPERATURES

Redrawing our thermal circuit, we have the diagram of Fig. 2.8. According to Ohm's law, we have

$$P_{max} = \frac{T_J - T_A}{\theta_{JA}} = \frac{T_{Jmax} - T_{ambient}}{\theta_{JC} + \theta_{CS} + \theta_{SA}}$$

$$= \frac{200°C - 25°C}{75°C/W + 10°C/W + 50°C/W} \tag{2.4}$$

$$= \frac{175°C}{135°C/W}$$

$$P_{max} = 1.3 \text{ W}$$

$P_{max} = ?$ J   C   S   A

$\theta_{JC} = 75°$ C/W $\theta_{CS} = 10°$ C/W $\theta_{SA} = 50°C/W$

$T_J = T_{Jmax} = 200°C$ T_C T_S $T_A = 25°C$

Fig. 2.8

For a given transistor, the thermal resistance from junction to case is a fixed value. Generally, the thermal resistance of the insulator will not vary appreciably for a specific size of transistor case. However, there is an amazing material called berillia (berillium oxide), which has the thermal conductivity of aluminum and the electrical resistance of glass. However, it is considerably more expensive than mica and, therefore, mica is more commonly used as an insulator between transistor case and heat sink.

The one source of thermal resistance that we can largely control is the heat sink. If we omit the heat sink entirely, the thermal resistance from sink to ambient air (θ_{SA}) becomes a very large number. Thus, only a small amount of power can be dissipated without raising the junction temperature to prohibitive values. Transistor manufacturers will often specify two thermal resistances, θ_{JC} and θ_{JA} (with no heat sink). From this data it is easy to determine θ_{SA}.

$$\theta_{SA} \text{ (without heat sink)} = \theta_{JA} \text{ (without heat sink)} - \theta_{JC} \tag{2.5}$$

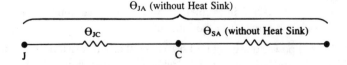

Θ_{JA} (without Heat Sink)

Θ_{JC} Θ_{SA} (without Heat Sink)

J   C

Fig. 2.9

The transistor manufacturer may also specify the maximum power that a transistor can dissipate with an infinite heat sink and without a heat sink. An infinite *anything* is difficult to come by in the real world, and, therefore, what the manufacturer is doing by specifying an infinite heat sink is saying that the transistor case is maintained at 25°C by some external cooling mechanism while temperature measurements are taken.

PROBLEMS

1. Heat is generated in a semiconductor device by the flow of electricity. A (back-biased or forward-biased) junction generates the most heat?

2. What thermal mechanism causes the destruction of a semiconductor device?

3. Heat flow analysis is similar to?

4. Which material can withstand a greater temperature before damage ensues, germanium or silicon?

5. Thermal resistance is proportional to what physical parameters?

6. Describe the heat flow path for the power dissipated in a semiconductor.

7. A heat sink should have what properties?

8. What is the most common construction material for commercially available heat sinks?

9. For the thermal circuit shown below, compute T_S, T_C, and T_J.

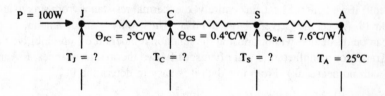

10. If the transistor in Problem 9 were made of germanium with a maximum junction temperature rating of 105°C, would the above operation be acceptable?

11. If $T_{Jmax} = 105°C$ in Problem 9, what is the maximum power that could be safely dissipated in the junction?

12. With a power dissipation of 50W, the thermal resistances given in Problem 9, and $T_{Jmax} = 105°C$, what is the maximum allowable ambient temperature?

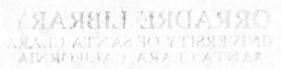

Chapter 3

Field-Effect Transistor Amplifier
DC Bias
Design and Analysis
by Graphical and Algebraic Techniques

3.1 INTRODUCTION

The design and analysis of the dc bias of field-effect transistors is the topic that will be investigated in this chapter. Because field-effect transistors are unilateral devices, their analysis, at least from a small-signal standpoint, is somewhat less involved than for bipolar transistors. In many textbooks, bipolar transistors are investigated after diodes and prior to FETs because the concepts of forward- and reversed-biased diode junctions are appropriate to the collector-to-base and emitter-to-base junctions of transistors. From a chronological standpoint, bipolar transistors also gained acceptance in commercial applications prior to field-effect transistors.

In Chapter 1 we learned that field-effect transistors could be thought of as having a very high input impedance due to a back-biased diode from source to gate, or due to an insulation layer between the gate and substrate. Thus, the source-to-gate voltage is the controlling parameter of the drain current. The question arises, "What voltage is necessary to obtain a desired drain current?" To answer this question, we may take two solution approaches: graphical and analytical (algebraic).

3.2 GRAPHICAL TECHNIQUES

A graphical approach requires that we have available to us a plot of input voltage, V_{sg} versus output current I_D. We may use a set of curves which are plots of V_{sg} versus I_D for specific V_{SD} voltages as shown in Fig. 3.1 to determine a desired or unknown operating point. Because all field-effect transistors have transfer characteristic curves with similar shapes, the curves may be "normalized," so that any transistor operating point may be determined for any transistor.

50

We will not pursue the analysis or design of FET circuits with normalized curves in our investigation because of the added complexity in using this solution approach. It should be pointed out that the effect of varying the source-to-drain voltage is to create a family of curves that are parallel to each other and quite close together. As an approximation, a single curve is often used instead of the family.

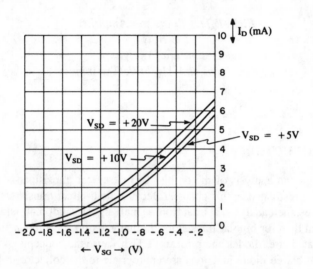

Fig. 3.1 Transfer Characteristic Curves of MPF-102 JFET from a Tektronix 575 Curve Tracer

Let us examine a transfer characteristic curve for a specific source to drain voltage (see Fig. 3.2). We note that as the source-to-gate voltage becomes increasingly negative, the drain current decreases until it becomes zero. The specific source-to-gate voltage at which drain current ceases to flow is referred to as the "pinch-off" voltage (V_P). We may remember this by analogy such that the effect of the increasing source-to-gate voltage is similar to the effect of squeezing a plastic tube through which water is flowing. If we squeeze the tube with sufficient force (pinch-off), we can constrict the tube so that the water is unable to flow.

Another key point on the transfer characteristic curve(s) concerns where the maximum drain current flows. Maximum current flows when the gate is no longer negative with respect to the source, i.e., no longer a back-biased diode. The electric field depleting the injection of majority carriers is thus no longer present. This point is indicated on the curves as I_{DSS}. The various considerations that must be kept in mind when performing a dc bias design of a JFET can include the following:

1. We must be certain that our chòice of the amount of drain current, source-to-drain voltage (referred to as the Q point currents and voltages or just the Q point), and transistor power dissipation does not exceed any of the manufacturer's ratings.
2. Generally, the greater the Q point drain current, the greater the voltage amplification of the transistor.
3. Approximately one-half of the power supply voltage should appear across the source-to-drain terminals of the transistor, approximately one-half of the power supply voltage should appear across the drain resistor, and a small voltage should appear across the source resistor (for a self-bias circuit). With the above division of the power supply voltages, a maximum signal swing can be anticipated.

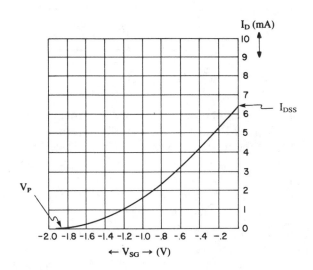

Fig. 3.2

Let us examine a fixed-bias JFET amplifier circuit. The name "*fixed* bias" is derived from the fact that the source-to-gate voltage, which controls both the drain current and source-to-drain voltage, is derived directly from a fixed power supply voltage. See Fig. 3.3. We note the symbols V_{DD} for the drain power supply and V_{GG} for the gate power supply. The double subscript, e.g., DD, has become a convention in electronics to designate power supply voltages connected between ground and a particular transistor terminal, i.e., ground-to-drain power supply is V_{DD}.

In the fixed-bias circuit of Fig. 3.3, we note a designation of v_{in} and v_{out}. For the purposes of this chapter we will not be interested in those ac (or signal) voltages. They are only shown in order to familiarize the student with the concept of an amplifier stage that has an impressed input signal and yields an amplified output signal.

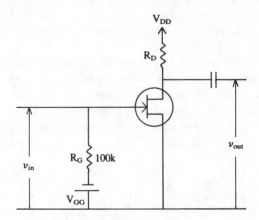

Fig. 3.3 Fixed Bias JFET Amplifier Circuit

We are more interested in developing design techniques that will enable us to specify the value of R_D and V_{GG}. R_G is usually chosen as a large value so that we do not "load" the previous amplifier circuit (or "stage" as it is sometimes called). The term "load" implies that the effect of one stage being connected to another will tend to reduce the voltage amplification of a previous stage. If R_G is chosen to be too large a value, leakage drain-to-gate currents could result in an unanticipated change in the operating point.

Example (3.1). Suppose that we wanted to design a JFET amplifier dc bias circuit to meet the following specifications: $V_{DD} = +20$ V, $V_{SDQ} = 1/2 \ V_{DD}$, and $I_{DQ} = 3.0$ mA (maximum current without exceeding maximum ratings).

Referring to the curves shown in Fig. 3.4, we note that a maximum power curve has been "sketched-in" and a dotted vertical line has been constructed through $V_{SDQ} = +10$ V. We arbitrarily choose the Q point on a $V_{SGQ} =$ constant curve ($= -0.6$ V) and read the value of $I_{DQ} = 3.0$ mA (see Fig. 3.5). Because we have 10 volts across the source to drain terminals, we also have 10 volts across R_D. Therefore, according to Ohm's law,

$$R_D = \frac{V_{RD}}{I_{DQ}} = \frac{10 \text{ V}}{3 \text{ mA}} = 3.33 \text{ k } \Omega \qquad (3.1)$$

We must choose a V_{GG} power supply equal to 0.6 V, since $|V_{GG}| = |V_{SG}|$. *Note*: There is no voltage drop across R_G since no current is flowing. Thus, we obtain the diagram of Fig. 3.6—and our design is complete!

OUTPUT
CURVES

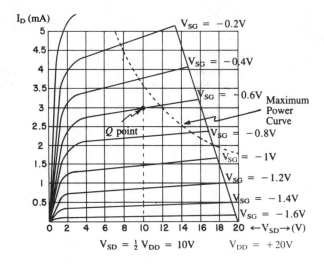

Fig. 3.4

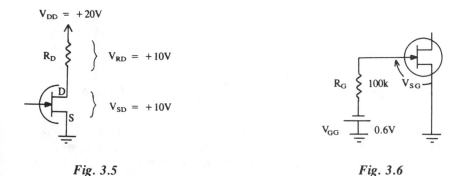

Fig. 3.5   Fig. 3.6

The previous design is unlikely to be used in a practical circuit because two power supplies are employed. If we could successfully design a dc bias circuit for a JFET by employing a single power supply V_{DD} and achieve essentially the same bias conditions, the expense of a second power supply can be avoided. A single power supply design is referred to as a self-bias design. The current flowing through the transistor I_D is employed to generate the required source-to-gate voltage at the Q point, V_{SGQ}. Thus, the transistor it*self* is developing its required source-to-gate voltage. Let us examine how this feat is accomplished. Referring to Fig. 3.7, we note that there is an extra resistor in series with the transistor

54

and drain resistor. Because this resistor is connected to the source terminal of the JFET, it is referred to as the source resistor, R_S. Referring to Fig. 3.8, we see that two "routes" may be taken in summing the total voltage, V_{SG}. We can start at the source and "jump" directly to the gate, but that would not enlighten us concerning the potential difference from source to gate. Another path to "travel" is from source to ground and then from ground to gate. If we "travel" this path, the first voltage summed is $-I_D R_S$, the voltage across the source resistor. If we travel from ground to gate, no additional voltage drops or rises are encountered because no current is flowing in R_G ($V_{RG} = I_G R_G$), where $I = 0$. We have "arrived" at the gate terminal. Thus,

$$V_{SG} = -I_D R_S + I_G R_G \tag{3.2}$$

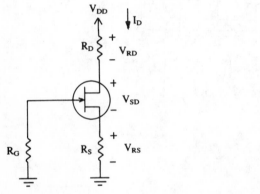

Fig. 3.7

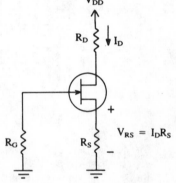

Fig. 3.8

Example (3.2). Referring to the output curves used in the previous example, we have the curves given in Fig. 3.9. If we place the Q point in the same location as in the previous example, we note that I_{DQ} is once again = 3.0 mA and $V_{SG} = -0.6$ V. Therefore,

$$R_S = \frac{|V_{SGQ}|}{I_D} = \frac{.6 \text{ V}}{3.0 \text{ mA}} = .2 \text{ k}$$

and R_D is computed as follows:

$$R_D = \frac{V_{RD}}{I_D} = \frac{?}{3.0 \text{ mA}} \tag{3.3}$$

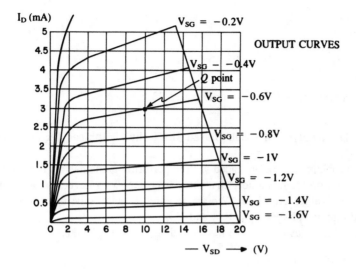

Fig. 3.9

However, V_{RD} is no longer 10 V as in the previous problem. Examining the circuit shown in Fig. 3.10, we see that

$$
\begin{aligned}
V_{RD} &= V_{DD} - V_{SD} - V_{RS} \\
&= 20 - 10 - 0.6 \\
&= 9.4 \text{ V}
\end{aligned}
\tag{3.4}
$$

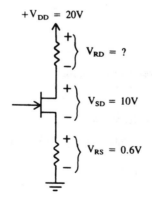

Fig. 3.10

Therefore,

$$R_D = \frac{V_{RD}}{I_{DQ}} = \frac{9.4 \text{ V}}{3 \text{ mA}} = 3.13 \text{ k}$$

(R_D in previous fixed bias example was 3.33 k.)

3.3.1 DC Analysis of a Self-Biased JFET (by Graphical Techniques)

The problem that we are faced with in performing an analysis is the opposite of a design problem, namely that the power supply voltage(s) along with the component values are given. We are required to determine the circuit current I_D, and circuit voltages V_{SD}, V_{RD} and V_{SG}.

For a fixed bias circuit, we have the diagram of Fig. 3.11.

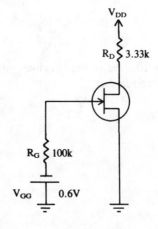

Fig. 3.11

Example (3.3). We can refer to the output curves and perform the following operations:

  1. Locate V_{DD} on the horizontal axis

  2. Draw a "dc load line" diagonally from V_{DD} to the left, up to I_y, where

$$I_y = \frac{V_{DD}}{R_D + R_S} \qquad \text{(for fixed bias, } R_S = 0\text{)}$$

  3. Locate the intersection of the dc load line and

$$V_{SGQ} = V_{GG} \hspace{5cm} \text{(for fixed bias)}$$

For our example,

$$V_{DD} = +20 \text{ V}, \; V_{GG} = -0.6 \text{ V}, \; R_D = 3.33 \text{ k}$$

Then

$$I_y = \frac{V_{DD}}{R_D + R_S} = \frac{20 \text{ V}}{3.33 \text{ k} + 0} = 6 \text{ mA} \hspace{3cm} (3.5)$$

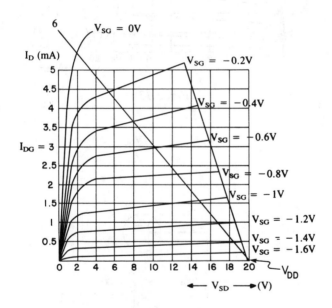

Fig. 3.12

Performing the necessary graphical operations, we have the results shown in Fig. 3.13. From our construction efforts, we read the values of $I_{DQ} = 3.0$ mA and $V_{SDQ} = 10$ V, which checks against our original design!

Example (3.4). To perform a graphical, dc analysis of a self-biased JFET, common source connected amplifier, we begin in the same manner as for a fixed-bias circuit. First, we construct the dc load line on the output curves. Referring to the circuit shown in Fig. 3.14, we plot the dc load line on the output curves. The slope of dc load line is

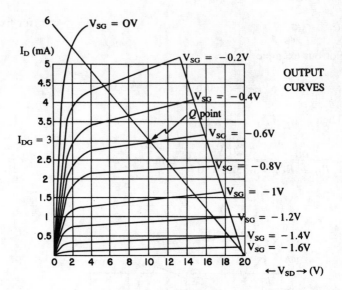

Fig. 3.13

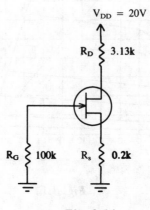

Fig. 3.14

$$\frac{1}{R_D + R_S} \tag{3.6}$$

Referring to the output curves (see Fig. 3.15), we have, according to Ohm's law:

$$I_y = \frac{V_{DD}}{R_S + R_D} = \frac{20}{.2k + 3.13k} = 6 \text{ mA}$$

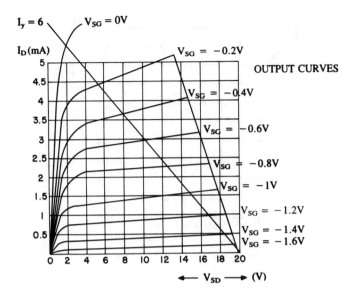

Fig. 3.15

The construction of the load line still has not helped to determine the operating point. However, if we refer to the transfer characteristic curve shown in Fig. 3.16, and plot R_S, where the line representing R_S crosses the transfer curve is the Q point (see Fig. 3.17).

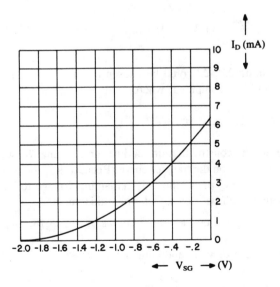

Fig. 3.16

60

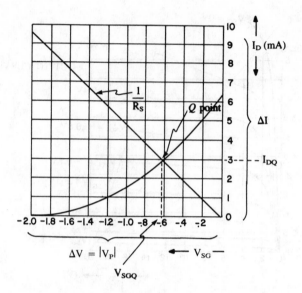

Fig. 3.17

To plot the line representing R_S, we may assume *any* ΔV and use Ohm's law to determine ΔI. Arbitrarily, ΔV was chosen equal to the pinch-off voltage, $|V_p|$. Thus,

$$\Delta I = \frac{\Delta V}{R_S} = \left|\frac{V_P}{R_S}\right| \quad \text{in this instance, } \Delta I = \left|\frac{-2}{.2}\right| = 10 \text{ mA}$$

Instead of assuming ΔV, ΔI could be chosen (such as $|I_{DSS}|$ and the corresponding ΔV could be computed, where

$$\Delta V = \Delta I \, R_S \quad \text{e.g., } \Delta V = I_{DSS} R_S$$

After completing the construction on the line representing R_S, we may read the value of I_{DQ} and V_{SDQ} on the transfer curve. Because V_{RS} is equal to the source-to-gate voltage at the Q point, the only remaining voltage to be determined is V_{RD}. Two methods can be used to determine this value:

1. Locate the Q point on the output curves of Fig. 3.18.

2. Read the value of V_{SDQ} and compute V_{RD}:

$$V_{RD} = V_{DD} - V_{SDQ} - V_{RS} \tag{3.7}$$

Or

3. $V_{RD} = I_{DQ}R_D$ and $V_{RS} = I_{DQ}R_S$ (3.8)

4. $V_{SDQ} = V_{DD} - V_{RD} - V_{RS}$ (3.9)

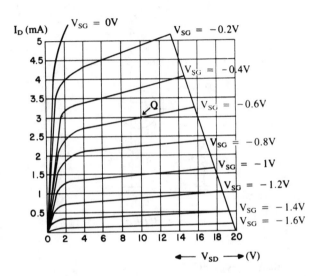

Fig. 3.18(a)

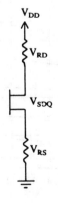

Fig. 3.18(b)

Example (3.5). Let us use an example to demonstrate the required steps to analyze the dc operating point of a self-biased JFET amplifier stage, given the following amplifier stage with characteristic curves of the transistor as shown in Fig. 3.19(a,b).

First, we construct the dc load line:

$$I_y = \frac{V_{DD}}{R_D + R_S} = \frac{20 \text{ V}}{3.13 \text{ k} + 0.2 \text{ k}} = 6 \text{ mA}$$

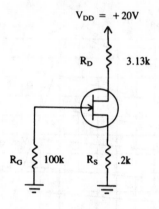

Fig. 3.19(a)

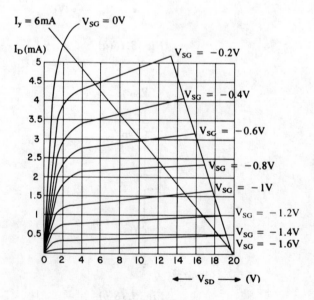

Fig. 3.19(b)

Second, we use the transfer characteristic curve(s) to locate the operating point using Fig. 3.20.

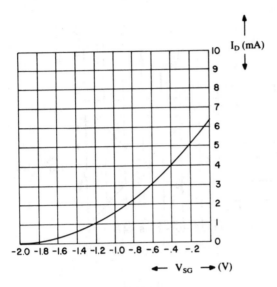

Fig. 3.20

To plot the line representing the value of R_S, we assume ΔI and determine the corresponding ΔV (by use of Ohm's law). Let

$$\Delta I = | I_{DSS} |$$
$$\Delta E = \Delta I R_S = I_{DSS} R_S$$
$$= 6.5(.2) = 1.3 \text{ V}$$

We see that the Q point is now determined, but in order to completely define the circuit, we must still determine two voltages, V_{RD} and V_{SDQ}.

Two approaches can be taken.

1. From the output curves:

$$V_{RD} = I_D R_D$$
$$= 3 \text{ mA } (3.13 \text{ k})$$
$$= 9.39 \text{ V}$$

2. Knowing I_{DQ}, we derive

$$V_{RD} = I_D R_D = 3 \text{ mA } (3.13 \text{ k})$$
$$= 9.39 \text{ V}$$

$$V_{RS} = |V_{SGQ}| = 0.6 \text{ V}$$
$$V_{SD} = V_{DD} - V_{RD} - V_{RS} = 20 - 9.4 - 0.6 = 10 \text{ V}$$

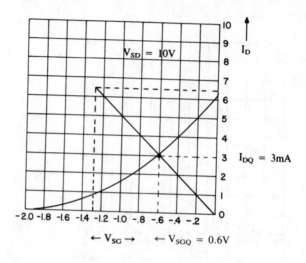

Fig. 3.21

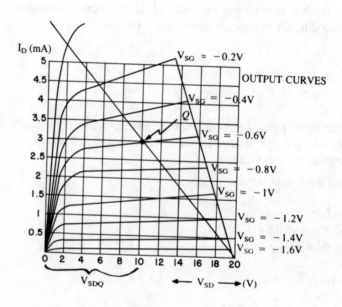

Fig. 3.22

At this point, it is appropriate to consider some of the practical considerations in designing a JFET common source connected amplifier. Examining the JFET transfer characteristic curve of Fig. 3.23, we note that if we want to use a large drain current, which will tend to maximize the output signal power and decrease the stage output impedance, the source-to-gate voltage becomes an insignificant value. Therefore, the source-to-gate voltage is the limiting factor of the allowable maximum amplitude of the input voltage.

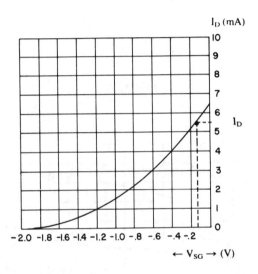

Fig. 3.23

Another consideration in our dc design is the stability of the Q point. As power levels are increased, heating of the transistor will cause an increase in the leakage current. Leakage currents require that the designer employ lower values for the gate resistor. One method of stabilizing the dc operating point is to increase the value of the source resistor, thereby increasing feedback of a compensating voltage. Increasing the value of the source resistor will tend to increase the source-to-gate voltage, which, in turn, tends to reduce the drain current. For a particular design application, reduced values of drain currents may yield undesirable circuit performance.

A method of overcoming the decrease in drain current resulting from an increase in the source resistance value is to bias the gate positive with respect to ground (*not* with respect to the source—for JFETs the source to gate junction must remain reverse biased). The circuit shown in Fig. 3.24 accomplishes this goal.

66

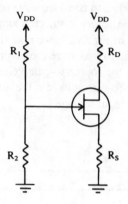

Fig. 3.24

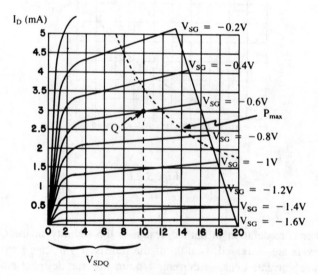

Fig. 3.25

In order to design to a particular Q point, we select the desired Q point drain current I_D, the source to drain voltage, and locate this point on the output curves (see Fig. 3.25). This selection automatically "nails down" the value of V_{SGQ} (for convenience the Q point can be selected on a $V_{SG} = $ const. curve). The Q point is then transferred to the transfer characteristic curve (see Fig. 3.26).

If we have chosen a value for R_S, to ensure a desired amount of feedback, we then plot this on the transfer characteristic curve as follows:

$$\Delta V = \Delta I R_S$$

or

$$\Delta V = I_{DQ} R_S$$

We thereby obtain the curves of Fig. 3.27.

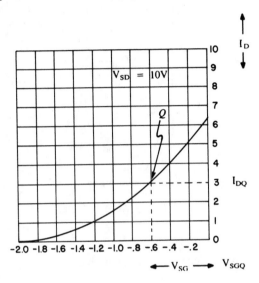

Fig. 3.26

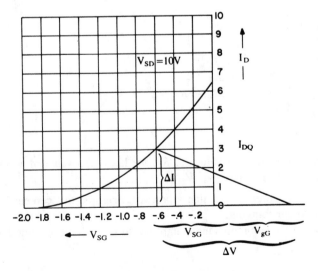

Fig. 3.27

Once the ground to gate voltage has been determined, we can establish resistor values for the voltage divider that are required in the gate circuit (see Fig. 3.28).

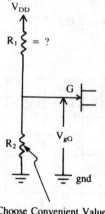

Choose Convenient Value

Fig. 3.28

By use of the voltage divider rule, we know that

$$V_{gG} = V_{DD} \frac{R_2}{R_1 + R_2}$$

solving for R_1:

$$V_{gG} (R_1 + R_2) = V_{DD}R_2$$
$$V_{gG}R_1 + V_{gG}R_2 = V_{DD}R_2$$
$$V_{gG}R_1 = V_{DD}R_2 - V_{gG}R_2 = R_2 (V_{DD} - V_{gG})$$
$$R_1 = R_2 \left(\frac{V_{DD} - V_{gG}}{V_{gG}} \right) \tag{3.10}$$

If we arbitrarily assume a value for R_2, we know the value of V_{DD} and have previously determined the value V_{gG}.

Example (3.6). Let us examine a practical example to ''cement'' the above concepts into our thinking.

Design a JFET common source amplifier to meet the following design characteristics:

$$I_{DQ} = 3 \text{ mA}, \qquad V_{SDQ} = 10 \text{ V}, \qquad R_S = 500 \text{ }\Omega$$

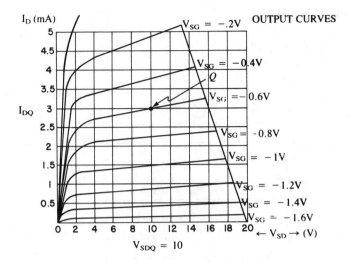

Fig. 3.29

Referring to the output curves given in Fig. 3.29, we locate the Q point. We then locate the Q point on the transfer characteristic curve of Fig. 3.30.

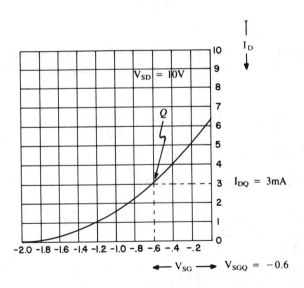

Fig. 3.30

By use of Ohm's law:

$$\Delta V = \Delta I\, R_3 = I_{DQ} R_S$$
$$\Delta V = 3(.5) = 1.5 \text{ V}$$

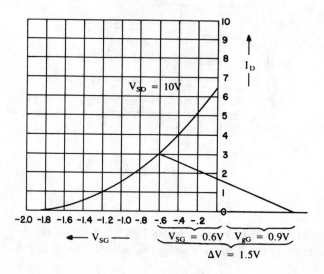

Fig. 3.31

We see from Fig. 3.31 that $V_{gG} = 0.9$ V. To determine a value for R_1, we assume a value of R_2 (conveniently, 100 k) and use the previously determined formula (3.10):

$$R_1 = R_2 \frac{(V_{DD} - V_{gG})}{V_{gG}} = 100 \text{ k} \frac{(20 \text{ V} - .9 \text{ V})}{.9 \text{ V}}$$
$$R_1 = 2.122 \text{ M}\Omega$$

All that remains is to determine the value of R_D:

$$V_{RD} = V_{DD} - V_{SDQ} - V_{RS} = V_{DD} - V_{SDQ} - I_D R_S$$
$$= 20 - 10 - 3(.5)$$
$$V_{RD} = 8.5 \text{ V}$$
$$R_D = \frac{V_{RD}}{I_{DQ}} = \frac{8.5 \text{ V}}{3 \text{ mA}} = 2.83 \text{ k}$$

Example (3.7). In order to *analyze* a JFET common source amplifier with a voltage divider in the gate circuit by graphical techniques, we reverse the procedure that we have developed for design. Given the circuit shown in Fig. 3.32, determine the circuit voltage and current.

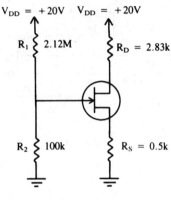

Fig. 3.32

To determine the Q point, we use the transfer characteristic curve, but first we must compute the ground to gate voltage by use of the voltage divider rule:

$$V_{gG} = V_{DD} \frac{R_2}{R_1 + R_2} = 20 \frac{.1M}{2.12M + .1M} = .9 \text{ V}$$

To plot the value of R_S on the transfer characteristic curve, we assume ΔV and use Ohm's law to compute ΔI. Let $\Delta V = 2$ volts.

$$\Delta I = \frac{\Delta V}{R_S} = \frac{2 \text{ V}}{0.5 \text{ k}} = 4 \text{ mA}$$

See Fig. 3.33. Thus, we read $I_{DQ} = 3.0$ mA, $V_{SGQ} = -0.6$ V. From this we calculate:

$$V_{RD} = I_{DQ} R_D = 3 \text{ mA} (2.83 \text{ k}) = 8.5 \text{ V}$$
$$V_{RS} = I_{DQ} R_S = 3 \text{ mA} (0.5 \text{ k}) = 1.5 \text{ V}$$
$$V_{SDQ} = V_{DD} - V_{RD} - V_{RS} = 20 - 8.5 - 1.5 = 10 \text{ V}$$

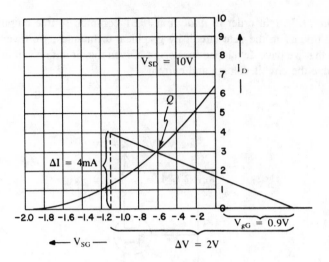

Fig. 3.33

3.3 ALGEBRAIC DESIGN AND ANALYSIS OF DC BIAS JFET CIRCUITS

All of our previous work in this chapter assumed that the characteristic curves were available for the particular transistor which we intended to use, either from manufacturer's specifications or by photographing them from a curve tracer. A complete design and analysis procedure may be developed whereby the only data required are the transistor short circuit, saturation current in the common source configuration I_{DSS}, and the pinch-off voltage V_P. These parameters may be determined if necessary with only a dc power supply and a dc voltmeter.

We can exploit the fact that the transfer characteristic curve for any JFET may be constructed by using the expression:

$$I_D = I_{DSS} \left(1 - \frac{V_{SG}}{V_P} \right)^2$$

Example (3.8). Suppose we wished to design a fixed bias, common source, JFET amplifier (dc design). Assume that we are given $V_{DD} = +20$ V and $I_{DQ} = 3$ mA, and that we must design for a maximum output signal voltage swing, i.e., $V_{SDQ} = V_{RD}$. Also $V_P = -1.87$ V and $I_{DSS} = 6.5$ mA.

Rewriting:

$$I_D = I_{DSS}\left(1 - \frac{V_{SG}}{V_P}\right)^2$$

$$\frac{I_D}{I_{DSS}} = \left(1 - \frac{V_{SG}}{V_P}\right)^2 \rightarrow \sqrt{\frac{I_D}{I_{DSS}}} = 1 - \frac{V_{SG}}{V_P}$$

$$\frac{V_{SG}}{V_P} = 1 - \sqrt{\frac{I_D}{I_{DSS}}} \rightarrow V_{SG} = V_P\left(1 - \sqrt{\frac{I_D}{I_{DSS}}}\right) \tag{3.11}$$

Examining our fixed bias circuit shown in Fig. 3.34, we can determine R_D:

$$R_D = \frac{V_{RD}}{I_{DQ}} = \frac{10 \text{ V}}{3 \text{ mA}} = 3.33 \text{ k}$$

$$V_{GG} = |V_{SGQ}| = V_P\left(1 - \sqrt{\frac{I_D}{I_{DSS}}}\right)$$

$$V_{GG} = 1.87\left(1 - \sqrt{\frac{3}{6.5}}\right)$$

$$V_{GG} = .6 \text{ V}$$

and our design is complete!

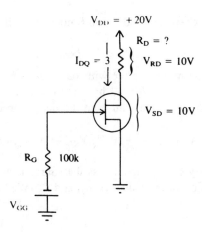

Fig. 3.34

74

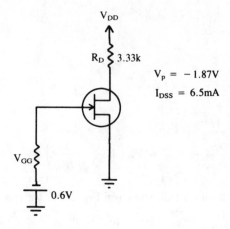

Fig. 3.35

Example (3.9). To perform an *analysis*, we reverse the process. For the circuit shown in Fig. 3.35, analyze the circuit voltages and current.

Since $V_{SDQ} = V_{GG} = 0.6$ V, we can use the formula

$$I_D = I_{DSS} \left(1 - \frac{V_{SG}}{V_P}\right)^2 = 6.5\left(1 - \frac{-.6}{-1.87}\right)^2 = 3 \text{ mA}$$

$$V_{RD} = I_D R_D = 3 \text{ mA } (3.33 \text{ k}) = 10 \text{ V}$$

3.3.1 Design of a Self-Biased Common Source Amplifier (by Analytical Techniques)

Example (3.10). Continuing further with our *algebraic dc bias design and analysis of common source amplifiers*, the next topic is the *design of self-biased common source amplifier*. Let us illustrate the design procedures with an example.

Given $I_{DQ} = 3$ mA, $V_{DD} = +20$ V, $V_{RD} = V_{SDQ}$, determine R_S and R_D (see Fig. 3.36).

If we do not specify the voltage across the source resistor, then the design entails the determination of the values of R_D and R_S. We know that

$$|V_{SGQ}| = V_{RS} = I_D R_S \quad \text{and} \quad |V_{SGQ}| = |V_P| \left(1 - \sqrt{\frac{I_D}{I_{DSS}}}\right)$$

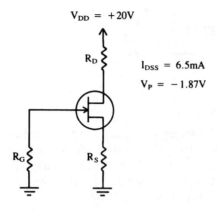

$V_{DD} = +20V$

R_D

$I_{DSS} = 6.5mA$

$V_P = -1.87V$

R_G R_S

Fig. 3.36

equating:

$$I_D R_S = |V_P| \left(1 - \sqrt{\frac{I_D}{I_{DSS}}}\right)$$

therefore,

$$R_S = \frac{|V_P|}{I_{DQ}} \left(1 - \sqrt{\frac{I_D}{I_{DSS}}}\right) = \frac{1.87}{3}\left(1 - \sqrt{\frac{3}{6.5}}\right) \qquad (3.12)$$

$$R_S = .2 \text{ k}$$

Employing Kirchoff's voltage law:

$V_{RS} + V_{SDQ} + V_{RD} = V_{DD}$ but: $V_{SDQ} = V_{RD}$
$V_{RS} + V_{RD} + V_{RD} = V_{DD}$
$V_{RS} + 2 V_{RD} = V_{DD}$
$$V_{RD} = \frac{V_{DD} - V_{RS}}{2}$$

$$V_{RD} = \frac{V_{DD} - |V_P|\left(1 - \sqrt{\frac{I_D}{I_{DSS}}}\right)}{2}$$

$$R_D = \frac{V_{RD}}{I_D} = \frac{V_{DD} - |V_P|\left(1 - \sqrt{\frac{I_D}{I_{DSS}}}\right)}{2 I_D} \qquad (3.13)$$

$$R_D = \frac{20 - 1.87 \left(1 - \sqrt{\dfrac{3}{6.5}}\right)}{2\,(3)}$$

$$R_D = 3.23 \text{ k}$$

and our design is complete!

Example (3.11).   To *design* the slightly more involved circuit for a *self-biased, common source amplifier with a voltage divider in the gate circuit,* we must assume a resistor value for R_S or an arbitrary voltage value across R_S. In the following design example a value for V_{RS} was chosen.

Given $I_{DQ} = 3$ mA, $I_{DSS} = 6.5$ mA, $V_P = -1.87$ V, $V_{RS} = 1.5$ V, $V_{RD} = V_{SDQ}$, determine R_D, R_S, and R_1 (see Fig. 3.37).

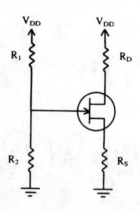

Fig. 3.37

According to Kirchoff's voltage law:

$$V_{RS} + V_{SDQ} + V_{RD} = V_{DD}$$

since $V_{RD} = V_{SDQ}$

$$V_{RS} + V_{RD} + V_{RD} = V_{DD}$$
$$V_{RS} + 2V_{RD} = V_{DD}$$
$$2V_{RD} = V_{DD} - V_{RS}$$
$$V_{RD} = \frac{V_{DD} - V_{RS}}{2}$$
$$R_D = \frac{V_{RD}}{I_D} = \frac{V_{DD} - V_{RS}}{2\,I_D} = \frac{20 - 1.5}{2(3)} = 3.08 \text{ k} \qquad (3.14)$$
$$R_S = \frac{V_{RS}}{I_D} = \frac{1.5}{3} = .5 \text{ k}$$

Two down, two to go!

Let us try to make life simple, let $R_2 = 100$ k and let us find R_1.

From our previous work, we know that

$$V_{SG} = V_P\left(1 - \sqrt{\frac{I_D}{I_{DSS}}}\right) \tag{3.11}$$

Examining the circuit shown below (Fig. 3.38) and using Kirchoff's voltage law:

$$-V_{RS} + V_{gG} = V_{SG} = V_P\left(1 - \sqrt{\frac{I_D}{I_{DSS}}}\right)$$

therefore,

$$V_{gG} = V_P\left(1 - \frac{I_D}{I_{DSS}}\right) + V_{RS} \tag{3.15}$$

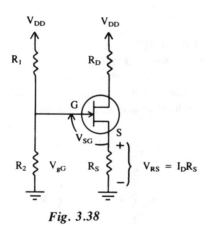

Fig. 3.38

By the voltage divider rule:

$$V_{gG} = V_{DD}\frac{R_2}{R_1 + R_2}$$

and from our previous manipulations:

$$R_1 = R_2\left(\frac{V_{DD} - V_{gG}}{V_{gG}}\right) \tag{3.10}$$

Rewriting:

$$R_1 = R_2\left(\frac{V_{DD}}{V_{SG}} - 1\right)$$

Substituting:

$$R_1 = R_2\left[\frac{V_{DD}}{V_P\left(1 - \sqrt{\frac{I_D}{I_{DSS}}}\right) + V_{RS}} - 1\right] \tag{3.16}$$

For our example:

$$R_1 = 100\text{ k}\left[\frac{20}{-1.87\left(1 - \sqrt{\frac{3}{6.5}}\right) + 1.5} - 1\right]$$

$$R_1 = 2.12\text{ M}\Omega$$

and we have completed our design!

To *analyze* a *common source amplifier*, if we have developed formulas that enable the analysis of the more complex circuit in the case, where there is a voltage divider in the gate circuit, it will also be possible to analyze the circuit for the case where a voltage divider does not exist.

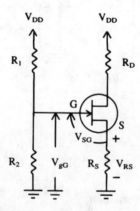

Fig. 3.39 Derivation of the Formula for Drain Current in a Self-Biased FET Common Source Amplifier (with Gate Circuit Voltage Divider)

According to Kirchoff's voltage law:

$$V_{SG} = -I_D R_S + V_{gG}$$

From Example (3.12):

$$V_{SG} = V_P \left(1 - \sqrt{\frac{I_D}{I_{DSS}}}\right)$$

Equating:

$$-I_D R_S + V_{gG} = V_P \left(1 - \sqrt{\frac{I_D}{I_{DSS}}}\right) = V_P - V_P \sqrt{\frac{I_D}{I_{DSS}}}$$

Therefore,

$$I_D R_S + V_P - V_{gG} = V_P \sqrt{\frac{I_D}{I_{DSS}}}$$

Squaring

$$I_D^2 R_S^2 + 2I_D R_S (V_P - V_{gG})^2 + (V_P - V_{gG})^2 = V_P^2 \left(\frac{I_D}{I_{DSS}}\right)$$

$$I_D^2 R_S^2 + I_D \left[2R_S(V_P - V_{gG}) - \frac{V_P^2}{I_{DSS}}\right] + (V_P - V_{gG})^2 = 0$$

$$I_D = \frac{1}{2R_S^2}\left(\frac{V_P^2}{I_{DSS}} - 2R_S(V_P - V_{gG})\right)$$

$$\pm \frac{1}{2R_S^2} \frac{\sqrt{4R_S^2(V_P - V_{gG})^2 - \dfrac{4R_S(V_P - V_{gG})V_P^2}{I_{DSS}}}}{\dfrac{V_P^4}{I_{DSS}} - 4R_S^2(V_P - V_{gG})}$$

$$= \frac{\dfrac{V_P^2}{I_{DSS}} - 2R_S(V_P - V_{gG}) \pm V_P \sqrt{\dfrac{V_P^2}{I_{DSS}} - 4\dfrac{R_S(V_P - V_{gG})}{I_{DSS}}}}{2R_S^2}$$

$$= \frac{\dfrac{V_P^2}{I_{DSS}} - \dfrac{2R_S(V_P - V_{gG})I_{DSS}}{I_{DSS}} \pm \dfrac{V_P}{I_{DSS}}\sqrt{V_P^2 - 4I_{DSS}R_S(V_P - V_{gG})}}{2R_S^2}$$

$$I_D = \frac{V_P{}^2 - 2I_{DSS}R_S(V_P - V_{gG}) \pm V_P\sqrt{V_P{}^2 - 2\,[2I_{DSS}R_S(V_P - V_{gG})]}}{2I_{DSS}R_S{}^2}$$

$$(3.17)$$

Formula (3.17) is well suited to computer or programmable calculator usage. Let us use formula (3.17) to analyze two of our previous design examples.

Example (3.13). Analyze the JFET circuit shown in Fig. 3.40.

Given $R_D = 3.23\,\text{k}$, $R_S = 0.2\,\text{k}$, $I_{DSS} = 6.5\,\text{mA}$, $V_P = -1.87\,\text{V}$, determine: I_{DQ}, V_{RD}, V_{RS}, V_{SDQ}.

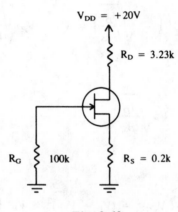

Fig. 3.40

Employing formula (3.17):

$$I_D = \frac{V_P{}^2 - 2I_{DSS}R_S(V_P - V_{gG})}{2I_{DSS}R_S{}^2}$$

$$+ \frac{V_P\sqrt{V_P{}^2 - 2[2I_{DSS}R_S(V_P - V_{gG})]}}{2I_{DSS}R_S{}^2}$$

$$= \frac{(-1.87)^2 - 2(6.5)(.2)(-1.87 - 0)}{2(6.5)(.2)^2}$$

$$+ \frac{(-1.87)\sqrt{(-1.87)^2 - 2[2(6.5)(.2)(-1.87 - 0)]}}{2(6.5)(.2)^2}$$

$$I_D = 3\,\text{mA}$$
$$V_{RD} = I_D R_D = 3\,\text{mA}\,(3.23\,\text{k}) = 9.69\,\text{V}$$

$$V_{RS} = I_D R_S = 3 \text{ mA } (.2 \text{ k}) = .6 \text{ V}$$

$$V_{SDQ} = V_{DD} - V_{RD} - V_{RS} = 20 - 9.69 - .6 = 9.71 \text{ V}$$

The result checks against design!

Example (3.14).   Analyze the JFET circuit shown in Fig. 3.41.

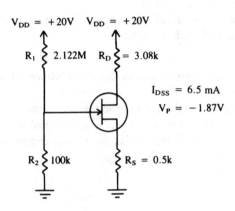

Fig. 3.41

Given $V_{DD} = +20$ V, $R_1 = 2.122$ M, $R_2 = 100$ k, $R_D = 3.08$ k, $R_S = 0.5$ k, determine V_{gG}, I_D, V_{RD}, V_{RS}, V_{SDQ}:

$$V_{gG} = V_{DD} \frac{R_2}{R_1 + R_2}$$

$$= 20 \left(\frac{0.1 \text{ M}}{2.122 \text{ M} + 0.1 \text{ M}} \right)$$

$$V_{gG} = 0.9 \text{ V}$$

Employing formula (3.17):

$$I_D = \frac{V_P^2 - 2I_{DSS}R_S(V_P - V_{gG})}{2I_{DSS}R_S^2}$$

$$+ \frac{V_P \sqrt{V_P^2 - 2[2I_{DSS}R_S(V_P - V_{gG})]}}{2I_{DSS}R_S^2}$$

$$= \frac{(-1.87)^2 - 2(6.5)(.5)(-1.87 - .9)}{2(6.5)(.5)^2}$$

$$+ \frac{(-1.87)\sqrt{(-1.87)^2 - 2[(6.5)(.5)(-1.87 - .9)}}{2(6.5)(.5)^2}$$

$$I_D = 3 \text{ mA}$$
$$V_{RD} = I_D R_D = 3 \text{ mA}(3.08\text{k}) = 9.24 \text{ V}$$
$$V_{RS} = I_D R_S = 3 \text{ mA}(.5\text{k}) = 1.5 \text{ V}$$
$$V_{SDQ} = V_{DD} - V_{RD} - V_{RS} = 20 - 9.24 - 1.5 = 9.26 \text{ V}$$
$$V_{SGQ} = -V_{RS} + V_{gG} = -1.5 + .9 = -.6 \text{ V}$$

The result checks against design!

This completes the graphical and analytical design and analysis of the common source connected FET dc bias circuit.

3.3.2 Design of a Common Drain (Source Follower) Circuit

Another frequently used circuit is the common drain connection. This configuration is more often referred to as the *source follower*. We note from the circuit shown in Fig. 3.42 that the drain to V_{DD} connection does not have a resistor located in the drain to V_{DD} "leg."

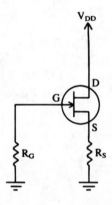

Fig. 3.42

Let us develop dc bias equations for the design of a source follower. In order to develop the appropriate formulas, let us define the percentage of the power supply voltage V_{DD} that appears across R_S as $\gamma \times 100$, i.e.,

$$\gamma = \frac{V_{RS}}{V_{DD}}$$

In most linear amplifier applications, $\gamma = 0.5$.

Example (3.15). For the circuit shown (Fig. 3.43), determine the value of R_S to achieve the given currents and voltages.

Given $I_{DQ} = 3$ mA, $V_{SDQ} = 12$ V, $V_{RS} = 8$ V, $I_{DSS} = 6.5$ mA, $V_P = -1.87$ V, determine R_S.

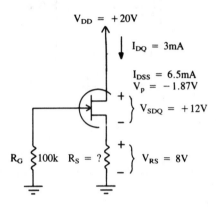

Fig. 3.43

We might be tempted to use Ohm's law as follows:

$$R_S = \frac{V_{RS}}{I_{DQ}} = \frac{8 \text{ V}}{3 \text{ mA}} = 2.67 \text{ k}\Omega$$

The fallacy in this approach is that $|V_{SGQ}| = V_{RS}$, and hence the source-to-gate voltage would exceed the pinch-off voltage and no current would flow. Thus, we see that, in most instances, when we design a source follower, a voltage divider is required in the gate circuit. Let us start over. See Fig. 3.44.

Given $I_{DQ} = 3$ mA, $V_{SDQ} = 12$ V, $V_{RS} = 8$ V, $I_{DSS} = 6.5$ mA, $V_P = -1.87$ V, $R_2 = 100$ k, determine R_S and R_1.

The above calculation for R_S is still valid:

$$R_S = 2.67 \text{ k}$$

Referring to formula (3.15):

$$V_{gG} = V_P \left(1 - \sqrt{\frac{I_D}{I_{DSS}}}\right) + V_{RS}$$

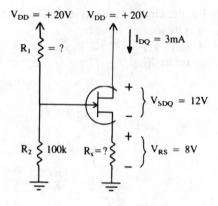

Fig. 3.44

The voltage divider rule when applied to the gate circuit yields:

$$V_{gG} = V_{DD} \frac{R_2}{R_1 + R_2}$$

and by manipulating we previously developed the following:

$$R_1 = R_2 \left(\frac{V_{DD} - V_{gG}}{V_{gG}} \right)$$

$$= R_2 \left(\frac{V_{DD}}{V_{gG}} - 1 \right)$$

$$(3.18)$$

Combining the formulas, we obtain

$$R_1 = \left[R_2 \frac{V_{DD}}{V_P \left(1 - \sqrt{\frac{I_D}{I_{DSS}}} \right) + V_{RS}} - 1 \right]$$

$$(3.19)$$

Substituting into the formula, we have

$$R_1 = 100 \text{ k} \left[\frac{20}{-1.87 \left(1 - \sqrt{\frac{3}{6.5}} \right) + 8} - 1 \right] = 170 \text{ k}$$

It is important to point out that for maximum sinusoidal signal swing, the voltages across the transistor and source resistor should be equal. In the above example, different voltages were chosen to demonstrate that the design formulas do not require equal values for V_{RS} and V_{SDQ}.

3.3.3 Analysis Formulas

Now that we have developed the source follower dc bias design formulas, next we need the *analysis formulas*. Fortunately, we are not required to "reinvent the wheel," since the formulas for the common source circuit are appropriate with the exception that $R_D = 0 \ \Omega$.

Let us check our last design example.

Example (3.16). Analyze the circuit shown in Fig. 3.45.

Given $R_1 = 170$ k, $R_2 = 100$ k, $R_S = 2.67$ k, $I_{DSS} = 6.5$ mA, $V_P = -1.87$ V, determine V_{gG}, I_{DQ}, V_{RS}.

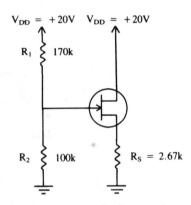

Fig. 3.45

$$V_{gG} = V_{DD} \frac{R_2}{R_1 + R_2} = 20 \frac{100 \text{ k}}{170 \text{ k} + 100 \text{ k}} = 7.4 \text{ V}$$

$$I_D = \frac{V_P^2 - 2I_{DSS}R_S(V_P - V_{gG})}{2I_{DSS}R_S^2}$$

$$+ \frac{V_P \sqrt{V_P^2 - 2[(2I_{DSS}R_S(V_P - V_{gG})]}}{2I_{DSS}R_S^2}$$

$$I_D = \frac{(-1.87)^2 - 2(6.5)(2.67)(-1.87 - 7.4)}{2(6.5)(2.67)^2}$$

$$+ \frac{(-1.87)\sqrt{(-1.87)^2 - 2[(2(6.5)(2.67)(-1.87 - 7.4)]}}{2(6.5)(2.67)^2}$$

$I_D = 3$ mA

$V_{RS} = I_D R_S = 3$ mA $(2.67$ k$) = 8$ V

The results check out against our design.

Note: A graphical solution would also be appropriate!

It is important to note that in the derivations and examples throughout this chapter, reference was made to JFETs. The same design and analysis techniques are also appropriate for depletion-mode MOSFETs. The depletion-mode MOSFET has an added advantage in that the gate may be driven positive with respect to the source without drawing current. Thus, a very simple bias circuit for an *N*-channel MOSFET might appear as shown in Fig. 3.46(a,b).

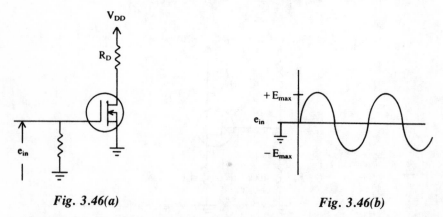

Fig. 3.46(a)   Fig. 3.46(b)

We note that a source resistor or gate circuit resistor voltage divider is not required. For the above circuit, the source-to-gate voltage is zero volts. Thus, an ac signal applied between ground and gate would drive the gate both positive and negative with respect to the source.

3.4 DESIGN AND ANALYSIS TECHNIQUES FOR THE DC BIAS OF ENHANCEMENT-MODE MOSFET COMMON-SOURCE AMPLIFIER CIRCUITS

Up to this point in the chapter, we have developed techniques for designing and analyzing dc bias circuits for JFETs and depletion-mode MOSFETs. There are some unique circumstances when an enhancement-mode MOSFET (EMOS-FET) is suitable for use in amplifier applications.

Recalling from Chapter 1, the EMOSFET will not conduct any drain current when the source-to-gate voltage is zero volts. Therefore, the gate must be connected to a positive voltage (through a resistor) for an *N*-channel EMOSFET. Two typical techniques for accomplishing this bias requirement are depicted by the schematics shown in Fig. 3.47(a,b).

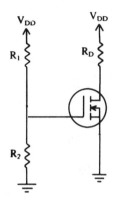

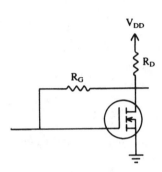

Fig. 3.47(a)   *Fig. 3.47(b)*

The advantage of utilizing the circuit shown in Fig. 3.47(b) is that dc operating point stabilization is achieved. To illustrate this, suppose that a small increase in drain current occurs. Perhaps this increase is due to the replacement of a defective transistor by a new unit with different characteristics. Since greater drain current is flowing, a greater voltage drop across R_D will occur, thus lowering the potential at drain and gate (as measured with respect to ground). The reduced gate potential will, in turn, tend to reduce the drain current, thus stabilizing the operating point. A potential disadvantage of the connection shown in Fig. 3.47(b) is the negative feedback of the signal voltage. Negative feedback, as we will learn in subsequent chapters (Ch. 8, 9, 16), will reduce the voltage amplification of an input signal. The circuit shown in Fig. 3.48 alleviates this potential problem.

The large gate resistors effectively isolate the input and output signals from being shorted to ground by the bypass capacitor, C_B. However, C_B effectively shorts to ground the small amount of the output signal that would appear at the left-most end of R_{G2}. The dc voltage that appears at the drain is still available at the gate for bias purposes. It may be unnecessary to employ the bypass capacitor if the source impedance is low compared with the gate bias resistor because, according to the voltage divider rule, a negligible feedback signal would appear at the gate.

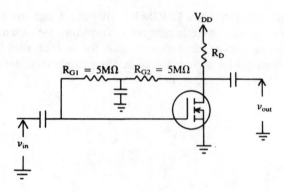

Fig. 3.48

Using *graphical techniques*, let us *design* an *N-channel EMOSFET dc bias circuit*.

Example (3.17). For the circuit shown in Fig. 3.49, design for $V_{SD} = \frac{1}{2} V_{DD}$.

Given $R_G = 5$ MΩ, $V_{DD} = +14$ V, $V_{SD} = \frac{1}{2} V_{DD} = +7$ V, determine R_D.

By projecting a line vertically upward from the point $V_{SDQ} = +7$ V until it intersects the transfer characteristic curve, we have located the Q point. By projecting horizontally to the left, we can determine the value of I_{DQ} (for this example, $I_{DQ} = 5.4$ mA), but this operation is not necessary for design purposes. To determine the value of R_D, we construct the dc load line by connecting a straight line from V_{DD} on the horizontal axis through the Q point. The slope of the load line represents the value of R_D. Maximum accuracy in determining the value of R_D can be obtained by employing the largest possible triangle (for the purpose of determining the slope of the hypotenuse), i.e.,

$$R_D = \frac{V_{DD}}{I_y} = \frac{14 \text{ V}}{10.8 \text{ mA}} = 1.3 \text{ k}\Omega$$

The design is complete!

To perform *graphical analysis* of an *EMOSFET common-source dc bias circuit*, the above procedure is reversed. First, the load line is plotted on the transfer characteristic curve, and then the Q point values are read from the curves. If only a single resistor is used in the gate circuit, then the source-to-gate voltage and source-to-drain voltage are equal, i.e., $V_{SGQ} = V_{SDQ}$.

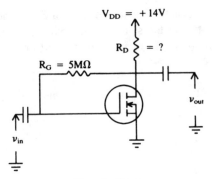

Fig. 3.49(a)

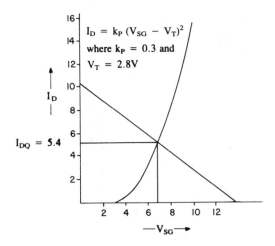

Fig. 3.49(b)

3.4.1 Algebraic Design Techniques for the DC Bias of an EMOSFET Common-Source Amplifier

In chapter 1, we noted that the algebraic relationship between drain current and source-to-gate voltage for a EMOSFET was given by:

$$I_D = k_P (V_{SG} - V_T)^2, \qquad (V_{SG} > V_T) \qquad (3.20)$$

where k_P and V_T are functions of the transistor manufacturing process. We also know from previous design investigations that for maximum signal swing,

$V_{SD} = \frac{1}{2} V_{DD}$. If we bias the N-channel EMOSFET as shown in the schematic of Fig. 3.50, we note that the source-to-gate voltage is equal to the source to drain voltage ($V_{SG} = V_{SD}$) since there is little or no voltage drop in R_G ($I_G = 0$). Therefore,

$$V_{SG} = V_{SD} = 1/2V_{DD} \qquad \text{(for maximum signal swing)}$$

or

$$I_D = k_P \left(\frac{V_{DD}}{2} - V_T \right)^2$$

thus,

$$R_D = \frac{V_{RD}}{I_D} = \frac{1/2V_{DD}}{k_P \left(\dfrac{V_{DD}}{2} - V_T \right)^2}$$

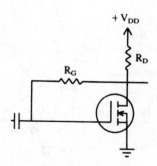

Fig. 3.50

Example (3.18). For the circuit shown in Example (3.17), design for maximum output signal swing.

Given $R_G = 5\ \text{M}\Omega$, $k_P = 0.3$, $V_T = 2.8\ \text{V}$, $V_{DD} = +14\ \text{V}$, determine R_D.

$$R_D = \frac{V_{DD}}{2k_P \left(\dfrac{V_{DD}}{2} - V_T \right)^2} \tag{3.21}$$

$$R_D = \frac{14}{2(0.3) \left(\dfrac{14}{2} - 2.8 \right)^2} = 1.32\ \text{k}$$

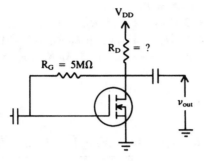

Fig. 3.51

To *analyze an EMOSFET dc bias circuit*, we proceed as follows: Given V_{DD}, R_D, R_G, k_P, V_T, $I_D = k_P (V_{SG} - V_T)^2$ $(V_{SG} > V_T)$, determine V_{SDQ}, I_{DQ}.

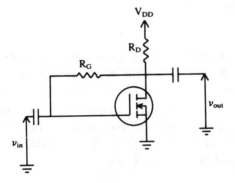

Fig. 3.52

According to Kirchoff's voltage law:

$$V_{SD} + V_{RD} = V_{DD}$$

also

$$V_{SD} = V_{SG} \text{ and } V_{RD} = I_D R_D$$

therefore,

$$V_{SD} + I_D R_D = V_{DD}$$
$$V_{SD} + k_P(V_{SG} - V_T)^2 R_D = V_{DD}$$

$$V_{SD} + k_P(V_{SD} - V_T)^2 = R_D V_{DD}$$
$$V_{SD} + k_P(V_{SD}^2 - 2 V_{SD}V_T + V_T^2) R_D = V_{DD}$$
$$V_{SD} + k_P R_D V_{SD}^2 - 2k_P R_D V_{SD} V_T + k_P R_D V_T^2 = V_{DD}$$
$$V_{SD}^2 (k_P R_D) + V_{SD}(1 - 2 k_P R_D V_T) + k_P R_D V_T^2 - V_{DD} = 0$$

$$V_{SD} = \frac{1}{2k_P R_D}$$

$$\times\ 2k_P R_D V_T - 1$$

$$\pm \sqrt{4k_P^2 R_D^2 V_T^2 - 4k_P R_D V_T + 1 - 4k_P R_D(k_P R_D V_T^2 - V_{DD})}/2k_P R_D$$

$$V_{SD} = \frac{(2k_P R_D)V_T - 1 \pm \sqrt{1 - 2(2k_P R_D)(V_T - V_{DD})}}{(2 k_P R_D)} \tag{3.22}$$

Also,

$$V_{SD} + I_D R_D = V_{DD}$$

therefore,

$$I_D = \frac{V_{DD} - V_{SD}}{R_D} \tag{3.23}$$

The evaluation of Equations (3.21) and (3.22) is well suited for solution with a programmable calculator. Where the $\pm$ sign appears, use a + sign for computational purposes.

Example (3.19). Compute the source-to-drain voltage and drain current for the EMOSFET common-source amplifier stage shown in Fig. 3.53.

Given $V_{DD} = +14$ V, $R_D = 1.32$ k, $R_G = 5$ MΩ, $k_P = 0.3$, $V_T = 2.8$ V, determine V_{SDQ}, I_{DQ}.

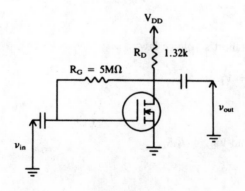

Fig. 3.53

Employing Equation (3.21), we have

$$V_{SD} = \frac{(2k_P R_D)\,V_T - 1 + \sqrt{1 - 2(2k_P R_D)(V_T - V_{DD})}}{(2k_P R_D)}$$

$$V_{SD} = \frac{[2(.3)(1.32)](2.8) - 1 + \sqrt{1 - 2\,[2(.3)(1.32)]\,(2.8 - 14)}}{[2(.3)(1.32)]}$$

$$V_{SD} = 7\ \text{V}$$

The result checks against our design!

$$I_D = \frac{V_{RD}}{R_D} = \frac{7\ \text{V}}{1.32\text{k}} = 5.3\ \text{mA}$$

The steps needed to compute V_{SD} using a Sharp 5813 programmable calculator are:

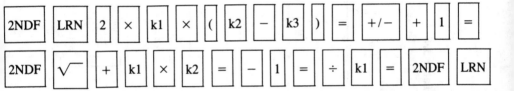

where: $k1 = 2\,k_P R_D\ (=.792)$, $k\,2 = V_T(=2.8)$, $k3 = V_{DD}(= +14\ \text{V})$ for this example.

Table 3.1
FET Amplifier DC Bias Design Formulas

COMMON SOURCE CONFIGURATION	Fixed Bias JFET or DMOSFET	$R_D = \dfrac{V_{RD}}{I_{DQ}}$ $\qquad V_{GG} = |V_P|\left(1 - \sqrt{\dfrac{I_D}{I_{DSS}}}\right)$
	Self Bias JFET or DMOSFET	$R_S = \dfrac{|V_P|}{I_{DQ}}\left(1 - \sqrt{\dfrac{I_D}{I_{DSS}}}\right)$ $\qquad R_D = \dfrac{V_{DD} - |V_P|\left(1 - \sqrt{\dfrac{I_D}{I_{DSS}}}\right)}{2I_D}$
	Self Bias JFET or DMOSFET with gate voltage divider	$R_S = \dfrac{V_{RS}}{I_D}, \; R_D = \dfrac{V_{DD} - V_{RS}}{2I_D}, \; R_1 = R_2\left[\dfrac{V_{DD}}{V_P\left(1 - \sqrt{\dfrac{I_D}{I_{DSS}}}\right) + V_{RS}} - 1\right]$
	Self Bias EMOSFET	$R_D = \dfrac{V_{DD}}{2k_P\left(\dfrac{V_{DD}}{2} - V_T\right)^2}$
COMMON DRAIN—SOURCE FOLLOWER	JFET or DMOSFET with gate voltage divider	$R_S = \dfrac{V_{RS}}{I_{DQ}}, \; R_1 = R_2\left[\dfrac{V_{DD}}{V_P\left(1 - \sqrt{\dfrac{I_D}{I_{DSS}}}\right) + V_{RS}} - 1\right]$

Table 3.2
FET Amplifier DC Bias Analysis Formulas

COMMON SOURCE CONFIGURATION	Fixed Bias JFET or DMOSFET	$I_D = I_{DSS}\left(1 - \dfrac{V_{SG}}{V_P}\right)^2$ $\quad V_{RD} = I_D R_D$
	Self Bias JFET or DMOSFET with or without gate voltage divider	$V_{gG} = V_{DD}\dfrac{R_2}{R_1 + R_2}$, $\quad I_D = \dfrac{V_P^2 - 2I_{DSS}R_S(V_P - V_{gG}) + \sqrt{V_P^2 - 2[2I_{DSS}R_S(V_P - V_{gG})]}}{2I_{DSS}R_S^2}$ $\quad V_{RD} = I_D R_D,\ V_{RS} = I_D R_S,\ V_{SD} = V_{DD} - V_{RD} - V_{RS}$
	Self Bias EMOSFET	$V_{SD} = \dfrac{(2k_P R_D)\,V_T - 1 + \sqrt{1 - 2\,(2k_P R_D)(V_T - V_{DD})}}{2k_P R_D},\ I_D = \dfrac{V_{DD} - V_{SD}}{R_D}$
COMMON DRAIN— SOURCE FOLLOWER	JFET or DMOSFET with or without gate voltage divider	$V_{gG} = V_{DD}\dfrac{R_2}{R_1 + R_2}$, $\quad I_D = \dfrac{V_P^2 - 2I_{DSS}R_S(V_P - V_{gG}) + \sqrt{V_P^2 - 2[2I_{DSS}R_S(V_P - V_{gG})]}}{2I_{DSS}R_S^2}$ $\quad V_{RS} = I_D R_S \qquad V_{SD} = V_{DD} - V_{RS}$

96

PROBLEMS

1. By using the output curves shown below (Fig. 3.54), design a common source FET amplifier to have a maximum output swing, i.e., $V_{SDQ} = V_{RD}$, and employ a quiescent drain current of 1.5 mA.

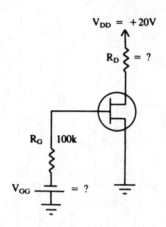

$V_{DD} = +20V$

$R_D \gtrless = ?$

R_G | 100k

$V_{GG} = ?$

Fig. 3.54(a)

OUTPUT CURVES

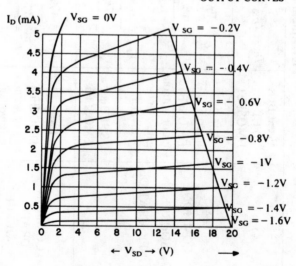

Fig. 3.54(b)

2. By using the output curves shown below (Fig. 3.55), design a common source FET amplifier to have a maximum output signal swing, i.e., $V_{SDQ} = V_{RD}$, and employ a quiescent drain current of 2.7 mA.

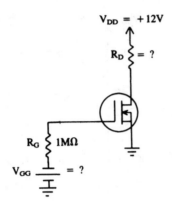

$V_{DD} = +12V$

$R_D = ?$

R_G 1MΩ

$V_{GG} = ?$

Fig. 3.55(a)

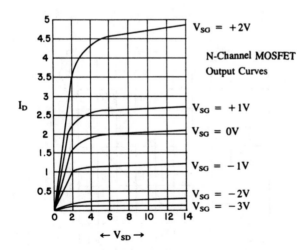

$V_{SG} = +2V$

N-Channel MOSFET
Output Curves

$V_{SG} = +1V$

$V_{SG} = 0V$

$V_{SG} = -1V$

$V_{SG} = -2V$
$V_{SG} = -3V$

I_D

$\leftarrow V_{SD} \rightarrow$

Fig. 3.55(b)

3. By using graphical techniques, analyze the circuit designed in Problem 1 and determine I_{DQ}, V_{GG}, V_{RD}, and V_{SDQ}.

4. Repeat Problem 3 for the results of Problem 2.

5. By using graphical techniques and the given output and transfer characteristic curves shown below (Fig. 3.56(b,c)), design a FET bias circuit to satisfy the conditions shown in Fig. 3.56(a).

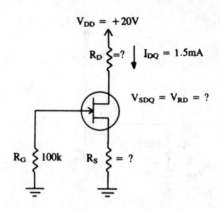

Fig. 3.56(a)

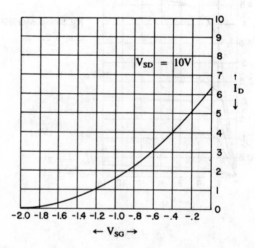

Fig. 3.56(b)

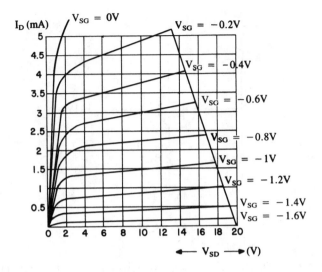

Fig. 3.56(c)

6. By using graphical techniques, analyze the circuit designed in Problem 5.

7. By using algebraic design techniques, repeat Problem 1 if $I_{DSS} = 6.5$ mA and $V_P = -1.87$ V.

8. By using algebraic analytical techniques, analyze the results of Problem 7 if $I_{DSS} = 6.5$ mA and $V_P = 1.87$ V.

9. By using algebraic design techniques, repear Problem 5 if $I_{DSS} = 6.5$ mA and $V_P = 1.87$ V.

10. By using algebraic analytical techniques, analyze the results of Problem 9 if $I_{DSS} = 6.5$ mA and $V_P = -1.87$ V.

11. Design a source follower to meet the specifications shown in Fig. 3.57. Given $I_{DQ} = 1.5$ mA, $V_{SDQ} = 11$ V, $I_{DSS} = 6.5$ mA, $V_P = -1.87$ V, $R_2 = 100$ k, determine R_1 and R_S.

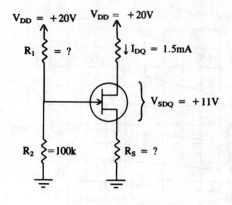

Fig. 3.57

12. Analyze the results of Problem 10. Given $I_{DSS} = 6.5$ mA and $V_P = -1.87$ V.

Chapter 4

Bipolar Transistor Amplifier Stage DC Bias Design and Analysis by Algebraic Techniques

4.1 DC BIAS DESIGN AND ANALYSIS

In a previous chapter (Ch. 1), the general characteristics of bipolar (sec. 1.2) and field-effect transistors (sec. 1.3) were presented. In this chapter we will discuss the methods by which a transistor may be connected in order to allow electricity to flow through it.* There is no magical process by which a transistor amplifies a weak electrical signal and increases the signal's power level. As we will see shortly, an amplifier of an alternating current (ac) signal is merely a device whose capabilities enable it to change a dc level. Let us review some dc electricity rules.

The voltage divider rule may be applied to determine the voltage across a particular resistor in a series "string." If two resistors are connected in series, and if they are of equal resistance values, then the voltage drop across each is the same value. (See Fig. 4.1.)

If resistor R_1 in Fig. 4.2 is a variable resistor, we can examine three different cases for values of R_1:

1. $R_1 = R_2 \, \Omega$
2. $R_1 = 0 \, \Omega$
3. $R_1 = \infty \, \Omega$

For case 1, the voltmeter will indicate 5 V as previously demonstrated. For case two (see Fig. 4.3(a)), the voltmeter will indicate zero volts because point A is essentially at ground potential, and therefore both voltmeter leads are connected together. Finally, for case three (see Fig. 4.3(b)), because no current is flowing through R_2, there will be no voltage drop across R_2 and the voltmeter will indicate the ground-to-power supply voltage ($+10V$).

---

*This process of "turning-on" a transistor is commonly referred to as "biasing."

102

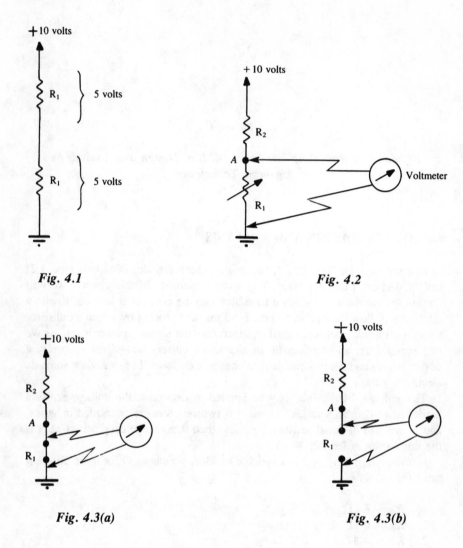

Fig. 4.1

Fig. 4.2

Fig. 4.3(a)

Fig. 4.3(b)

If we slowly vary the resistance of R_1 from a value equal to R_2 ohms, from infinity to zero ohms in a cyclic fashion, the voltage from ground to the top of R_1 would vary as shown in Fig. 4.4.

For a bipolar transistor, if we connect the emitter to ground and the collector to a resistor, the transistor could be thought of as a replacement for resistor R_1 in Fig. 4.1 (see Fig.4.5). A small amount of voltage from ground to base in addition to a small base current can control large collector voltage and current.

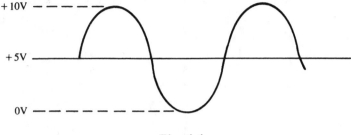

Fig. 4.4

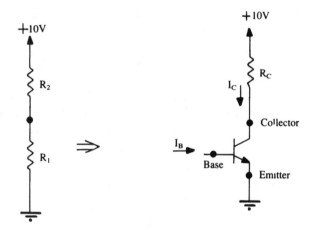

Fig. 4.5

$\sqrt{c_E}$ $\sqrt{B E}$

The collector-to-emitter voltage is typically many times the value of the base-to-emitter voltage, and the collector current is many times the base current. Thus, a small change in the base circuit power level can control or greatly change the power level in the collector circuit.

Let us further examine how the transistor amplifies a weak electrical signal and generates a strong signal.

Referring to Fig. 4.6, if the base current is at a level such that the collector current yields exactly a 5V drop across the collector resistor, by varying the base current, the collector voltage will vary proportionately. Let us see what happens if the base current increases from some *rest* value (referred to as the quiescent value). If I_B increases, then I_C increases. The increase in I_C results in a larger value for the voltage drop across R_C. Let us say that the voltage has increased

from a rest value of 5 V to 6 V. This leaves only 4 V across the transistor
$(10 - 6 = 4)$.

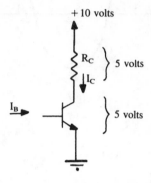

Fig. 4.6

If we next adjust the base current back to the rest value, the collector current
returns to the quiescent value. Sketching the points for the voltage appearing
across the transistor, we would see the curve outlined in Fig. 4.7.

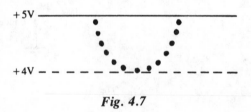

Fig. 4.7

If we then gradually reduce the base current below the rest value, the collector
current will vary proportionately. (It will also decrease.)

If the collector current is reduced below the rest value, the voltage drop across
R_C decreases below the rest value. Let us say that V_{RC} decreases from 5 V to 4
V. See Fig. 4.8.

Thus, if there are 4 V across the collector resistor R_C, then there must be
6 V across the collector-to-emitter junction of the transistor $(10 - 4 = 6)$.
Hence, if we return the base current to the rest value, I_C will return to its rest
value and V_{RC} will return to its rest value (5 V). A curve for V_{EC} can be traced
out as shown in Fig. 4.9.

If we use the curves shown in Figs. 4.7 and 4.9 and connect them together, we
see the curve shown in Fig. 4.10, which can be readily recognized as a sine
wave.

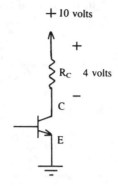

Fig. 4.8

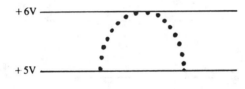

Fig. 4.9

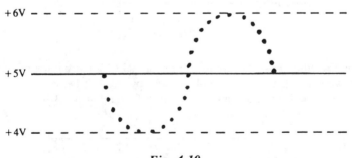

Fig. 4.10

One of the questions that arises is: "What is the maximum voltage sine wave that could be generated across the emitter-to-collector junction of a transistor connected to a series resistor and then connected to a 10 V power supply?"

To answer this question, let us examine a physical analogy. Consider a weight suspended by a spring connected to a 10 ft high ceiling. Further, suppose that the spring is stretched 5 ft due to the weight (see Fig. 4.11). If we pull the weight all the way down to the floor, then release it, the weight can conceivably oscillate all the way from the floor to the ceiling and back. From the rest position it will oscillate up 5 ft and down 5 ft, or we could say that it has a peak-to-peak oscillation of 10 ft (see Fig. 4.12).

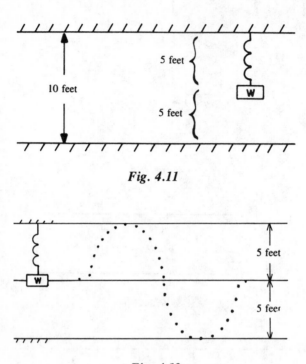

Fig. 4.11

Fig. 4.12

If on the other hand, the spring is only stretched 2 ft from the rest position by the weight, the weight can only move upward 2 ft before contacting the ceiling, but could move downward the full 5 ft (see Fig. 4.13).

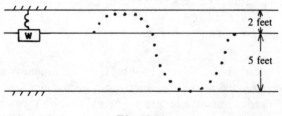

Fig. 4.13

Thus, we can see that for maximum excursion in both directions the weight should be extended halfway between the ceiling and floor. The same analogy can be applied to the transistor and series resistor combination.

Example (4.1). In the rest position, one-half of the power supply voltage should be across the transistor and the other half of the power supply voltage should be across the series-connected collector resistor. Under these circumstances, the output voltage may rise and fall by an equal amount. If, however, the rest voltage is not centered, the maximum swing is reduced.

Exceeding the maximum allowable swing would result in a distorted waveform (see Fig. 4.14).

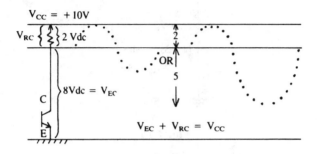

$$V_{EC} + V_{RC} = V_{CC}$$

Fig. 4.14

4.2 THE COMMON EMITTER CIRCUIT

Although this chapter is devoted to the simple task of devising methods by which the transistor can be connected to enable it to conduct an unvarying amount of current (dc), we have gone to great lengths to describe the desirability of having one-half of the power supply voltage across the transistor and the other half of the power supply voltage across the series connected collector resistor. We will now make use of this design criterion.

4.2.1 DC Design of an Unstabilized Common Emitter Stage

Before we proceed with the design, let us develop a somewhat formal definition of what we mean (throughout this textbook) by the word "design."

Definition—Given an existing set of electrical conditions that must be met, such as desired electrical currents and voltages, design is determining what component values must be employed in a circuit in order to achieve the desired electrical circuit parameters.

One additional point of clarification is in order. Wherever a schematic is drawn and dotted lines with a capacitor are shown connected to a transistor, the dotted lines imply that if an alternating electrical current was to be amplified by the transistor being studied, the connections would be made to the terminals shown.

Example (4.2). Suppose that we are given a power supply voltage of $V_{CC} = 10$ V and the steady-state (quiescent) collector must be 5 mA. From our previous discussions, we know that the emitter-to-collector voltage should be equal to the voltage across the collector resistor, i.e., $V_{EC} = V_{RC} = \frac{1}{2} V_{CC}$. Therefore, $V_{RC} = 5$ V.

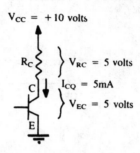

$V_{CC} = +10$ volts

R_C } $V_{RC} = 5$ volts

C   } $I_{CQ} = 5$mA

   } $V_{EC} = 5$ volts

E

Fig. 4.15

According to Ohm's law we may now calculate the value of R_C:

$$R_C = \frac{V_{RC}}{I_{CQ}} = \frac{5\ \text{V}}{5\text{mA}} = 1\ \text{k}\,\Omega \tag{4.1}$$

The above formula completes the design of the output portion of the unstabilized, common emitter stage. The collector and its associated resistor are referred to as the output circuit because the transistor, when used as an amplifier of weak ac electrical signals, has the signal "fed" to a succeeding electronic device by way of a connection to the collector terminal.

We must now investigate the electrical connections that are necessary to "turn-on" the common emitter stage. From the Chapter 3, we learned that one of the methods by which a transistor could increase the electrical power level consists of supplying more output current (collector current) as compared with the input current (base current). We also learned that there is a (somewhat) linear relationship between the collector and base currents. This proportionality constant was referred to as the transistor *beta* (β).

Thus,

$$\beta = \frac{I_C}{I_B}$$

For this design example, let us assume that we measure either the beta of the transistor, or determine the beta from the manufacturers specifications and determine it to have a value of 100 (β = 100). By employing this relationship we may determine the base current required to obtain a collector current of 5 mA for our design.

$$\beta = \frac{I_C}{I_B}$$

Therefore,

$$I_B = \frac{I_C}{\beta} = \frac{5 \text{ mA}}{100} = .05 \text{ mA}$$

The question now arises, "How do we enable .05 mA of current to flow into the base of the transistor?" Redrawing the common emitter amplifier stage, we have the schematic of Fig. 4.16. Also, we recall from the previous chapter that the base-to-emitter junction is a forward-biased diode with a voltage drop of approximately 0.6 V (for silicon).

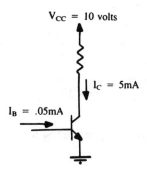

Fig. 4.16

If we were to connect the base terminal directly to the positive terminal of the 10 V power supply (see Fig. 4.17), an unlimited amount of current would flow into the base because there would be nothing in the circuit to limit the current flow (neglecting the low forward resistance of the base-to-emitter junction).

Therefore, the design problem becomes, "How can the current into the base-to-emitter junction be adjusted to precisely .05 mA, if we have a 10 V power supply available?" We might suspect that a resistor in the circuit would limit the base current. That is a correct assumption. Then, the question arises, "What value must the base resistor be?" Referring to Fig. 4.18, we see that there will be a total of 9.4 V across R_B.

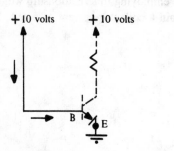

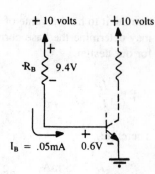

Fig. 4.17

Fig. 4.18

Using Ohm's law, we can now determine the resistance value of the base resistor:

$$R_B = \frac{V_{RB}}{I_B} = \frac{V_{CC} - V_{EB}}{I_B}$$

Therefore,

$$R_B = \frac{10 - 0.6 \text{ V}}{0.05 \text{ mA}} = \frac{9.4}{0.05} = 188 \text{ k } \Omega$$

The design is completed. Redrawing the circuit we have, we have the schematic of Fig. 4.19.

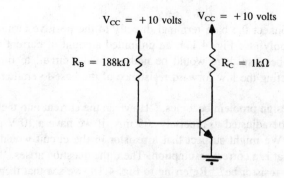

Fig. 4.19

4.2.2 DC Analysis of an Unstabilized Common Emitter Stage

The inverse operation of the design function is referred to as "analysis." Let us define analysis as follows:

Definition—Given an existing electrical circuit design, with known power supply voltage, other fixed circuit voltage drops, and known component values, analysis is determining the remaining circuit currents and voltages.

For the design function, we started at the ouptut (collector circuit) and worked our way to the input (base circuit) by determining the necessary component values. The analysis function requires that we follow an inverse procedure.

Example (4.3) Starting at the input (base circuit), let us determine the amount of base current that is flowing (see Fig. 4.20).

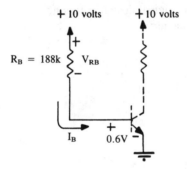

Fig. 4.20

Because we know that base current is flowing, since the base-to-emitter junction is a forward-biased diode junction, the emitter-to-base voltage drop may be thought of as a constant 0.6 V drop, with the polarity signs as shown in Fig. 4.20.

According to Kirchoff's voltage law, we have

$$V_{EB} + V_{RB} = V_{CC}$$

therefore,

$$V_{RB} = V_{CC} - V_{EB}$$

However,

$$V_{RB} = I_B R_B$$

therefore,

$$I_B R_B = V_{CC} - V_{EB}$$

thus,

$$I_B = \frac{V_{CC} - V_{EB}}{R_B} \tag{4.3}$$

Substituting the numerical data, we have

$$I_B = \frac{10 - 0.6}{188} = 0.05 \text{ mA}$$

This checks against our original design criterion.

From the previous chapter:

$$\beta = \frac{I_C}{I_B}$$

therefore,

$$I_C = \beta I_B \tag{4.4}$$

Substituting, we obtain

$$I_C = 100(0.05) = 5 \text{ mA}$$

This agrees with the original design value from the previous example.

According to Ohm's law, the voltage across the collector resistor is equal to the current passing through it (the collector current I_C) times the resistance value of the collector resistor (R_C). Thus,

$$V_{RC} = I_C R_C$$

therefore,

$$V_{RC} = 5 \text{ mA } (1\text{k}) = 5 \text{ V}$$

Referring to Fig. 4.21, by Kirchoff's voltage law, the emitter-to-collector voltage is the total voltage (power supply voltage V_{CC}) minus the voltage drop across the collector resistor V_{RC}:

$$V_{EC} = V_{CC} - V_{RC} \qquad\qquad (4.5)$$

Therefore,

$$V_{EC} = 10 - 5 = 5 \text{ V}$$

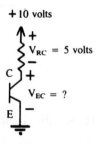

+10 volts

$V_{RC} = 5$ volts

$V_{EC} = ?$

C

E

Fig. 4.21

Example (4.4). In the case where the transistor under consideration is of the *pnp* type, we learned that all voltages and currents are reversed in polarity and direction. Let us examine an example with a *pnp* transistor employed as the active device shown in Fig. 4.21(a).

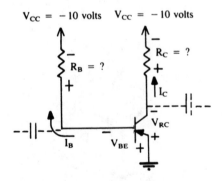

$V_{CC} = -10$ volts $V_{CC} = -10$ volts

$R_B = ?$

$R_C = ?$

I_C

V_{RC}

I_B V_{BE}

Fig. 4.21(a)

Given $V_{CC} = -10$ V, $V_{BE} = 0.3$ V, $I_C = 5$ mA, $\beta = 100$, determine R_C and R_B. Assume: $V_{RC} = V_{CE}$. Starting with an examination of the output circuit, according to Kirchoff's voltage law:

$$-V_{CC} + V_{RC} + V_{CE} = 0$$

or

$$V_{CC} = V_{RC} + V_{CE}$$

However, V_{RC} was assumed to equal to V_{CE}. Therefore,

$$V_{CC} = 2V_{RC}$$

then

$$V_{RC} = \frac{V_{CC}}{2} = \frac{10}{2} = 5 \text{ V}$$

with the polarity as shown in Fig. 4.21(b). Then,

$$R_C = \frac{V_{RC}}{I_{CQ}} = \frac{5 \text{ V}}{5 \text{ mA}} = 1 \text{ k}\Omega \qquad (4.6)$$

Examining the input circuit:

$$-V_{CC} + V_{RB} + V_{BE} = 0$$

or

$$V_{RB} = V_{CC} - V_{BE}$$

therefore,

$$V_{RB} = 10 - 0.3 = 9.7 \text{ V}$$

However, $\beta = 100$:

$$\beta = \frac{I_C}{I_B}$$

therefore,

$$I_B = \frac{I_C}{\beta} = \frac{5}{100} = 0.05 \text{ mA}$$

Also,

$$R_B = \frac{V_{RB}}{I_B} = \frac{9.7}{0.05} = 198 \text{ k }\Omega$$

This completes our design.

Example (4.5).   Proceeding with the analysis as a check on our design work, we start with the input circuit (see Fig. 4.21(b)).

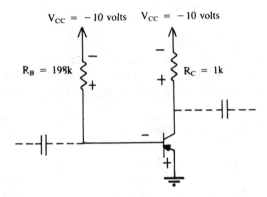

$V_{CC} = -10$ volts $V_{CC} = -10$ volts

$R_B = 198k$ $R_C = 1k$

Fig. 4.21(b)

Given $V_{CC} = -10$ V, $V_{BE} = 0.3$ V, $\beta = 100$, determine V_{RB}, I_B, I_C, V_{RC}. We can write

$$-V_{CC} + V_{RB} + V_{BE} = 0$$

Thus,

$$V_{RB} = V_{CC} - V_{BE}$$

Therefore,

$$V_{RB} = 10 - 0.3 = 9.7 \text{ V}$$

Then,

$$I_B = \frac{V_{RB}}{R_B} = \frac{9.7}{198 \text{ k}} = 0.5 \text{ mA}$$

Also,

$$\beta = I_C/I_B$$

Thus,

$$I_C = \beta I_B = 100(.05) = 5 \text{ mA}$$

Then,

$$V_{RC} = I_C R_C = 5 \text{ mA } (1 \text{ k}) = 5 \text{ V}$$

which checks against our design!

The value that we have determined for the emitter to collector voltage checks correctly with the original design value of the previous example. This completes the dc bias design and analysis of the unstabilized common emitter amplifier.

4.3 THE COMMON BASE CIRCUIT

4.3.1 Common Base DC Bias Design

Another method of connecting a bipolar transistor is the common base configuration. Referring to Fig. 4.22, we note that the transistor has been "tilted on its side," so that the base lead is common to both the input circuit on the left and the output circuit on the right. In contrast with the common emitter amplifier, the input signal is applied to the emitter terminal, rather than the base terminal as in the common emitter amplifier configuration.

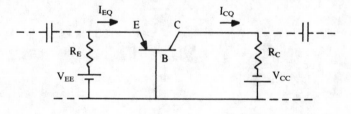

Fig. 4.22

Example (4.5). In all designs we must have some given information in addition to the parameters that must be determined. Referring to Fig. 4.23(a), given will be the output power supply $V_{CC} = 10$ V, the input power supply $V_{EE} = 3$ V, the base-to-emitter voltage drop $V_{BE} = 0.3$ V, the steady-state (or quiescent) collector current that is flowing in the output circuit $I_{CQ} = 5$ mA. We must determine the unknowns, which are I_{EQ}, R_C, and R_E.

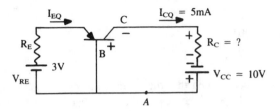

Fig. 4.23(a)

Before we proceed with the design, two points should be made. First, the transistor is of the *pnp* variety and the voltage drop from base to emitter is only 0.3 V. This tells us that the unit is a germanium transistor. Second, for maximum output signal, it would be appropriate for the voltage across the collector-to-base junction to be equal to the voltage drop across the collector resistor R_C, i.e., $V_{CB} = V_{RC}$. The reason for this was discussed in the beginning of this chapter.

Proceeding with the design, let us use Kirchoff's voltage law around the output loop in Fig. 4.23. starting at point A and proceeding counterclockwise, we have

$$-V_{CC} + V_{RC} + V_{CB} = 0$$

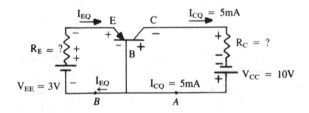

Fig. 4.23(b)

For maximum signal swing, using the same reasoning as in the common emitter stage,

$$V_{RC} = V_{CB}$$

therefore,

$$V_{CC} = V_{RC} + V_{CB}$$
$$= 2\,V_{RC}$$

therefore,

$$V_{RC} = \frac{V_{CC}}{2} = \frac{10}{2} = 5 \text{ V}$$

Since we have established that the voltage across the collector load resistor R_C is five volts, and one of the design criterion that we started with was that a steady-state collector current I_{CQ} of 5 mA would flow, we are now in a position to calculate the value of the collector resistor:

$$R_C = \frac{V_{RC}}{I_{CQ}} \tag{4.7}$$

$$R_C = \frac{5 \text{ V}}{5 \text{ mA}} = 1 \text{ k}\Omega$$

In the previous chapter we learned that there was a relationship between the collector and emitter currents,

$$\alpha = \frac{I_C}{I_E}$$

Solving the equation for I_E,

$$I_E = \frac{I_C}{\alpha}$$

Note: A determination of a bipolar transistor dc α is not as accurately obtained from the transistor curves as is the dc β. The dc β of a transistor may be measured quite accurately by way of a photograph of the transistor characteristic curves. The curves may be obtained by use of a commercially available transistor curve tracer. Once the β has been determined, the relation developed in Chapter 1 may be used to determine α:

$$\alpha = \frac{\beta}{\beta + 1}$$

If α is given as .98 (from measurements or manufacturers specifications), we then have

$$I_E = \frac{I_C}{\alpha} = \frac{5}{.98} = 5.1 \text{ mA}$$

We have thus related the input circuit to the output circuit. Since we know the current that must flow in the input circuit, if we can determine the voltage across the emitter resistor, we can determine the resistor value. Using Kirchoff's voltage law starting at point B (see Fig. 4.23(b)) and summing the voltages in a counterclockwise direction we get:

$$V_{BE} + V_{RE} - V_{EE} = 0$$

therefore,

$$V_{RE} = V_{EE} - V_{BE}$$

or

$$V_{RE} = 3 - 0.3 = 2.7 \text{ V}$$

Using Ohm's law:

$$R_E = \frac{V_{RE}}{I_E} = \frac{2.7 \text{ V}}{5.1 \text{ mA}} = 0.529 \text{ k } \Omega \tag{4.8}$$

This completes the design of the common base transistor.

4.3.2   Analysis of Common Base, Bipolar Transistor Stage

Now that we have completed the design of the common base stage, we need to know if our design is correct. Let us develop the procedures for the analysis of the common base transistor circuit.

When analyzing the common emitter stage, we noted that the analysis procedure was initiated at the input circuit. Before proceeding further, we should list the given parameters and those parameters that must be determined.

Example (4.6).   Refer to the circuit diagram of Fig. 4.24. Given V_{CC}, V_{EE}, R_C, R_E, V_{BE} and α, determine V_{RE}, I_E, I_C and V_{RC}.

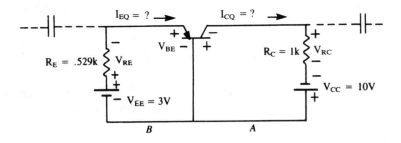

Fig. 4.24

Using Kirchoff's voltage law at node B and summing the voltages in a counter-clockwise direction, we have

$$V_{BE} + V_{RE} - V_{EE} = 0 \qquad (4.9)$$

Therefore,

$$V_{RE} = V_{EE} - V_{BE}$$

or

$$V_{RE} = 3 - 0.3 = 2.7 \text{ V}$$

Using Ohm's law, we can now calculate the value of the emitter current:

$$I_{EQ} = \frac{V_{REQ}}{R_E} \qquad (4.10)$$

Therefore,

$$I_{EQ} = \frac{2.7 \text{ V}}{0.529 \text{ k}} = 5.1 \text{ mA}$$

Knowing the emitter current and the α of the transistor, we may calculate I_C:

$$\alpha = \frac{I_C}{I_E} \qquad (4.11)$$

Therefore,

$$I_C = \alpha I_E$$
$$I_C = 0.98(5.1) = 5 \text{ mA}$$

Now that we know the value of I_C, by using Ohm's law we may determine the value of V_{RC}:

$$V_{RC} = I_C R_C = 5 \text{ mA } (1 \text{ k}) = 5 \text{ V (with polarity as shown in Fig. 4.24)}$$

Referring to Fig. 4.24, we may employ Kirchoff's voltage around the output loop. Starting at point A and proceeding in a counterclockwise direction:

$$-V_{CC} + V_{RC} + V_{CB} = 0$$

Therefore,

$$V_{CB} = V_{CC} - V_{RC} \quad \text{or} \quad V_{CB} = 10 - 5 = 5 \text{ V} \tag{4.12}$$

This value is in agreement with our original design requirement.

4.3.3 Comon Base Approximation

At this point it is appropriate to mention that if we did not know the α for a transistor, we could assume that the α value was equal to one with little loss in accuracy. For a transistor with a β approaching 100, let us say 99, the α value would be

$$\alpha = \frac{\beta}{\beta + 1} = \frac{99}{99 + 1} = \frac{99}{100} = 0.99$$

If I_E was equal to 1 mA in a transistor, the I_C would be .99 mA. Assuming that $I_C = 1$ mA results in an error of

$$\% \text{ Error} = \left| \frac{\text{correct value } - \text{ approximate value}}{\text{correct value}} \right| \times 100$$

$$= \left| \frac{0.99 - 1}{0.99} \right| \times 100 = \left| \frac{- 0.01}{0.99} \right| \times 100 \doteq 1\%$$

For design purposes, we may assume that $I_C = I_E$ in both common base circuit design and analysis. This would result in little error and some simplification of the computation.

An additional assumption which is sometimes made is that the base-to-emitter voltage drop is negligible, and therefore may be ignored. The larger the value of the emitter power supply V_{EE}, the more legitimate the assumption that V_{BE} is negligible. Let us examine the previous common base design example using the above approximation.

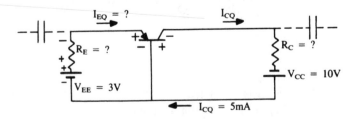

Fig. 4.25

Example (4.7). Refer to the diagram of Fig. 4.25. Given $V_{RC} = V_{CB} = 5$ V, $V_{BE} = 0.3$ V. In a manner similar to the previous design example:

$$R_C = \frac{V_{RC}}{I_{CQ}} = \frac{5 \text{ V}}{5 \text{ mA}} = 1 \text{ k}$$

Approximation: $I_E \doteq I_C = 5$ mA.

$$V_{RE} = V_{EE} - V_{BE} = 3 - 0.3 = 2.7 \text{ V}$$

therefore,

$$R_E = \frac{V_{RE}}{I_E} = \frac{2.7}{5} = 540 \text{ } \Omega$$

Comparing emitter resistors for the exact and approximate designs, we have

$$\% \text{ difference} = \left| \frac{529 - 540}{529} \right| \times 100 = 2.1\%$$

Let us examine the previous common base analysis example using the above approximation (see Fig. 4.26).

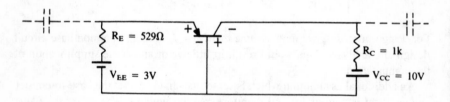

Fig. 4.26

$$I_E = \frac{V_{EE} - V_{BE}}{R_E} = \frac{3 - 0.3}{529} = \frac{2.7}{529} = 5.104 \text{ mA}$$

Assume:

$$I_C \doteq I_E = 5.104 \text{ mA}$$
$$V_{RC} = I_C R_C = 5.104(1) = 5.104 \cdot \text{V}$$
$$V_{CB} = V_{CC} - V_{RC} = 10 - 5.104 = 4.896 \text{ V}$$

$$\% \text{ error in } V_{CB} = \left| \frac{5 - 4.896}{5} \right| \times 100 = 2.08\%$$

4.3.4 Stability

The stability of a transistor amplifier is a measurement "yardstick" used to gauge the changes in the dc operating voltages and currents caused by changes in the internal electrical parameters of the transistor. The lower the stability factor, the greater the resistance to change of the dc operating point. Several stability factors are usually considered when examining the changes in a transistor amplifier operating point. The individual stability factors most commonly considered are:

1. The change in the collector current caused by changes in the leakage current from either the collector-to-base junction or collector-to-emitter junction;
2. The change in the collector current caused by changes in the forward-biased base-to-emitter junction;
3. The change in the collector current caused by changes in the β of the transistors.

There are other transistor parameters, such as the static base-to-emitter forward-biased junction resistance, which will also affect the transistor operating point. These second-order effects are not usually considered. A convenient mathematical expression that may be used to predict the overall change in the collector current is expressed as a partial differential equation:

$$dI_C = \frac{\partial I_C}{\partial I_{CBO}} \cdot dI_{CBO} + \frac{\partial I_C}{\partial V_{EB}} \cdot dV_{EB} + \frac{\partial I_C}{\partial \beta} \cdot d\beta \qquad (4.12)$$

Each of the partial derivative terms is referred to as a stability factor. Thus,

$$S_{ICBO} = \frac{\partial I_C}{\partial I_{CBO}} \,, \; S_{VEB} = \frac{\partial I_C}{\partial V_{EB}} \,, \; S_{\beta} = \frac{\partial I_C}{\partial \beta} \qquad (4.13)$$

The evolution of transistor technology from germanium transistors to silicon transistors and the evolution in manufacturing techniques from the point-contact, to the junction alloy, to the mesa, and finally to the planar process, have progressively lowered the quantity of undesirable leakage current that the circuit designer must contend with. A contemporary silicon transistor may exhibit less than a nanoamp of leakage current at room temperature.

The second stability factor listed above is a predictor of collector current changes caused by changes in the voltage drop of the forward-biased base-to-emitter junction. In a typical common emitter amplifier stage design, the base current is derived from the collector power supply. Because the collector power supply voltage is usually appreciably greater than that required by the base circuit, the combination of V_{CC} and R_B begins to function in a manner similar to a constant current source. The constant current source analogy is particularly valid because changes in V_{EB} (the voltage V_{EB} is in series opposition to V_{CC}) are inconsequential compared to V_{CC}. Thus, with I_B derived from a relatively constant current source, I_C is relatively unaffected by changes in V_{EB}.

The third stability factor listed above concerns changes in the transistor β and is discussed in detail in the next section (sec. 4.3.5).

4.3.5 DC Stabilization of a Bipolar Transistor Connected in the Common Emitter Configuration

In the beginning of this chapter, we subjectively examined the importance of establishing the dc conditions (operating point), such that approximately one-half of the power supply voltage appeared across the collector resistor (sometimes referred to as the collector-load resistor) and the other half of the power supply voltage appeared across the transistor collector-to-emitter leads. In the previous chapter, we also noted that variations occur in bipolar transistor electrical parameters as a result of variations in certain parameters, which are sometimes uncontrollable, such as temperature, age, collector-to-emitter dc voltage, *et cetera*.

Although each of these parameter variations should be of concern to the designer, a large variation in bipolar transistor β will be experienced during manufacturing as a result of variations in β from one transistor to another. We might do well to discuss one economic issue of concern to any designer. As might be anticipated, the closer the manufacturer is required to hold the "spread" of the tolerances of devices that are being supplied to an "end user," the more expensive the parts will be to the manufacturer. These costs must be passed along to the end user. Therefore, the end user, whether he is the manufacturer of a system employing electronic parts or the user of transistors in a piece of electronic equipment, must design and use equipment to "tolerate" as wide a spread as possible in the supplied parts' tolerances.

In the beginning of this chapter, we noted that the maximum signal swing in the output of the common-emitter connected transistor depended upon the power supply voltage being *shared equally* between the collector load resistor and the collector-to-emitter junction of the transistor. Let us see what happens to the dc bias conditions of the transistor if we allow the β to "stray" from its original value. For convenience, let us use the common emitter stage that we designed and analyzed earlier.

Original Problem. Given $V_{CC} = +10$ V, $V_{EB} = 0.6$ V, $I_{CQ} = 5$ mA, $\beta = 100$, determine R_C and R_B. Assume: $V_{RC} = V_{EC}$.

Example (4.8). For the circuit shown in Fig. 4.27, we determined that $R_B = 188$ k and $R_C = 1$ k. This design was predicated upon $V_{RC} = V_{EC} = 5$ V. Let β decrease by 30%, and determine the new V_{RC} and V_{EC}:

$$\beta_2 = 0.7\beta_1 = 0.7 \, (100) = 70$$
$$V_{EB} + V_{RB} = V_{CC}$$

Therefore, $V_{RB} = V_{CC} - V_{EB} = 10 - 0.6 = 9.4$ V

$$I_B = \frac{V_{RB}}{R_B} = \frac{9.4 \text{ V}}{188 \text{ k}} = 0.05 \text{ mA}$$

Note that the base current has not changed, although the transistor β has varied.

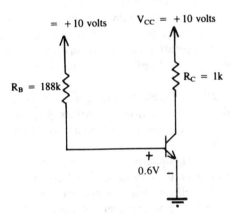

Fig. 4.27

Now, let us calculate I_{CQ}:

$$I_C = \beta I_B = 70(0.05) = 3.5 \text{ mA}$$

Then,

$$V_{RC} = I_{CQ} R_C = 3.5 \text{ mA} (1 \text{ k}) = 3.5 \text{ V}$$

and

$$V_{EC} = V_{CC} - V_{RC} = 10 - 3.5 = 6.5 \text{ V}$$

The maximum signal swing is determined by the dc voltage across either the collector load resistor or the emitter-to-collector junction. The lesser value determines the maximum signal swing, i.e., 3.5 V peak-to-peak. Thus, we can see that the maximum signal swing has decreased from 5 V (β = 100) to 3.5 V (β = 70).

Example (4.9). Let us examine the effect of a 30% increase in the value of the transistor β:

$$\beta_2 = 1.3\beta_1 = 1.3(100) = 130$$
$$I_C = \beta I_B = 130(0.05) = 6.5 \text{ mA}$$
$$V_{RC} = I_C R_C = 6.5 \text{ mA (1 k)} = 6.5 \text{ V}$$
$$V_{EC} = V_{CC} - V_{RC} = 10 - 6.5 \text{ V}$$

Again, the maximum signal swing is now limited to 3.5 V peak-to-peak.

In a supplier's "lot" of transistors, a β spread of 70 to 130 would be entirely reasonable, if the supplier's specifications were not "held tighter."

As previously noted, parts costs are inversely proportional to parameter spreads. However, we are faced with a situation where a wide spread in the transistor beta could result in the operating conditions (operating point) of the transistor amplifier bias changing considerably. This could result in "clipping" of the output waveform.

If we employ a method whereby the change in the operating point caused by variations in β could be minimized, wide transistor β spreads could be tolerated (thus, reducing parts costs). Methods have been devised for this purpose, called *dc bias stabilization techniques*.

Let us examine one technique, referred to as *emitter stabilization*. Suppose we insert a resistor between the emitter lead and ground. The collector current flows down through the collector load resistor into the transistor and joins with the base current to form the emitter current (see Fig. 4.28).

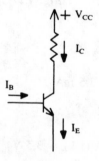

Fig. 4.28

By far the greatest portion of the emitter current is composed of the collector current (98–100%). For a specific value of the base current, the collector current is equal to β times the base current. Therefore, the emitter current is approximately equal to β times the base current. Let us see what happens if the beta increases (with an emitter resistor in the circuit), as shown in Fig. 4.29.

If β increases, then I_E increases to $I_{E2} = I_{E1} + \Delta I_E$. The increase in I_E results in an increase in voltage across R_E; or $V_{RE2} = V_{RE1} + \Delta V_{RE}$. See Fig. 4.30.

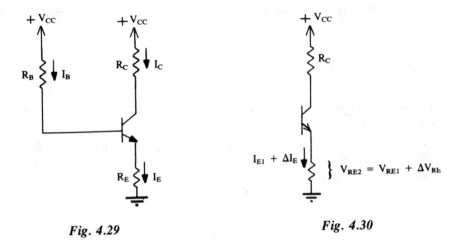

Fig. 4.29

Fig. 4.30

If the voltage across R_E has increased, the voltage from ground to base has also increased. See Fig. 4.31.

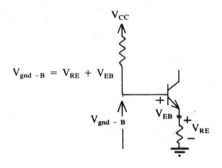

Fig. 4.31

If the voltage across the ground-to-base points has increased, then the voltage across the base resistor has decreased. See Fig. 4.32.

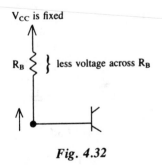

Fig. 4.32

If the voltage across R_B has decreased, then by Ohm's law I_B decreases.

$$\downarrow \quad \downarrow$$
$$I_B = \frac{V_{RB}}{R_B}$$

The decrease in I_B causes a comparable decrease in I_C and the circuit tends to return to its initial condition.

Example (4.10). Let us demonstrate with an example based on Fig. 4.33, from the first common emitter design problem. When beta increased to 130, V_{RC} increased from 5 V to 6.5 V.

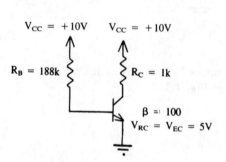

Fig. 4.33

With an emitter resistor, let $V_{RC} = 4.5$ V, $V_{EC} = 4.5$ V and $V_{RE} = 1$ V, $I_{CQ} = 5$ mA, $V_{EB} = 0.6$ V, $\beta = 100$.

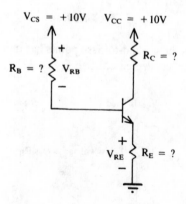

Fig. 4.34

Notice in Fig. 4.34 that we have arbitrarily allowed one volt to appear across the emitter resistor, which leaves 9 V (of the 10 V power supply) to be divided equally among the collector-load resistors and the collector-to-emitter junction of the transistor. Later in this chapter, we will investigate the effects of various percentages of the power supply voltage appearing across the emitter resistor.

$$R_C = \frac{V_{RC}}{I_{CQ}} = \frac{4.5 \text{ V}}{5 \text{ mA}} = 0.9 \text{ k}$$

$$I_E = I_B + I_C = 0.05 + 5 = 5.05 \text{ mA}$$

$$R_E = \frac{V_{RE}}{I_E} = \frac{1}{5.05} = 0.19802 \text{ k}$$

The voltage from ground to base is

$$V_{gnd-B} = V_{RE} + V_{EB} = 1 + 0.6 = 1.6 \text{ V}$$

Therefore, $V_{RB} = V_{CC} - V_{gnd-B} = 10 - 1.6 = 8.4 \text{ V}$

$$R_B = \frac{V_{RB}}{I_B} = \frac{8.4 \text{ V}}{0.05 \text{ mA}} = 168 \text{ k} \, \Omega$$

The design is completed, so let us check the results.

4.3.6 Development of Emitter Stabilized Common Emitter DC Bias, Analysis Formulas

According to Kirchoff's voltage law,

$$V_{RE} + V_{EB} + V_{RB} = V_{CC}$$

However,

$$V_{RE} = I_E R_E = (I_C + I_B)R_E = (\beta I_B + I_B)R_E = (\beta + 1)I_B R_B R_E$$

Also,

$$V_{RB} = I_B R_B$$

Thus,

$$V_{RE} + V_{EB} + V_{RB} = V_{CC} \quad \rightarrow \quad V_{RB} + V_{RE} = V_{CC} - V_{EB}$$

Substituting

$$I_B R_B + (\beta + 1) I_B R_B = V_{CC} - V_{EB}$$

Factoring

$$I_B [R_B + (\beta + 1) R_E] = V_{CC} - V_{EB}$$

Thus,

$$I_B = R_B + \frac{V_{CC} - V_{EB}}{(\beta + 1)R_E} \tag{4.14}$$

Also,

$$I_C = \beta I_B \tag{4.15}$$

and

$$V_{RC} = I_C R_C \tag{4.16}$$

and

$$V_{RE} = I_E R_E = (I_C + I_B)R_E \tag{4.17}$$

and according to Kirchoff's voltage law:

$$V_{EC} = V_{CC} - V_{RE} - V_{RC}$$

Example (4.11). Checking our design example:

$$I_B = \frac{V_{CC} - V_{EB}}{R_B + (\beta + 1)R_E} = \frac{10 - 0.6}{168 + 101(0.19802)}$$

$$= \frac{9.4}{168 + 20} = \frac{9.4}{188} = 0.05 \text{ mA}$$

$$I_C = \beta I_B = 100(0.05) = 5 \text{ mA}$$

$$V_{RC} = I_C R_C = 5(0.9) = 4.5 \text{ V}$$

$$V_{RE} = (I_C + I_B)R_E = (5.05)(0.19802) = 1 \text{ V}$$

$$V_{EC} = V_{CC} - V_{RE} - V_{RC} = 10 - 1 - 4.5 = 4.5 \text{ V}$$

which checks!

Example (4.12). Let us see what happens to V_{RC} and V_{EC} if the β increases by 30%.

$$I_B = \frac{V_{CC} - V_{EB}}{R_B + (\beta + 1)R_B} = \frac{10 - 0.6}{168 + (131)(0.19802)} = \frac{9.4}{168 + 25.94}$$

$$= \frac{9.4}{193.94}$$

$$I_B = 0.04874 \text{ mA}$$

Therefore, $I_C = \beta I_B = 131(0.04874) = 6.349$ mA
$V_{RC} = I_C R_C = 6.349 \ (0.9) = 5.714$ V
$V_{RE} = (I_C + I_B) \ R_E = 6.349 \ (0.19802) = 1.257$ V
$V_{EC} = V_{CC} - V_{RE} - V_{RE} = 10 - 1.257 - 5.714 = 3.0283$ V

In Example (4.8) without stabilization, V_{EC} varied from 5 V to 3.5 V, for a 30% increase in β. This results in a change of:

$$5 - 3.5 \text{ V} = 1.5 \text{ V}$$

Example (4.13). For emitter stabilization with a 30% increase in β, we have

$$4.5 - 3.028 = 1.472 \text{ volts } (= -32.7\%)$$

Let us see what happens to V_{RC} and V_{EC} if the β decreases by 30%.

$$I_B = \frac{V_{CC} - V_{EB}}{R_B + (\beta + 1)R_E} = \frac{10 - 0.6}{168 + 71(0.19802)} = 0.05163 \text{ mA}$$
$$I_C = \beta \ I_B = 70(0.05163) = 3.614 \text{ mA}$$
$$V_{RC} = I_C \ R_C = 3.614(0.9) = 3.2527 \text{ V}$$
$$V_{RE} = (I_C + I_B) = (3.614 + 0.0516)(0.19802) = 0.7259 \text{ V}$$
$$V_{EC} = V_{CC} - V_{RE} - V_{RC} = 10 - 0.7259 - 3.253 = 6.021 \text{ V } (= +33\%)$$

For the unstabilized cases, V_{EC} increased to 6.5 volts.

We can see that emitter stabilization has a questionable beneficial effect on stabilizing the change in voltage across a transistor, and the beneficial effect is noticeable only when the transistor β increases from the nominal design value.

Stabilization against changes in the transistor operating point are compensated by the emitter resistor as a result of leakage current increases. Emitter stabilization is more effective when leakage current is a factor in the transistor operating environment.

A more effective method of providing stabilization against change in the transistor β is by use of an additional resistor from base to ground. Additionally, the resistor in the emitter-to-ground path is usually included. This bias arrangement is given a special name: *universal stabilization* (see Fig. 4.35).

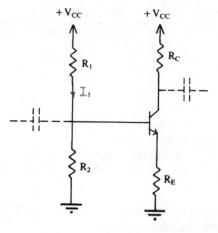

Fig. 4.35

Let us see how this circuit stabilizes the output voltages. Assume that the β of the transistor is increased above the nominal value for some arbitrary reason, such as the replacement of the transistor. With increased β, the collector current will attempt to increase, as will the emitter current. With the forecasted increase in emitter current, a corresponding increase in voltage across the emitter resistor will attempt to take place. The resistors R_1 and R_2 form a voltage divider. Their effectiveness as an unvarying voltage divider will be a function of the base current as a small percentage of the "bleeder" current through R_1 and R_2. Assuming that the voltage across R_2 remains relatively constant, as the voltage across R_E tends to increase, V_{EB} is reduced. If V_{EB} is reduced, then I_B and I_C are reduced. This effect tends to "cancel out" the undesirable effects of the increased β.

Let us investigate the design of the universal bias circuit. Before we can proceed, however, we must mention that there are two variables which will arise in this design. The first is the ratio of the bleeder current flowing down through R_1 *versus* the base current. Let us define this ratio, δ, where

$$\delta = \frac{I_1}{I_B}$$

The larger the value of δ, the more stable will be the voltage divider formed by R_1 and R_2. This should aid in circuit stability. However, a stable voltage divider implies that I_B is negligible compared to I_1 and I_2. (See Fig. 3.46.)

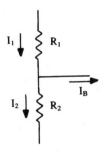

Fig. 4.36

If the currents I_1 and I_2 are large (compared to I_B), then the resistors R_1 and R_2 must be small. A low resistance has two disadvantages:

1. A large dc power is dissipated within these resistors;
2. The load resistance that a previous stage would "see" would be quite low, thereby reducing the previous stage voltage amplification (gain).

The second variable is the percentage of the dc power supply voltage that appears across the emitter resistor. Let us define that ratio as γ.

$$\gamma = \frac{V_{RE}}{V_{CC}} \qquad (4.20)$$

Therefore, $V_{RE} = \gamma V_{CC}$

The development of the general design equations and an example is as follows in section 4.4.

4.4 GENERAL DESIGN EQUATIONS FOR A UNIVERSAL STABILIZED COMMON EMITTER AMPLIFIER

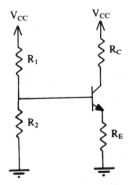

Fig. 4.37

134

Let, $I_1 = \delta I_B$, $V_{RE} = \gamma V_{CC}$, $I_C = \beta I_B$, $V_{CE} = V_{RC}$. Choose V_{CC}, I_{CQ}, Given V_{EB}.

$$V_{RE} + V_{EC} + V_{RC} = V_{CC}$$

Therefore,

$$V_{RC} = V_{CC} - V_{EC} - V_{RE} = V_{CC} - V_{RC} - V_{RE}$$

or

$$2V_{RC} = V_{CC} - V_{RE} = V_{CC} - \gamma V_{CC} = V_{CC}(1 - \gamma)$$

Therefore,

$$V_{RC} = \frac{V_{CC}(1 - \gamma)}{2}$$

Also,

$$R_C = \frac{V_{RC}}{I_C} = \frac{V_{CC}(1 - \gamma)}{2I_C} \tag{4.21}$$

$$V_{RE} = \gamma V_{CC} = I_E R_E$$

Therefore,

$$R_E = \frac{\gamma V_{CC}}{I_E} = \frac{\gamma V_{CC}}{I_C + I_B} = \frac{\gamma V_{CC}}{I_C + I_C/\beta}$$

$$= \frac{\gamma V_{CC}}{I_C} \frac{1}{\left(\dfrac{1 + \beta}{\beta}\right)}$$

$$R_E = \frac{\gamma V_{CC}}{I_C}\left(\frac{\beta}{\beta + 1}\right) \tag{4.22}$$

$$V_{R2} = V_{RE} + V_{EB} = \gamma V_{CC} + V_{EB}$$

Also,

$$V_{R2} + V_{R1} = V_{CC}$$

Therefore,

$$V_{R1} = V_{CC} - V_{R2} = V_{CC} - (\gamma V_{CC} + V_{EB})$$

$$= V_{CC} - \gamma V_{CC} - V_B = V_{CC}(1 - \gamma) - V_{EB}$$

but

$$R_1 = \frac{V_{R1}}{I_1} = \frac{V_{CC}(1 - \gamma) - V_{EB}}{I_1} = \frac{V_{CC}(1 - \gamma) - V_{EB}}{\delta I_B}$$

$$= \frac{V_{CC}(1 - \gamma) - V_{EB}}{\delta I_C/\beta}$$

Therefore,

$$R_1 = \frac{\beta}{\delta I_C} [V_{CC} (1 - \gamma) - V_{EB}] \qquad (4.23)$$

$$I_1 = I_2 + I_B \text{ or } \delta I_B = I_2 + I_B \rightarrow I_2 = \delta I_B - I_B$$

$$I_2 = I_B (\delta - 1) = \frac{I_C}{\beta} (\delta - 1)$$

$$R_2 = \frac{V_{R2}}{I_2} = \frac{V_{RE} + V_{EB}}{I_2} = \frac{\gamma V_{CC} + V_{EB}}{I_2}$$

$$= \frac{\gamma V_{CC} + V_{EB}}{I_C/\beta (\delta - 1)}$$

$$R_2 = \frac{\beta (\gamma V_{CC} + V_{EB})}{I_C (\delta - 1)} \qquad (4.24)$$

Now that we have developed the design formulas, an illustrative example is in order.

Example (4.14). Let $I_1 = 4I_B$.
 Therefore,

$$\delta = \frac{I_1}{I_B} = 4$$

Also, let

$$V_{RE} = 1 \text{ V}$$

Thus,

$$\gamma = \frac{V_{RE}}{V_{CC}} = 0.1$$

Also,

$$V_{CC} = +10 \text{ V}, \beta = 100, V_{EB} = 0.6, I_{CA} = 5 \text{ mA, and } V_{RC}$$
$$= V_{EC} = 4.5 \text{ V}$$

$$R_C = \frac{V_{CC}}{I_C} \quad \frac{1 - \gamma}{2} = \frac{10}{5} \quad \left(\frac{1 - 0.1}{2}\right) = 0.9 \text{ k}$$

$$R_E = \frac{\delta V_{CC}}{I_C} \left(\frac{\beta}{\beta + 1}\right) = \frac{0.1 (10)}{5} \left(\frac{100}{101}\right) = 0.19802 \text{ k}$$

$$R_1 = \frac{\beta}{\delta I_C} [V_{CC} (1 - \gamma) - V_{EB}] = \frac{100}{4(5)} [10 (1 - 0.1) - 0.6] = 42 \text{ k}$$

$$R_2 = \frac{\beta (\gamma V_{CC} + V_{EB})}{I_C (\delta - 1)} = \frac{100 [0.1(10) + 0.6]}{5(4 - 1)} = 10.67 \text{ k}$$

As we can see the computations are not laborious, but if repetitive calculations must be made, this undertaking could be time consuming. Bearing this in mind, a computer program has been written in BASIC (see Appendix A) to assist in the component calculations for the universal stabilized common emitter stage.

4.4.1 Approximations for the Universal Stabilized Common Emitter Stage

Two assumptions are occasionally made when dealing with the design and analysis of the common emitter stage with universal stabilization. They are:

1. The emitter-to-base voltage drop is negligible;

2. The voltage developed across R_2 in the base voltage divider is not affected by the base current (i.e., $I_B \doteq 0$).

The effect of neglecting V_{EB} is a decision that must be made as a function of the power supply voltage value. The greater the value of the power supply voltage, the more legitimate it is to assume that V_{EB} can be neglected. An example investigating the resulting error in neglecting V_{EB} is left as a problem for the student.

The assumption that I_B does not affect the voltage, which is developed in the voltage divider, will now be investigated (see Fig. 4.38).

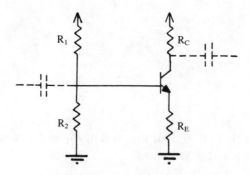

Fig. 4.38

Example (4.15). Given V_{CC}, V_{EB}, I_C, assume I_B does not affect the voltage divider, V_{RC}, V_{EC}, R_1, and determine R_C, R_E, R_2.

$$V_{RE} + V_{EC} + V_{RC} = V_{CC}$$

Therefore,

$$\gamma V_{CC} + V_{EC} + V_{RC} = V_{CC} \rightarrow V_{EC} + V_{RC} = V_{CC} - \gamma V_{CC}$$

Therefore,

$$2V_{RC} = V_{CC} - \gamma V_{CC}$$

Therefore,

$$V_{RC} = \frac{V_{CC}(1 - \gamma)}{2}, R_C = \frac{V_{RC}}{I_C} = \frac{V_{CC}(-\gamma)}{2I_C} \tag{4.25}$$

$$R_E = \frac{V_{RE}}{I_E} = \frac{\gamma V_{CC}}{I_E} = \frac{\gamma V_{CC}}{I_C + I_B} = \frac{\gamma V_{CC}}{I_C + I_C/\beta} = \frac{\gamma V_{CC}}{I_C(1 + 1/\beta)}$$

$$R_E = \frac{\beta\,\gamma V_{CC}}{I_C(\beta + 1)} \tag{4.26}$$

$$V_{gnd-B} = V_{RE} + V_{EB} = \gamma V_{CC} + V_{EB}$$

$$V_{gnd-B} = V_{R2} = V_{CC}\frac{R_2}{R_1 + R_2}$$

$$\gamma V_{CC} + V_{EB} = V_{CC}\frac{R_2}{R_1 + R_2}$$

Choosing a value for R_1, we have

$$(R_1 + R_2)(\gamma V_{CC} + V_{EB}) = V_{CC}R_2$$
$$R_1(\gamma V_{CC} + V_{EB}) + R_2(\gamma V_{CC} + V_{EB}) = V_{CC}R_2$$
$$R_1(\gamma V_{CC} + V_{EB}) = V_{CC}R_2 - R_2(\gamma V_{CC} + V_{EB})$$
$$R_1(\gamma V_{CC} + V_{EB}) = R_2(V_{CC} - (\gamma V_{CC} + V_{EB}))$$
$$= R_2(V_{CC} - \gamma V_{CC} - V_{EB})$$
$$R_1(\gamma V_{CC} + V_{EB}) = R_2(V_{CC}(1 - \gamma) - V_{EB})$$

$$R_2 = \frac{R_1(\gamma V_{CC} + V_{EB})}{V_{CC}(1 - \gamma) - V_{EB}} \tag{4.27}$$

Let us use these formulas and compare the results against the previous design example.

Example (4.16). Let

$$V_{RE} = 1\ V$$

Therefore,

$$\gamma = \frac{V_{RE}}{V_{CC}} = 0.1$$

if $V_{CC} = +10$ V, $\beta = 100$, $V_{EB} = 0.6$ V, $I_{CQ} = 5$ mA, $R_1 = 42$ k, then

$$R_C = \frac{V_{CC}(1 - \gamma)}{2I_C} = \frac{10(1 - 0.1)}{2(5)} = 0.9\ k$$

$$R_E = \frac{\beta\,\gamma V_{CC}}{(\beta + 1)\,I_C} = \frac{100(0.1)(10)}{101(5)} = \frac{100}{505} = 0.19802 \text{ k}$$

$$R_2 = \frac{R_1\,(\gamma V_{CC} + V_{EB})}{V_{CC}\,(1 - \gamma) - V_{EB}} = \frac{42(0.1(10) + 0.6)}{10(1. - 0.1) - 0.6} = \frac{42(1.6)}{8.4} = 8 \text{ k}$$

By assuming a value of 42k for R_1, the computations for R_2 yields a value of 8k. This differs from the exact equations where R_2 is calculated to be 10.67 k.

We also note that the values for R_C and R_E are identical to the values that were computed using the exact formulas.

If we use exact analysis formulas, which will be developed later in this chapter, we can check on the inaccuracies that result in using the above approximations:

$$I_B = \frac{V_{CC} - V_{EB}\left(1 + \dfrac{R_1}{R_2}\right)}{R_1 + (\beta + 1)\,R_E\left(1 + \dfrac{R_1}{R_2}\right)}$$

$$I_B = \frac{10 - 0.6\,(1 + 42/8)}{42 + (101)\,(0.19802)\,(1 + 42/8)}$$

$$= \frac{10 - 3.75}{42 + 125} = \frac{6.25}{167} = 0.03743 \text{ mA}$$

$$I_C = \beta I_B = 100(0.03743) = 3.743 \text{ mA}$$

$$V_{RC} = I_C R_C = (3.743)(0.9) = 3.37 \text{ V}$$

$$I_E = (\beta + 1)I_B = 101(0.03743) = 3.78 \text{ mA},$$

$$V_{RE} = I_E R_E = 3.78(0.198) = 0.749 \text{ V}$$

$$V_{EC} = V_{CC} - V_{RC} - V_{RE} = 10 - 3.37 - 0.749 = 5.88 \text{ V}$$

$$V_{gnd\text{-}B} = V_{RE} + V_{EB} = 0.749 + 0.6 = 1.349 \text{ V}$$

$$I_2 = \frac{V_{R2}}{R_2} = \frac{1.349}{8} = 0.169 \text{ mA}$$

$$I_1 = I_B + I_2 = 0.0374 + 0.169 = 0.206 \text{ mA}$$

Check:

$$I_1 = \frac{V_{R1}}{R_1} = \frac{V_{CC} - V_{R2}}{R_1} = \frac{10 - 1.349}{42} = \frac{8.65}{42} = 0.206 \text{ mA}$$

which checks!

Let us compare the results using the approximation against "designed-for" parameters by referring to Table 4.1.

We can see that there is a considerable loss in accuracy by employing the approximation formulas. However, the greater the bleeder current flowing down through R_1 and R_2, the less significant will be the base current compared to the

Table 4.1

Parameter	Designed-for Parameter Value	Approximation Results	% Difference— Approximation Results Compared to Design Parameter Value
I_B	0 mA	0.0374 mA	∞
I_C	5.0 mA	3.743 mA	25.2%
I_E	5.0 mA	3.78 mA	24.4%
V_{RC}	4.5 V	3.37 V	25.2%
V_{RE}	1.0 V	0.749 V	25.2%
V_{EC}	4.5 V	5.88 V	30.7%

bleeder current. Thus, by reducing the values of R_1 and R_2, design accuracy should improve.

4.4.2 Analysis of the Universal Stabilized Common Emitter Stage

With the design formulas developed, it is desirable to be able to determine if our computed results for the four resistors are, in fact, the correct values.

Example (4.17). Refer to the diagram of Fig. 4.39. Given β, V_{CC}, V_{EB}, R_1, R_2, R_C, R_E, determine I_B, I_C, V_{RC}, I_E, V_{RE}, V_{R2}, I_2, I_1.

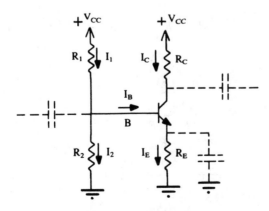

Fig. 4.39

$$V_{gnd\text{-}B} = V_{R2} = V_{CC} - V_{R1} = V_{CC} - I_1 R_1 \qquad (4.28)$$
$$V_{gnd\text{-}B} = I_2 R_2$$
$$V_{gnd\text{-}B} = V_{RE} + V_{EB} = I_E R_E + V_{EB} = (I_C + I_B)\, R_E + V_{EB}$$
$$= (\beta\, I_B + I_B)\, R_E + V_{EB}$$
$$= I_B\, (\beta + 1) R_E + V_{EB}$$

Also,

$$I_1 = I_2 + I_B \qquad (4.29)$$

Thus,

$$I_B = I_1 - I_2$$
$$= \frac{V_{CC} - V_{gnd\text{-}B}}{R_1} - \frac{V_{gnd\text{-}B}}{R_2}$$
$$= \frac{V_{CC}}{R_1} - \frac{V_{gnd\text{-}B}}{R_1} - \frac{V_{gnd\text{-}B}}{R_2} = \frac{V_{CC}}{R_1} - V_{gnd\text{-}B}\left(\frac{1}{R_1} + \frac{1}{R_2}\right)$$

From (4.28) and (4.29):

$$I_B = \frac{V_{CC}}{R_1} - [I_B\, (\beta + 1)\, R_E + V_{EB}]\left(\frac{1}{R_1} + \frac{1}{R_2}\right)$$
$$I_B = \frac{V_{CC}}{R_1} - I_B\, (\beta + 1)\, R_E \left(\frac{1}{R_1} + \frac{1}{R_2}\right) - V_{EB}\left(\frac{1}{R_1} + \frac{1}{R_2}\right)$$

Therefore,

$$I_B + I_B\, (\beta + 1)\, R_E \left(\frac{1}{R_1} + \frac{1}{R_2}\right) = \frac{V_{CC}}{R_1} - V_{EB}\left(\frac{1}{R_1} + \frac{1}{R_2}\right)$$
$$I_B \left[1 + (\beta + 1) R_E \left(\frac{1}{R_1} + \frac{1}{R_2}\right)\right] = \frac{V_{CC}}{R_1} - V_{EB}\left(\frac{1}{R_1} + \frac{1}{R_2}\right)$$

Therefore,

$$I_B = \frac{\dfrac{V_{CC}}{R_1} - V_{EB}\left(\dfrac{1}{R_1} + \dfrac{1}{R_2}\right)}{1 + (\beta + 1)\, R_E \left(\dfrac{1}{R_1} + \dfrac{1}{R_2}\right)}$$
$$= \frac{\dfrac{V_{CC}}{R_1} - V_{EB}\left(\dfrac{R_1 + R_2}{R_1 R_2}\right)}{\dfrac{R_1}{R_1} + (\beta + 1)\, R_E \left(\dfrac{R_1 + R_2}{R_1 R_2}\right)}$$

$$I_B = \dfrac{V_{CC} - V_{EB} \left(\dfrac{R_1 + R_2}{R_2} \right)}{R_1 + (\beta + 1) R_E \left(\dfrac{R_1 + R_2}{R_2} \right)}$$

Therefore, we have

$$I_B = \dfrac{V_{CC} - V_{EB} \left(1 + \dfrac{R_1}{R_2} \right)}{R_1 + (\beta + 1) R_E \left(1 + \dfrac{R_1}{R_2} \right)} \qquad (4.30)$$

If the derivation of the formula for I_B is correct, the formula for I_B for the emitter stabilized common emitter stage should be a simpler version of formula (4.30). For the emitter stabilized common emitter stage, R_2 is not present, i.e., its value is infinite. If $R_2 = \infty$ is substituted into formula (4.30), the result is

$$I_B = \dfrac{V_{CC} - V_{EB}}{R_1 + (\beta + 1) R_E}$$

This is the formula that we developed for the emitter stabilized common emitter stage.

Carrying our check on the universal bias formula further, the unstabilized common emitter stage formula for I_B should evolve out of the above formula. In the unstabilized case $R_E = 0$, and thus we have

$$I_B = \dfrac{V_{CC} - V_{EB}}{R_1}$$

which checks!

Developing the analysis formulas further,

$$I_C = \beta I_B$$

and

$$V_{RC} = I_C R_C$$
$$V_{EC} = V_{CC} - V_{RC} - V_{RE}$$
$$I_E = I_C + I_B$$
$$V_{RE} = I_E R_E$$

also

$$V_{gnd-B} = V_{RE} + V_{EB} = V_{R2}$$

also

$$I_2 = \frac{V_{R2}}{R_2}$$

and

$$I_1 = I_2 + I_B$$
$$V_{R1} = I_1 R_1$$

or

$$V_{R1} = V_{CC} - V_{R2}$$

which should check on each other.

Example (4.18). Let us check our design Example (4.14) for correctness by employing the above analysis formulas:

$$I_B = \frac{V_{CC} - V_{EB}\left(1 + \dfrac{R_1}{R_2}\right)}{R_1 + (\beta + 1) R_E \left(1 + \dfrac{R_1}{R_2}\right)} = \frac{10 - 0.6\left(1 + \dfrac{42}{10.67}\right)}{42 + (101)(0.198)\left(1 + \dfrac{42}{10.67}\right)}$$

$$= \frac{10 - 0.6(4.936)}{42 + (101)(0.198)(4.936)} = \frac{10 - 2.96}{42 + 98.7} = \frac{7.038}{140.7} = 0.05 \text{ mA}$$

$I_C = \beta I_B = 100(0.05) = 5 \text{ mA}, \ V_{RC} = I_C R_C = 5(0.9) = 4.5 \text{ V}$
$I_E = I_C + I_B = 5.05 \text{ mA}, \ V_{RE} = I_E R_E = 5.05(0.198) = 1 \text{ V}$
$V_{R2} = V_{RE} + V_{EB} = 1 + 0.6 = 1.6 \text{ V}$

$$I_2 = \frac{V_{R2}}{R_2} = \frac{1.6}{10.67} = 0.15 \text{ mA}$$

$I_1 = I_2 + I_B = 0.15 + 0.05 = 0.2 \text{ mA}$
$V_{R1} = I_1 R_1 = 0.2 \ (42) = 8.4 \text{ V}$
$V_{EC} = V_{CC} - V_{RC} - V_{RE} = 10 - 4.5 - 1$
$V_{EC} = 4.5 \text{ V}$

which checks against our design!

In a manner similar to the development of design approximation formulas for the universal stabilized common emitter stage, we can develop approximate analysis formulas. Referring to the input resistor voltage divider, assume that the base current flow is a negligible portion of the bleeder current (see Fig. 4.40 (a,b)). Thus,

$$V_{R2} = V_{CC} \frac{R_2}{R_1 + R_2}$$

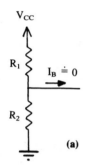

Fig. 4.40(a)

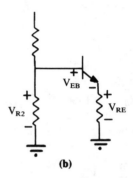

Fig. 4.40(b)

Also,

$$V_{R2} = V_{RE} + V_{EB}$$

thus,

$$V_{RE} = V_{R2} - V_{EB}$$

Therefore,

$$I_E = \frac{V_{RE}}{R_E}$$

Also

$$I_C \doteq I_E$$

thus,

$$V_{RC} = I_C R_E \quad \text{and} \quad V_{EC} = V_{CC} - V_{RC} - V_{RE}$$

Example (4.19). Let us use the original design that we developed for the universal stabilized common emitter stage (Example (4.14)). The design values were:

$V_{RC} = V_{EC} = 4.5$ V
$I_{CQ} = 5$ mA
$V_{RE} = 1$ V
$R_1 = 42$ k
$R_2 = 10.67$ k
$R_C = 0.9$ k
$R_E = 0.19802$ k

Substituting into the approximate analysis formulas:

$$V_{R2} = V_{CC} \frac{R_2}{R_1 + R_2} = 10 \frac{10.67}{42 + 10.67} = 2.026 \text{ V} \qquad (4.31)$$
$$V_{RE} = V_{R2} - V_{EB} = 2.026 - 0.6 = 1.426 \text{ V}$$
$$I_E = \frac{V_{RE}}{R_E} = \frac{1.426 \text{ V}}{0.198 \text{ k}} = 7.2 \text{ mA}$$
$$I_C = I_E = 7.2 \text{ mA}$$
$$V_{RC} = I_C R_C = 7.2(0.9) = 6.48 \text{ V}$$
$$V_{EC} = V_{CC} - V_{RC} - V_{RE} = 10 - 6.48 - 1.426 = 2.094 \text{ V}$$

A comparison of the correct values and approximate analysis values is given in Table 4.2.

Again we can see that unless the bleeder current flowing through R_1 and R_2 is appreciably larger than the base current ($> 10 I_B$), significant inaccuracies will be experienced in computing circuit parameters by assuming that I_B is negligible.

How much (if at all) has the use of universal stabilization improved the stability of the operating point? Let us check this by allowing the β to increase by 30%.

Example (4.20). Assuming the same increase as in the unstabilized case, we write

$$I_B = \frac{V_{CC} - V_{EB} \left(1 + \frac{R_1}{R_2} \right)}{R_1 + (\beta + 1) R_E (1 + R_1/R_2)}$$

Table 4.2

Parameter	Circuit Parameter Values	Result by use of Approximation Formulae	% Difference
I_B	0.05 mA	Not Calculated	—
I_C	5.0 mA	7.2 mA	44.0%
I_E	5.05 mA	7.2 mA	42.6%
V_{RC}	4.5 V	6.48 V	44.0%
V_{RE}	1.0 V	1.426 V	42.6%
V_{EC}	4.5 V	2.094 V	53.5%

$$I_B = \frac{10 - 0.6 (1 + 42/10.67)}{42 + 131 (0.19802) (1 + 42/10.67)}$$

$$= \frac{10 - 2.962}{42 + 128.05} = \frac{7.038}{170.05} = 0.04139 \text{ mA}$$

$I_C = 5.381$ mA

$V_{RC} = I_C R_C = 5.381 (0.9) = 4.843$ V

$I_E = (\beta + 1)I_B = 5.463$ mA, $V_{RE} = I_E R_E = 5.463(0.19802) = 1.082$ V

$V_{EC} = V_{CC} - V_{RC} - V_{RE} = 10 - 4.843 - 1.082 = 4.076$ V

$V_{gnd\text{-}B} = V_{RE} + V_{EB} = 1.082 + 0.6 = 1.682$V $(= V_{R2})$

$$I_2 = \frac{V_{R2}}{R_2} = \frac{1.682}{10.67} = 0.1576 \text{ mA}$$

$V_{R1} = V_{CC} - V_{R2} = 10 - 1.682 = 8.318$ V

$$I_1 = \frac{V_{R1}}{R_1} = \frac{8.318}{42} = 0.19804 \text{ mA}$$

Check $I_1 = I_2 + I_B = .15763 + .041389 = .0199$, which checks as before.

The emitter-to-collector voltage changed from 4.5 to 4.076 V $(= -9.4\%)$ for a change in β of 30% (100–130). For the unstabilized case where β increased from 100 to 130, V_{EC} decreased from 5 to 3.5 V $(= -30\%)$.

Example (4.21). Let us see what happens if β decreases by 30%. We write

$$I_B = \frac{V_{CC} - V_{EB} (1 + R_1/R_2)}{R_1 + (\beta + 1) R_E (1 + R_1/R_2)}$$

$$= \frac{10 - 0.6 (1 + 42/10.67)}{42 + 71(0.19802) (1 + 42/10.67)}$$

$$I_B = \frac{10 - 0.6\,(4.936)}{42 + 71(0.19802)\,(4.936)} = \frac{10 - 2.962}{42 + 69.4} = \frac{7.038}{111.4}$$

$$= 0.0632 \text{ mA}$$

$$I_C = \beta I_B = 70(0.0632) = 4.422 \text{ mA}$$

$$V_{RC} = I_C R_C = 4.422\,(0.9) = 3.98 \text{ V}$$

$$I_E = (\beta + 1)I_B = 4.486 \text{ mA}$$

$$V_{RE} = I_E R_E = 4.486\,(0.19802) = 0.888 \text{ V}$$

$$V_{EC} = V_{CC} - V_{RC} - V_{RE} = 10 - 3.98 - 0.888 = 5.132 \text{ V}$$

Thus, we obtain

$$\% \text{ change in } V_{EC} = \left[\frac{4.5 - 5.132}{4.5}\right] \times 100 = 14\%$$

which compares to a 30% change for the unstabilized case.

Of interest at this point is the rate of change of the collector voltage as a function of changes in β. By simply taking the first derivative of the expression for V_{RC} as a function of β, we arrive at the following:

$$I_B = \frac{V_{CC} - V_{EB}\,(1 + R_1/R_2)}{R_1 + (\beta + 1)\,R_E\,(1 + R_1/R_2)} \quad \begin{array}{l}\text{(universal stabilized} \\ \text{common emitter)}\end{array}$$

$$= \frac{T}{R_1 + (\beta + 1)\,K}$$

where $T = V_{CC} - V_{EB}\,(1 + R_1/R_2)$
      $K = R_E\,(1 + R_1/R_2)$

$$I_C = \beta I_B = \frac{\beta T}{R_1 + (\beta + 1)K}$$

$$V_{RC} = I_C R_C = \frac{\beta Q}{R_1 + (\beta + 1)K}$$

where $Q = TR_C$

$$V_{RC} = \frac{\beta Q}{R_1 + \beta K + K} = \frac{\beta Q}{\beta k + R_1 + K} = \frac{\beta Q}{\beta K + M}$$

where $M = R_1 + K$

$$V_{RC} = \frac{\beta\,Q/K}{\beta + M/K} = \frac{A\,\beta}{\beta + C}$$

where $A = Q/K,\ C = M/K$

$$\frac{dV_{RC}}{d\beta} = A\,[\beta\,(-(\beta + C)^{-2} + (\beta + C)^{-1}]$$

$$\frac{dV_{RC}}{d\beta} = A\left[\frac{-\beta}{(\beta + C)^2} + \frac{1}{\beta + C}\right]$$

$$= A\left[\frac{-\beta}{(\beta + C)^2} + \frac{\beta + C}{(\beta + C)^2}\right]$$

$$= A\frac{C}{(\beta + C)^2} = \frac{Q}{K}\frac{M/K}{(\beta + M/K)^2} = \frac{QM}{K^2(\beta + M/K)^2}$$

$$= \frac{QM}{(K\beta + M)^2}$$

$$= \frac{TR_C(R_1 + R_E(1 + R_1/R_2))}{[\beta R_E(1 + R_1/R_2) + R_1 + R_E(1 + R/R_2)]^2}$$

$$= \frac{[V_{CC} - V_{EB}(1 + R_1/R_2)]R_C[R_1 + R_E(1 + R_1/R_2)]}{[(\beta + 1)R_E(1 + R_1/R_2) + R_1]^2} \tag{4.32}$$

This computation for the rate of change of V_{RC} with respect to β for the universally stabilized common-emitter stage is quite involved, particularly if the computation must be repeated many times. Therefore, the program written in BASIC, which computes V_{RC} and V_{EC} *versus* β, also includes a computation for the rate of change of V_{RC}. It will be noted in the various computations for different combinations of δ and γ that in all cases the rate of change of V_{RC} with respect to β decreases as β increases (see Appendix A).

We can now see the value of employing universal stabilization to a common emitter stage. The charts shown in Appendix A are the results of employing the BASIC program, and Table A.1 is a summary of all of the computer runs for various combinations of δ and γ.

4.4.3 Common Emitter with Collector Feedback

Another approach to stabilizing the common emitter stage is by connecting a resistor from the collector to the base. Let us examine how this circuit operates. Referring to Fig. 4.41, let us assume that the transistor β has increased above its nominal value. If β increases then I_C tends to increase. The increase in I_C causes an increase in the voltage drop across R_C, with a consequent lowering of the potential at the transistor collector. With the tendency to reduce the collector potential, less voltage will tend to be across R_F, thereby reducing the current in R_F and the base current. The reduction in the base current tends to reduce the collector current increase (due to the increased β). This phenomenon is referred to as negative feedback.

Now that we have an overall concept of the operation of the common emitter stage with collector feedback, let us develop the design formulas.

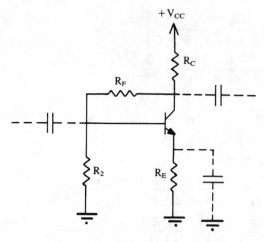

Fig. 4.41

4.5 DESIGN FORMULAS FOR A COMMON EMITTER STAGE WITH COLLECTOR FEEDBACK

Referring to Fig. 4.42, let $V_{RC} = V_{EC}$, $I_6 = \delta I_B$, $V_{RE} = \gamma V_{CC}$. Given V_{CC}, β, I_C, δ, γ, V_{EB}, determine R_C, R_F, R_E, R_2.

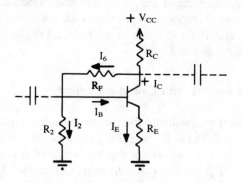

Fig. 4.42

$$V_{RE} = I_E R_E = (I_C + I_B)\, R_E = (I_C + I_C/\beta)\, R_E$$

$$= I_C R_E\,(1 + 1/\beta) = \frac{\beta + 1}{\beta}\, I_C R_E$$

$$\tag{4.33}$$

$$V_{RE} = \gamma V_{CC} = \frac{\beta + 1}{\beta} I_C R_E$$

Therefore,

$$R_E = \frac{\beta \gamma V_{CC}}{(\beta + 1) I_C}$$
$$V_{RE} + V_{EC} + V_{RC} = V_{CC}$$
$$V_{RE} + 2V_{RC} = V_{CC}$$
$$V_{RE} + 2(I_C + I_6) R_C = V_{CC}$$

$$\gamma V_{CC} + 2(I_C + \delta I_B)R_C = V_{CC}$$
$$\gamma V_{CC} + 2(I_C + \delta I_C/\beta)R_C = V_{CC}$$
$$2(I_C + \delta I_C/\beta)R_C = V_{CC} - \delta V_{CC}$$
$$2I_C \left(\frac{\beta + \delta}{\beta} \right) R_C = V_{CC} (1 - \delta)$$

Therefore,

$$R_C = \frac{\beta V_{CC} (1 - \gamma)}{2I_C (\beta + \delta)} \qquad\qquad (4.34)$$
$$I_6 = \frac{V_{RE} + V_{EC} - V_{gnd\text{-}B}}{R_F} = \frac{V_{RE} + V_{RC} - (V_{RE} + V_{EB})}{R_F}$$
$$= \frac{V_{RE} + V_{RC} - V_{RE} - V_{EB}}{R_F}$$
$$I_6 = \frac{V_{RC} - V_{EB}}{R_F} \Rightarrow R_F = \frac{V_{RC} - V_{EB}}{I_6} = \frac{V_{RC} - V_{EB}}{\delta I_B} = \frac{V_{RC} - V_{EB}}{\delta I_C/\beta}$$
$$R_F = \frac{\beta (V_{RC} - V_{EB})}{\delta I_C}$$

However,

$$V_{CC} = 2V_{RC} + V_{RE} = 2V_{RC} + \gamma V_{CC}$$

Therefore,

$$2V_{RC} = V_{CC} (1 - \gamma) \rightarrow V_{RC} = \frac{V_{CC}}{2} (1 - \gamma)$$
$$R_F = \frac{\beta}{\delta I_C} \left[\frac{V_{CC}}{2} (1 - \gamma) - V_{EB} \right]$$

$$= \frac{\beta}{\delta I_C} \left[\frac{V_{CC}(1 - \gamma) - 2V_{EB}}{2} \right]$$

$$R_F = \frac{\beta}{2\delta I_C} [V_{CC}(1 - \gamma) - 2V_{EB}] \tag{4.35}$$

$$V_{gnd\text{-}B} = I_2 R_2$$

Therefore,

$$R_2 = \frac{V_{gnd\text{-}B}}{I_2} = \frac{V_{RE} + V_{EB}}{I_6 - I_B} = \frac{\gamma V_{CC} + V_{EB}}{\delta I_B - I_B}$$

$$= \frac{\gamma V_{CC} + V_{EB}}{I_B (\delta - 1)} = \frac{\gamma V_{CC} + V_{EB}}{(I_C/\beta)(\delta - 1)}$$

$$R_2 = \frac{\beta (\gamma V_{CC} + V_{EB})}{I_C (\delta - 1)} \tag{4.36}$$

Now that we have developed the design formulas, an illustrative example is in order.

Example (4.22).

Let $I_6 = 4I_B$,

thus,

$$\delta = \frac{I_6}{I_B} = 4, \quad V_{RE} = 1 \text{ V}$$

Also,

$$V_{CC} = 10 \text{ V}, \quad \beta = 100, \quad V_{EB} = 0.6 \quad I_{CQ} = 5 \text{ mA}, \quad \gamma = \frac{V_{RE}}{V_{CC}} = 0.1$$

$$R_C = \frac{V_{CC}(1 - \gamma)\beta}{2I_C (\beta + \delta)} = \frac{10(1 - 0.1)100}{2(5)(100 + 4)} = \frac{90}{104} = 0.8654 \text{ k}$$

$$R_E = \frac{\gamma V_{CC}}{I_C}\left(\frac{\beta}{\beta + 1}\right) = \frac{0.1(10)}{5}\left(\frac{100}{101}\right) = \frac{20}{101} = 0.19802 \text{ k}$$

$$R_F = \frac{\beta}{2\delta I_C} [V_{CC}(1 - \gamma) - 2V_{EB}] = \frac{100}{2(4)(5)} [10(1 - 0.1) - 2(0.6)]$$

$$= 2.5 [9 - 1.2] = 2.5 [7.8] = 19.5 \text{ k}$$

$$R_2 = \frac{\beta (\gamma V_{CC} + V_{EB})}{I_C (\gamma - 1)} = \frac{100 [0.1(10) + 0.6]}{5(4 - 1)}$$

$$= \frac{100(1.6)}{15} = \frac{160}{15} = 10.67 \text{ k}$$

With the design formulas developed, it is appropriate to develop the corresponding analysis equations for the common emitter stage with collector feedback.

4.5.1 Analysis Equations for a Common Emitter Stage with Collector Feedback and an Emitter Resistor

Refer to Fig. 4.43. Given V_{CC}, V_{EB}, R_C, R_2, R_G, R_E, determine I_C, V_{RC}, V_{EC}.

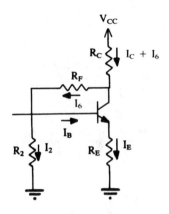

Fig. 4.43

$$I_6 = I_2 + I_B$$

Therefore,

$$I_B = I_6 - I_2$$

$$I_6 = \frac{V_{CC} - V_{RC} - V_{gnd\text{-}B}}{R_F}$$

$$= \frac{V_{CC} - (I_C + I_6)\,R_C - (V_{RE} + V_{EB})}{R_F}$$

$$= \frac{V_{CC} - I_C R_C - I_6 R_C - V_{RE} - V_{EB}}{R_F}$$

$$= \frac{V_{CC} - V_{RE} - V_{EB}}{R_F} - \frac{I_C R_C}{R_F} - \frac{I_6 R_C}{R_F}$$

$$I_6 + \frac{I_6 R_C}{R_F} = \frac{V_{CC} - V_{RE} - V_{EB}}{R_F} - \frac{\beta I_B R_C}{R_F}$$

$$I_6 \left(1 + \frac{R_C}{R_F}\right) = \frac{V_{CC} - V_{RE} - V_{EB} - \beta\,I_B R_C}{R_F}$$

Also,

$$I_2 = \frac{V_{R2}}{R_2} = \frac{V_{RE} + V_{EB}}{R_2} = \frac{(\beta + 1)\,I_B R_E + V_{EB}}{R_2}$$

$$I_6 \left(1 + \frac{R_C}{R_F}\right) = \frac{V_{CC} - (\beta + 1) I_B R_E - V_{EB} - \beta I_B R_C}{R_F}$$

$$I_6 \left(\frac{R_F + R_C}{R_F}\right) = \frac{V_{CC} - V_{EB} - I_B [(\beta + 1) R_E + \beta R_C]}{R_F}$$

$$I_6 = \frac{V_{CC} - V_{EB} - I_B [(\beta + 1) R_E + \beta R_C]}{R_F + R_C}$$

$$I_B = I_6 - I_2$$

$$= \frac{V_{CC} - V_{EB} - I_B [(\beta + 1) R_E + \beta R_C]}{R_F + R_C}$$

$$- \left[\frac{(\beta + 1) I_E R_E + V_{EB}}{R_2}\right]$$

$$= \frac{V_{CC} - V_{EB}}{R_F + R_C} - \frac{V_{EB}}{R_2} - \frac{I_B [(\beta + 1) R_E + \beta R_C}{R_F + R_C}$$

$$- \frac{I_B (\beta + 1) R_E}{R_2}$$

$$I_B + \frac{I_B [(\beta + 1) R_E + \beta R_C]}{R_F + R_C} + \frac{I_B (\beta + 1) R_E}{R_2} = \frac{V_{CC} - V_{EB}}{R_F + R_C} - \frac{V_{EB}}{R_2}$$

$$I_B \left[1 + \frac{(\beta + 1) R_E + \beta R_C}{R_F + R_C} + \frac{(\beta + 1) R_E}{R_2}\right] = \frac{V_{CC} - V_{EB}}{R_F + R_C} - \frac{V_{EB}}{R_2}$$

$$I_B [R_2 (R_F + R_C) + R_2 [(\beta + 1) R_E + \beta R_C] + (\beta + 1) R_E (R_F + R_C)] = (V_{CC} - V_{EB}) R_2 - V_{EB} (R_F + R_C)$$

$$I_B = \frac{(V_{CC} - V_{EB}) R_2 - V_{EB} (R_F + R_C)}{R_2 (R_F + R_C) + R_2 [(\beta + 1) R_E + \beta R_C] + (\beta + 1) R_E (R_F + R_C)}$$

$$I_B = \frac{(V_{CC} - V_{EB}) R_2 - V_{EB} (R_F + R_C)}{R_2 (R_F + R_C) + R_2 [(\beta + 1) R_E + \beta R_C] + (\beta + 1) R_E (R_F + R_C)}$$

$$= \frac{V_{CC} R_2 - V_{EB} R_2 - V_{EB} R_F - V_{EB} R_C}{R_2 R_F + R_2 R_C + R_2 \beta R_E + R_2 R_E + R_2 \beta R_C}$$

$$+ \frac{V_{CC} R_2 - V_{EB} R_2 - V_{EB} R_F - V_{EB} R_C}{(\beta + 1) R_E R_F + (\beta + 1) R_E R_C}$$

$$= \frac{V_{CC} R_2 - V_{EB} (R_2 + R_F + R_C)}{R_2 R_F + R_2 R_C (\beta + 1) + R_E}$$

$$\times \frac{V_{CC} R_2 - V_{EB} (R_2 + R_F + R_C)}{[(\beta + 1) R_2 + (\beta + 1) R_F + (\beta + 1) R_C]}$$

$$= \frac{V_{CC} R_2 - V_{EB} (R_2 + R_F + R_C)}{R_2 R_F + R_2 R_C (\beta + 1) + R_E (\beta + 1) (R_2 + R_F + R_C)}$$

$$I_B = \dfrac{V_{CC} - V_{EB}\left(1 + \dfrac{R_F + R_C}{R_2}\right)}{R_F + R_C\,(\beta + 1) + R_E\,(\beta + 1)\left(1 + \dfrac{R_F + R_C}{R_2}\right)} \qquad (4.37)$$

$I_C = \beta\, I_B$

$I_E = (\beta + 1)\, I_B$

$V_{RE} = I_E R_E$

$V_{R2} = V_{gnd\text{-}B} = V_{RE} + V_{EB}$

Therefore,

$$I_2 = \frac{V_{R2}}{R_2}$$

$$I_6 = I_2 + I_B \qquad (4.38)$$

$$V_{RC} = (I_6 + I_C)\,R_C \qquad (4.39)$$

or, for the purposes of a computer program (see Appendix A), the various circuit currents were not computed within the program. Therefore, the following derivation (Example (4.23)) was performed.

Example (4.23). Determine V_{RC} in terms of known quantities for a common emitter stage with collector feedback and an emitter resistor.

Given V_{EB}, R_2, R_C, R_E, β, I_B:

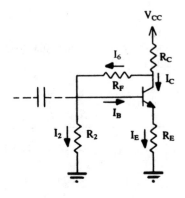

Fig. 4.44

$$V_{RC} = (\overset{I_6}{I_G} + I_C)R_C$$

$$I_6 = I_2 + I_B$$

154

Thus,

$$V_{RC} = (I_2 + I_B + I_C)R_C$$
$$= I_2R_C + (\beta + 1) I_ER_C$$

However,

$$I_2 = \frac{V_{RE} + V_{EB}}{R_2}$$

Thus,

$$V_{RC} = \frac{(V_{RE} + V_{EB}) R_C}{R_2} + (\beta + 1) I_BR_C$$

$$= \frac{V_{RE}R_C}{R_2} + \frac{V_{EB}R_C}{R_2} + (\beta + 1) I_BR_C$$

$$V_{RC} = (\beta + 1) I_B \frac{R_ER_C}{R_2} + \frac{V_{EB}R_C}{R_2} + (\beta + 1) I_BR_C$$

$$= (\beta + 1) I_BR_C (1 + R_E/R_2) + V_{EB}R_C/R_2 \tag{4.40}$$

Example (4.24). Let $\beta = 100$, $I_B = .05$ mA, $R_C = .8654$ k, $R_E = .19802$ k, $R_2 = 10.67$ k:

$$V_{RC} = (\beta + 1) I_BR_C (1 + R_E/R_2) + V_{EB}R_C/R_2$$
$$= (101) (0.05) (0.8654) (1 + 0.19802/10.67) + 0.6 (0.8654)/10.67$$
$$= 4.451 + 0.0487$$
$$V_{RC} = 4.5 \text{ V}$$

Example (4.25). Let us use these analysis formulas to check the design example component values (Example (4.22)):

$$I_B = \frac{V_{CC} - V_{EB} \left(1 + \dfrac{R_F + R_2}{R_2}\right)}{R_F + (\beta + 1) \left[R_C + R_E \left(1 + \dfrac{R_F + R_C}{R_2}\right)\right]}$$

$$= \frac{10 - 0.6 \left(1 + \dfrac{19.5 + 0.8654}{10.67}\right)}{19.5 + 101 \left[0.8654 + 0.19802 \left(1 + \dfrac{19.5 + 0.8654}{10.67}\right)\right]}$$

$$I_B = \frac{10 - 0.6\,(2.909)}{19.5 + 101\,[0.8654 + 0.576]} = \frac{10 - 1.745}{19.5 + 101\,(1.441)}$$

$$= \frac{8.255}{19.5 + 145.6} = \frac{8.255}{165.1} = 0.05 \text{ mA}$$

$I_C = \beta I_B = 5 \text{ mA}$, $I_E = (\beta + 1)I_B = 101(0.05) = 5.051 \text{ mA}$

$V_{RE} = I_E R_E = 5.051\,(0.19802) = 1 \text{ V}$

$V_{gnd\text{-}B} = V_{RE} + V_{EB} = 1.6 \text{ V}$

$I_{R2} = \dfrac{V_{R2}}{R_2} = \dfrac{1.6}{10.67} = 0.15 \text{ mA}$

$I_6 = I_2 + I_B = 0.15 + 0.05 = 0.2 \text{ mA}$

$V_{RC} = (I_6 + I_C)\,R_C = (0.2 + 5)\,(0.8654) = 4.5 \text{ V}$

which checks!

Therefore, $I_6 = 4I_B$ checks.

4.5.2 Effect of Variations of β

Of interest at this point, would be the effect of variations of β on the circuit stability.

Example (4.26). Let us determine what happens if β increases from a nominal value of 100 to 130 (30% increase).

$$I_B = \frac{V_{CC} - V_{EB}\left(1 + \dfrac{R_F + R_C}{R_2}\right)}{R_F + (\beta + 1)\left[R_C + R_E\left(1 + \dfrac{R_F + R_C}{R_2}\right)\right]}$$

$$= \frac{10 - 0.6\left(1 + \dfrac{19.5 + 0.8635}{10.67}\right)}{19.5 + 131\left[0.8635 + 0.19802\left(1 + \dfrac{19.5 + 0.8635}{10.67}\right)\right]}$$

$$= \frac{10 - 0.6\,(2.908)}{19.5 + 131\,[0.8635 + 0.19802\,(2.908)]}$$

$$= \frac{10 - 1.745}{19.5 + 131\,(0.8635 + 0.5759)}$$

$$= \frac{8.255}{19.5 + 131\,(1.439)} = \frac{8.255}{19.5 + 188.6}$$

$$I_B = \frac{8.255}{208.1} = 0.0397 \text{ mA}$$

$$I_C = \beta I_B = 130\,(0.0397) = 5.158 \text{ mA}$$

$$I_E = (\beta + 1)I_B = 131\,(0.0397) = 5.197 \text{ mA}$$

$$V_{RE} = I_E R_E = 5.197(0.19802) = 1.029 \text{ V}$$

$$V_{gnd\text{-}B} = V_{RE} + V_{EB} = 1.029 + 0.6 = 1.629 \text{ V}$$

$$I_2 = \frac{V_{R2}}{R_2} = \frac{1.623}{10.67} = 0.1527 \text{ mA}$$

$$I_6 = I_2 + I_B = 0.1527 + 0.0397 = 0.1924 \text{ mA}$$

$$V_{RC} = (I_6 + I_C)R_C = (0.1924 + 5.158)\,(0.8635) = 4.62 \text{ V}$$

$$V_{RF} = I_6 R_F = 0.1924\,(19.5) = 3.75 \text{ V}$$

Check:

$$V_{RC} = (\beta + 1)\,I_B R_C \left(1 + \frac{R_E}{R_2}\right) + V_{EB}\frac{R_C}{R_2}$$

$$= 131\,(0.0397)\,(0.8635)\left(1 + \frac{0.19802}{10.67}\right) + 0.6\left(\frac{0.8635}{10.67}\right)$$

$$V_{RC} = 4.571 + 0.0486 = 4.62 \text{ V}$$

$$V_{EC} = V_{CC} - V_{RC} - V_{RE} = 10 - 4.62 - 1.029 = 4.35 \text{ V}$$

which checks!

The percentage change in V_{EC} due to a $+30\%$ change in β is

$$\% \text{ change} = \frac{4.5 - 4.35}{4.5} \times 100 = -3.3\%$$

Example (4.27). For comparison purposes, we should also investigate the effects on circuit conditions caused by a 30% decrease in the transistor β (away from a nominal value of 100).

$$I_B = \frac{V_{CC} - V_{EB}\left(1 + \dfrac{R_F + R_C}{R_2}\right)}{R_F + (\beta + 1)\left[R_C + R_E\left(1 + \dfrac{R_F + R_C}{R_2}\right)\right]}$$

$$I_B = \frac{10 - 0.6\left(1 + \dfrac{19.5 + 0.8635}{10.67}\right)}{19.5 + 71\left[0.8635 + 0.19802\left(1 + \dfrac{19.5 + 0.8635}{10.67}\right)\right]}$$

$$= 0.06783 \text{ mA}$$

$$I_C = \beta I_B = 70(0.06783) \text{ mA} = 4.748 \text{ mA}$$

$I_E = (\beta + 1)I_B = 71(0.06783) = 4.816$ mA

$V_{RE} = I_E R_E = 4.816$ mA (0.19802) k $= 0.9537$ V

$V_{gnd\text{-}B} = V_{RE} + V_{EB} = 0.9537 + 0.6 = 1.554$ V

$I_2 = \dfrac{V_{R2}}{R_2} = \dfrac{1.554}{10.67} = 0.1456$ mA

$I_6 = I_2 + I_B = 0.1456 + 0.0678 = 0.2134$ mA

$V_{RF} = I_6 R_F = 0.2134(19.5) = 4.162$ V

$V_{RC} = (I_6 + I_C)R_C = (0.2134 + 4.478)0.8635 = 4.284$ V

Check:

$$V_{RC} = (\beta + 1)I_B R_C \left(1 + \frac{R_E}{R_2}\right) + V_{EB}R_C/R_2$$

$$= 71(0.06783)(0.8635)\left(1 + \frac{0.10802}{10.67}\right) + 0.6\left(\frac{0.8635}{10.67}\right)$$

$$V_{RC} = 4.236 + 0.0486 = 4.284 \text{ V}$$

which checks!

The percentage change in V_{EC} due to a -30% change in β is

$$\% \text{ change} = \left|\frac{4.5 - 4.762}{4.5}\right| \times 100 = +5.8\%$$

Let us compare these bias arrangement conditions for the common emitter (see Table 4.3):

β (Nominal) $= 100$, $V_{CC} = 10$ V, $I_{CQ} = 5$ mA

We can see the advantage of the use of stabilization, but in the next chapter we will see that there are certain changes in the ac circuit operation which must be allowed for when we use various stabilization techniques.

Table 4.3
($\delta = 4$, $\gamma = 0.1$ where appropriate)

Configuration	$\beta = 100$	$\beta = 130$	$\beta = 70$
Unstabilized	$V_{EC} = 5$ V	3.5 V (-30%)	6.5 V ($+30\%$)
Universal Stabilized	4.5 V	4.076 V (-9.4%)	5.132 V ($+14\%$)
Collector Feedback Stabilized	4.5 V	4.35 V (-3.3%)	4.762 V ($+5.8\%$)

158

In a manner similar to the analysis of the universal bias arrangement, a derivation of the rate of change of V_{RC} with respect to β has been developed for incorporation into the computer analysis program (see Appendix A).

Example (4.28). Derive the rate of change of V_{RC} with respect to β for a common emitter stage with collector feedback.

$$V_{RC} = (I_C + I_6)R_C = (\beta I_B + \delta I_B)R_C$$
$$= \beta I_B R_C + \delta I_B R_C = T + M$$

where

$$T = \beta I_B R_C$$
$$M = \delta I_B R_C$$

$$T = \beta \left\{ \dfrac{R_C \left[V_{CC} - V_{EB} \left(1 + \dfrac{R_F + R_C}{R_2} \right) \right]}{R_F + (\beta + 1) \left[R_C + R_E \left(1 + \dfrac{R_F + R_C}{R_2} \right) \right]} \right\}$$

Let

$$\Delta = R_C \left[V_{CC} - V_{EB} \left(1 + \dfrac{R_F + R_C}{R_2} \right) \right]$$

$$\theta = R_C + R_E \left(1 + \dfrac{R_F + R_C}{R_2} \right)$$

$$T = \dfrac{\beta \theta \Delta}{R_F + \beta \theta + \theta} = \dfrac{\beta \theta \Delta}{\beta \theta + \theta + R_F}$$

$$= \Delta \dfrac{\beta}{\beta \theta + \theta + R_F}$$

$$= \dfrac{\Delta}{\theta} \dfrac{\beta}{\beta + 1 + R_F/\theta}$$

Let

$$K = 1 + R_F/\theta$$

$$T = \dfrac{\Delta}{\theta} \dfrac{\beta}{\beta + K}$$

$$\dfrac{dT}{d\beta} = \dfrac{\Delta}{\theta} \left[\dfrac{-\beta}{(\beta + K)^2} + \dfrac{1}{\beta + K} \right]$$

$$= \dfrac{\Delta}{\theta} \left[\dfrac{-\beta}{(\beta + K)^2} + \dfrac{\beta + K}{(\beta + K)^2} \right]$$

$$\frac{dT}{d\beta} = \frac{\Delta}{\theta}\left[\frac{K}{(\beta + K)^2}\right]$$

$$M = \delta I_B R_C = \delta R_C I_B$$

$$= \frac{\delta R_C\left[V_{CC} - V_{EB}\left(1 + \frac{R_F + R_C}{R_2}\right)\right]}{R_F + (\beta + 1)\left[R_C + R_E\left(1 + \frac{R_F + R_C}{R_2}\right)\right]}$$

$$= \frac{\delta\Delta}{R_F + \beta\theta + \theta} = \frac{\delta\Delta}{\beta\theta + \theta + R_F}$$

$$= \delta\Delta\frac{1}{\beta\theta + \theta + \beta R_F}$$

$$= \frac{\delta\Delta}{\theta}\frac{1}{\beta + \frac{1 + \beta R_F}{\theta}} \cdot \frac{\delta\Delta}{\theta}\frac{1}{\beta + K}$$

$$\frac{dM}{d\beta} = \beta\frac{\theta\Delta}{\theta}\left[\frac{1}{(\beta + K^2}\right] \cdot - \frac{\delta\Delta}{\theta}\cdot\frac{1}{(\beta+K)^2}$$

$$\frac{dV_{RC}}{d\beta} = \frac{dT}{d\beta} + \frac{dM}{d\beta} \mp \frac{\Delta}{\theta}\left[\frac{k}{(\beta + K^2}\right]$$

$$- \frac{\delta\Delta}{\theta}\left[\frac{1}{(\beta + K)^2}\right]$$

$$\frac{dV_{RC}}{d\beta} = \frac{\Delta}{\theta}\left[\frac{K - \theta}{(\beta + K)^2}\right]$$

$$= \frac{\Delta}{\theta}\left\{\frac{\frac{\theta + R_F}{\theta} - \delta}{\left[\beta + \left(\frac{\theta + R_F}{\theta}\right)\right]^2}\right\}$$

$$= \frac{\Delta}{\delta}\left\{\frac{\frac{\theta + R_F - \delta\theta}{\theta}}{\left(\frac{\beta\theta + \theta + R_F}{\theta}\right)^2}\right\}$$

$$\frac{dV_{RC}}{d\beta} = \frac{\Delta}{\delta} \frac{\dfrac{\theta + R_F - \delta\theta}{\theta}}{\dfrac{(\beta\theta + \theta + R_F)^2}{\theta^2}}$$

$$= \frac{\Delta}{\theta} \left[\frac{\theta + R_F - \delta\theta \cdot \theta^2}{\theta (\beta\theta + \theta + R_F)^2} \right]$$

$$= \frac{\Delta (\theta + R_F - \delta\theta)}{(\beta\theta + \theta + R_F)^2}$$

$$= \frac{\Delta [\theta (1 - \delta) + R_F]}{[\theta (\beta + 1) + R_F]^2}$$

$$\frac{dV_{RC}}{d\beta} = \frac{R_C \left[V_{CC} - V_{EB} \left(1 + \dfrac{R_F + R_C}{R_2} \right) \right]}{\left\{ (\beta + 1) \left[R_C + R_E \left(1 + \dfrac{R_F + R_C}{R_2} \right) \right] + R_F \right\}^2}$$

$$\times \frac{\left\{ (1 - \delta) \left[R_C + R_E \left(1 + \dfrac{R_F + R_C}{R_2} \right) \right] + R_F \right\}}{\left\{ (\beta + 1) \left[R_C + R_E \left(1 + \dfrac{R_F + R_C}{R_2} \right) \right] + R_F \right\}^2}$$

(4.41)

4.5.3 Percentage Change in V_{EC} Due to Changes in β to Values of 10 and 100

To this point we have derived both exact and approximate formulas for the common emitter and common base configurations of the bipolar transistor. Tables 4.4 and 4.5 contain data which enable the reader to compare the effects of changes in β on the collector-to-emitter voltage. The β is varied from a nominal value of 100 to a low value of 10 and a high value of 1000. In addition to the variation in β, two other parameters are varied, δ (the ratio of the current through R_1 to the base current) and γ (the percentage of the power supply voltage V_{CC} to the emitter voltage V_{RE}).

It can be noted that, for universal stabilization (Table 4.4), increasing the ratio of the current through the bleeder to the base current, δ, moderately improves the stability. In a similar manner, increasing the percentage of the power supply voltage that appears across the emitter resistor, γ moderately improves

Table 4.4
Common Emitter with Universal Bias
$V_{CC} = 10$ V, $V_{EB} = .6$ V, $I_{CQ} = 5$ mA, $\beta_{Nominal} = 100$

	No Emitter Resistor $\gamma = 0$	$\gamma = .05$ (5%)	$\gamma = .1$ (10%)	$\gamma = .2$ (20%)
Emitter Stabilized— No gnd-base resistor $\delta = 1$	+90%	+98.7%	+108.4%	+131%
$\delta = 2$	+90%	+94%	+101.1%	+118.8%
$\delta = 4$	+90%	+85.9% / −69.9%	+89.9% / −52.7%	+100.5% / −45%
$\delta = 8$	+90%	+73.5% / −31.2%	+72.4% / −23.8%	+77.5% / −20.8%

Note: Shaded areas denote transistor in saturation at $\beta = 1000$.

Table 4.5
Common Emitter with Collector Feedback
$V_{CC} = 10$ V, $V_{EB} = .6$ V, $I_{CQ} = 5$ mA, $\beta_{Nominal} = 100$

	No Emitter Resistor $\gamma = 0$	$\gamma = .05$ (5%)	$\gamma = .1$ (10%)	$\gamma = .2$ (20%)
Emitter Stabilized— No gnd to base resistor $\delta = 1$	+79.2% / −72.6%	+86.5% / −72.8%	+94.4% / −72.2%	+112.3% / −72.3%
$\delta = 2$	+70.8% / −37.8%	+74.1% / −32.2%	+79.1% / −30.9%	+97% / −30.8%
$\delta = 4$	+58.4% / −19.4%	+57.5% / −15.2%	+59.6% / −14.4%	+66% / −14.3%
$\delta = 8$	+43.4% / −9.6%	+39.6% / −7.4%	+39.8% / −6.9%	+45.2% / −6.8%

the circuit stability for increasing β (β = 1000), but is detrimental for decreasing β (β = 10).

For the common emitter circuit with collector feedback (Table 4.5), improvements in stability can be noted as the bleeder current is increased (δ increases), but increases in the percentage of the power supply voltage that appear across the emitter resistor (γ increases) have a detrimental effect for decreasing β (β = 10) and only a marginally beneficial effect for increasing β (β = 1000).

4.6 THE COMMON COLLECTOR CIRCUIT (EMITTER FOLLOWER)

4.6.1 Common Collector DC Design

The common collector circuit is the third configuration of the bipolar transistor to be studied. The circuit advantages will be discussed in a later chapter (Ch. 9), but at this point we will concentrate on the circuit design for proper dc biasing (see Fig. 4.45).

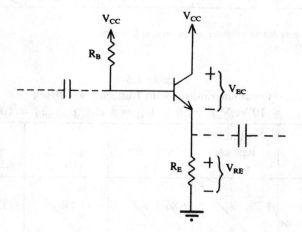

Fig. 4.45

For maximum signal swing, we will design for one-half of the power supply voltage (V_{CC}) to appear across the emitter resistor and the other half across the transistor emitter-to-collector junction. This rationale for dividing the power supply voltage is similar to the logic that was used with the common emitter stage.

If we are given certain design parameters for desired voltage and currents, we may thus develop the appropriate formulas, which will enable us to compute the values of the required resistances (see. Fig. 4.46).

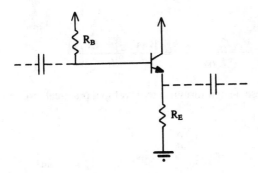

Fig. 4.46

Example (4.29). Given V_{CC}, I_{CQ}, β, V_{EB}, assume $V_{RE} = V_{EC}$, and determine R_E, R_B.

$$I_E = I_C + I_B = I_C + I_C/\beta \neq I_C (1 + 1/\beta) = I_C \left(\frac{\beta + 1}{\beta}\right)$$

$$V_{RE} = \frac{V_{CC}}{2}$$

Therefore,

$$R_E = \frac{V_{RE}}{I_E} = \frac{V_{CC}/2}{I_C \left(\frac{\beta + 1}{\beta}\right)} = \frac{\beta \, V_{CC}}{2I_C (\beta + 1)} \tag{4.42}$$

$$V_{gnd\text{-}B} = V_{RE} + V_{EB} = \frac{V_{CC}}{2} + V_{EB}$$

$$V_{RB} = V_{CC} - V_{gnd\text{-}B}$$

$$= V_{CC} - (\frac{V_{CC}}{2} + V_{EB})$$

$$= V_{CC} - \frac{V_{CC}}{2} - V_{EB}$$

$$V_{RB} = \frac{V_{CC}}{2} - V_{EB}$$

Also,

$$I_B = I_C/\beta$$

Thus,

$$R_B = \frac{V_{RB}}{I_B} = \frac{V_{CC}/2 - V_{EB}}{I_C/\beta} = \frac{\beta\,(V_{CC} - 2V_{EB})}{2I_C} \tag{4.43}$$

Let us use these design formulas to develop a practical circuit (see Fig. 4.47).

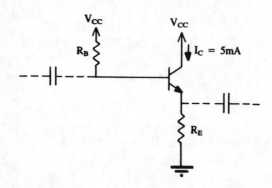

Fig. 4.47

Example (4.30). Given $V_{CC} = 10$ V, $V_{EB} = 0.6$ V, $I_{CQ} = 5$ mA, assume $V_{RE} = V_{EC}$, and determine R_E, R_B.

$$R_E = \frac{\beta V_{CC}}{2I_C\,(\beta + 1)} = \frac{100(10)}{2(5)(101)} = \frac{100}{101} = 0.99 \text{ k}$$

$$R_B = \frac{\beta\,(V_{CC} - 2V_{EB})}{2I_C} = \frac{100(10 - 2(0.6))}{2(5)} = 10(10 - 1.2) = 10(8.8)$$

$$R_B = 88 \text{ k}$$

With the design formulas developed, it is appropriate to determine their accuracy by applying the corresponding analysis formulas.

4.6.2 Common Collector DC Analysis

For the common collector dc analysis, the power supply voltage, β, base-to-emitter voltage drop, and resistance values are known and we wish to determine the other circuit voltages and currents. The formulas are developed as follows in Example (4.31). (See Fig. 4.48.)

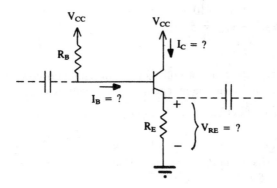

Fig. 4.48

Example (4.31). Given V_{CC}, V_{EB}, β, R_B, R_C, determine I_B, I_C, I_E, V_{RE}, V_{EC}, $V_{gnd\text{-}B}$.

$$V_{RE} + V_{EC} = V_{CC}$$
$$V_{RE} = I_E R_E = (I_C + I_B) R_E = (\beta I_B + I_B) R_E$$
$$V_{RE} = (\beta + 1) I_B R_E$$
$$V_{RE} + V_{EB} + V_{RB} = V_{CC}$$
$$I_B (\beta + 1) R_E + V_{EB} + I_B R_B = V_{CC}$$
$$I_B (\beta + 1) R_E + I_B R_B = V_{CC} - V_{EB}$$
$$I_B [(\beta + 1) R_E + R_B] = V_{CC} - V_{EB}$$
$$I_B = \frac{V_{CC} - V_{EB}}{R_B + (\beta + 1)R_E} \tag{4.44}$$

Also,

$$I_C = \beta I_B$$
$$I_E = (\beta + 1) I_B \tag{4.45}$$
$$V_{RE} = I_E R_E$$
$$V_{EC} = V_{CC} - V_{RE}$$
$$V_{gnd\text{-}B} = V_{RE} + V_{EB} \tag{4.46}$$

Using these formulas to check our previous designs, we have

$$I_B = \frac{V_{CC} - V_{EB}}{R_B + (\beta + 1) R_E} = \frac{10 - 0.6}{88 + (101)(0.99)} = \frac{9.4}{188} = 0.05 \text{ mA}$$
$$I_E = (\beta + 1) I_B = (101)(0.05) = 5.05 \text{ mA}$$
$$V_{RE} = I_E R_E = 5.05 (0.99) = 5 \text{ V}$$

$$V_{EC} = V_{CC} - V_{RE} = 10 - 5 = 5 \text{ V}$$
$$V_{gnd\text{-}B} = V_{RE} + V_{EB} = 5 + 0.6 = 5.6 \text{ V}$$

which checks!

With respect to the common collector circuit (emitter follower), an interesting point to investigate would be its circuit stability with respect to changes in β.

Example (4.32). Let us see what happens if the β increases by 30%.

$$I_B = \frac{V_{CC} - V_{EB}}{R_B + (\beta + 1) R_E} = \frac{10 - 0.6}{88 + (131)0.99} = \frac{9.4}{88 + 129.7} = \frac{9.4}{217.7}$$

$$I_B = 0.04318 \text{ mA}$$

$$I_E = (\beta + 1) I_B = 131(0.04318) = 5.67 \text{ mA}$$

$$V_{RE} = I_{E1} R_E = 5.657(0.99) = 5.6 \text{ V}$$

$$V_{EC} = V_{CC} - V_{RE} = 10 - 5.6 = 4.4 \text{ V}$$

$$\% \text{ Change in } V_{EC} = \left| \frac{5 - 4.4}{5} \right| \times 100 = \frac{60}{5} = -12\%$$

Example (4.33). Similarly, if the β decreases by 30%:

$$I_B = \frac{V_{CC} - V_{EB}}{R_B + (\beta + 1) R_E} = \frac{10 - 0.6}{88 + 71(0.99)}$$

$$= \frac{9.4}{88 + 70.29} = \frac{9.4}{150.29} = 0.06254 \text{ mA}$$

$$I_E = (\beta + 1) I_B = 71(0.06254) = 4.441 \text{ mA}$$

$$V_{RE} = I_E R_E = 4.441(0.99) = 4.396 \text{ V}$$

$$V_{EC} = V_{CC} - V_{RE} = 10 - 4.396 = 5.6 \text{ V}$$

$$\% \text{ change in } V_{EC} = \left| \frac{5 - 5.6}{5} \right| \times 100 = +12\%$$

4.6.3 Common Collector DC Design (with a Base Circuit Voltage Divider)

Additional stability may be obtained by employing a voltage divider in the base circuit in a manner similar to that employed in the common emitter circuit with universal stabilization. See Fig. 4.49. (Note that R_2 is an addition to the previously analyzed circuit (see Fig. 4.45).)

Proceeding with the development of the design equations: given V_{CC}, V_{EB}, I_{CQ}, β, assume $V_{RE} = V_{EC}$, $I_1 = \delta I_B$, and determine R_E, R_1, R_2. See Fig. 4.50.

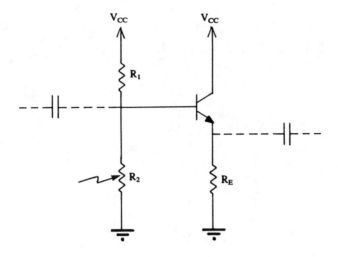

Fig. 4.49

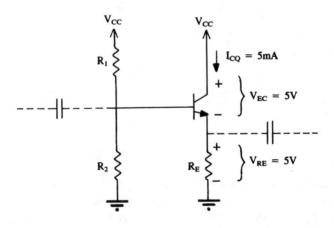

Fig. 4.50

$$I_E = I_C + I_B = I_C + I_C/\beta = I_C (1 + 1/\beta) = I_C \frac{(\beta + 1)}{\beta} \qquad (4.47)$$

$$V_{RE} = \frac{V_{CC}}{2}$$

Therefore,

$$R_E = \frac{V_{RE}}{R_E} = \frac{V_{CC}/2}{I_C\left(\frac{\beta+1}{\beta}\right)} = \frac{\beta V_{CC}}{2I_C\,(\beta+1)} \tag{4.48}$$

$$V_{gnd\text{-}B} = V_{RE} + V_{EB} = \frac{V_{CC}}{2} + V_{EB}$$

$$I_B = \frac{I_C}{\beta}$$

Also, $I_1 = \delta\,I_B = \delta\frac{I_C}{\beta}$

$$I_2 = I_1 - I_B = \frac{\delta I_C}{\beta} - \frac{I_C}{\beta} = \frac{I_C}{\beta}\,(\delta - 1)$$

$$R_2 = \frac{V_{R2}}{I_2} = \frac{V_{gnd\text{-}B}}{I_2} = \frac{\dfrac{V_{CC}}{2} + V_{EB}}{\dfrac{I_C\,(\delta-1)}{\beta\,(\delta-1)}} = \frac{\beta\,(V_{CC} + 2V_{EB})}{2I_C\,(\delta - 1)} \tag{4.49}$$

$$R_1 = V_{R1} = \frac{V_{CC} - V_{R2}}{I_1} = \frac{V_{CC} - \left(\dfrac{V_{CC}}{2} + V_{EB}\right)}{\delta\,I_C/\beta}$$

$$= \frac{V_{CC} - \dfrac{V_{CC}}{2} - V_{EB}}{\delta\,I_C/\beta}$$

$$R_1 = \frac{V_{CC}/2 - V_{EB}}{\delta\,I_C/\beta} = \frac{\beta\,(V_{CC} - 2V_{EB})}{2\delta\,I_C} \tag{4.50}$$

Let us use the formulas developed above and design an emitter follower where the current in resistor R_1 is four times the base current.

Given $V_{CC} = 10$, $I_C = 5$, $V_{EB} = 0.6$, $\beta = 100$, assume $V_{RE} = V_{EC}$, $\delta = 4$, and determine R_E, R_1, R_2. See Fig. 4.51.

$$R_E = \frac{\beta V_{CC}}{2I_C\,(\beta+1)} = \frac{100(10)}{2(5)(101)} = \frac{100}{101} = 0.99\text{ k}$$

$$R_2 = \frac{\beta\,(V_{CC} + 2V_{EB})}{2I_C\,(\delta - 1)} = \frac{100(10 + 2(0.6))}{2(5)\,(4 - 1)} = \frac{10(10 + 1.2)}{3}$$

$$R_2 = \frac{10(11.2)}{3} = \frac{112}{3} = 37.33\text{ k}$$

$$R_1 = \frac{\beta(V_{CC} - 2V_{EB})}{2\delta I_C} = \frac{100(10 - 2(0.6))}{2(4)(5)} = \frac{10(10 - 1.2)}{4}$$

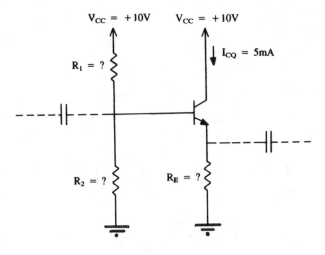

$$\text{Fig. 4.51}$$

$$R_1 = \frac{10(8.8)}{4} = 10(2.2) = 22 \text{ k}$$

4.6.4 Common Collector DC Analysis (with a Base Circuit Voltage Divider)

Now that we have developed the design formulas for the emitter-follower circuit with a voltage divider in the base circuit, it is appropriate to develop the corresponding analysis formulas so that we may check our design results.

Given V_{CC}, V_{EB}, β, R_1, R_2, R_E, determine I_B, I_E, V_{RE}, V_{EC}, V_{R1}, I_2, V_{R2}, I_1. See Fig. 4.52.

The development of the analysis formulas for the emitter follower is identical to the analysis of the common emitter circuit with universal stabilization (sec. 4.4.2) and is left to the student as an exercise.

The formulas are restated here:

$$I_B = \frac{V_{CC} - V_{EB}\,(1 + R_1/R_2)}{R_1 + (\beta + 1)\,R_E\,(1 + R_1/R_2)}$$

$$I_E = (\beta + 1)\,I_B$$

$$V_{RE} = I_E R_E$$

$$V_{EC} = V_{CC} - V_{RE}$$

$$V_{gnd\text{-}B} = V_{R2} = V_{RE} + V_{EB}$$

$$I_2 = \frac{V_{R2}}{R_2} = \frac{V_{RE} + V_{EB}}{R_2}$$

$$I_1 = \frac{V_{R1}}{R_1} = \frac{V_{CC} - V_{R2}}{R_1}$$

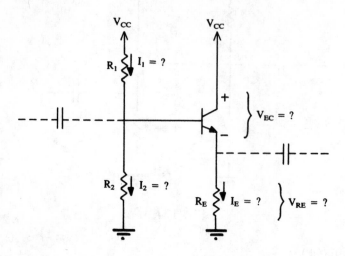

Fig. 4.52

Example (4.34).   Employing these analysis formulas as a check on the previous design example, we have:

$$I_B = \frac{V_{CC} - V_{EB}\left(1 + \dfrac{R_1}{R_2}\right)}{R_1 + (\beta + 1)\,R_E\left(1 + \dfrac{R_1}{R_2}\right)} = \frac{10 - 0.6\left(1 + \dfrac{22}{37.3}\right)}{22 + 101(0.99)\left(1 + \dfrac{22}{37.3}\right)}$$

$$I_B = \frac{10 - 0.6(1.59)}{22 + 101(0.99)\,(1.59)} = \frac{10 - 0.954}{22 + 159} = \frac{9.046}{181} = 0.05 \text{ mA}$$

$$I_E = (\beta + 1)\,I_B = 101(0.05) = 5.05 \text{ mA}$$

$$V_{RE} = I_E R_E = 5.05\,(0.99)$$

Thus,

$$V_{RE} = 5 \text{ V}$$

$$V_{EC} = V_{CC} - V_{RE} = 10 - 5 = 5 \text{ V}$$

$$I_2 = \frac{V_{R2}}{R_2} = \frac{V_{RE} + V_{EB}}{R_2} = \frac{5 + 0.6}{37.33} = \frac{5.6}{37.33} = 0.1501 \text{ mA}$$

$$I_1 = \frac{V_{CC} - V_{R2}}{R_1} = \frac{10}{22} = \frac{4.4}{22} = 0.2 \text{ mA}$$

$I_2 = 4I_B = 4.(0.05) = 0.2$ mA

which checks!

Example (4.35). Let us see how much the circuit voltages and currents change with a 30% increase in β:

$$I_B = \frac{V_{CC} - V_{EB}\left(1 + \dfrac{R_1}{R_2}\right)}{R_1 + (\beta + 1) R_E \left(1 + \dfrac{R_1}{R_2}\right)} = \frac{10 - 0.6\left(1 + \dfrac{22}{37.33}\right)}{22 + (131)(0.99)\left(1 + \dfrac{22}{37.33}\right)}$$

$$I_B = \frac{10 - 0.6(1.59)}{22 + 131(0.99)(1.59)} = \frac{10 - 0.954}{22 + 206.1} = \frac{9.046}{228.1} = 0.397 \text{ mA}$$

$I_E (\beta + 1) I_B = 131(0.0397) = 5.195$ mA

$V_{RE} = I_E R_E = 5.195(0.99) = 5.143$ V

$V_{EC} = V_{CC} - V_{RE} = 10 - 5.143 = 4.857$ V

% change in $V_{EC} = \left|\dfrac{5 - 4.857}{5}\right| \times 100 = \dfrac{14.3}{5} = -2.86\%$

This change is an improvement compared with the unstabilized emitter follower.

Example (4.36). Checking for a 30% decrease in β, we have

$$I_B + \frac{V_{CC} - V_{EB}\left(1 + \dfrac{R_1}{R_2}\right)}{R_1 + (\beta + 1) R_E \left(1 + \dfrac{R_1}{R_2}\right)} = \frac{10 - 0.6\left(1 + \dfrac{22}{37.33}\right)}{22 + 71(0.99)\left(1 + \dfrac{22}{37.33}\right)}$$

$$I_B = \frac{10 - 0.6(1.59)}{22 + 71(0.99)(1.59)} = \frac{10 - 0.954}{22 + 111.7} = \frac{9.05}{133.7} = 0.0677 \text{ mA}$$

$I_E = (\beta + 1) I_B = 71(0.0677) = 4.804$ mA

$V_{RE} = I_E R_E = 4.804(0.99) = 4.76$ mA

$V_{EC} = V_{CC} - V_{RE} = 10 - 4.76 = 5.24$ V

% change in $V_{EC} = \left|\dfrac{5 - 5.24}{5}\right| \times 100 = +4.89\%$

We can see that there is a distinct improvement in stability by use of the voltage divider, as detailed in Table 4.6.

What happens if β should change more drastically, such as a 100% increase or decrease? This question is left as an exercise for the student to solve (see Problem 25).

There is always a "trade-off" in making an engineering decision. By using the voltage divider circuit in the base circuit of the emitter follower, we have improved the stability with respect to changes in beta. The "trade-off" that we have made is that the total imput impedance that the previous stage (signal source) "sees" is a lesser value (than without the divider). The use of the voltage divider may reduce the amplification (gain) of the previous stage.

Table 4.6

% Change in β	% Change in V_{EC}	
	Unstabilized Emitter Follower	Emitter Follower with Base Circuit Voltage Divider
+30%	−12%	−2.86%
−30%	+12%	+4.89%

PROBLEMS

1. Design an unstabilized common emitter stage with the following criteria: $V_{CC} = 10$ V, $I_C = 20$ mA, $V_{EB} = 0.6$ V, $\beta = 100$, $V_{RC} = V_{EC}$.

2. For the design completed in Problem 1, perform an analysis to determine the correctness of your design. Allow β to increase by 30%. What is the new V_{EC}? What is the percent change in V_{EC} from V_{EC} at $\beta = 100$? Allow β to decrease by 30%. What is the new V_{EC}? What is the percent change from V_{EC} at $\beta = 100$?

3. Design an unstabilized common emitter stage with the following criteria: $V_{CC} = 30$ V, $I_C = 20$ mA. $V_{EB} = 0.6$ V, $\beta = 100$, $V_{RC} = V_{EC}$.

4. For the design completed in Problem 3, perform an analysis to determine the correctness of your design. Allow β to increase by 30%. What is the new V_{EC}? What is the percent change in V_{EC} from V_{EC} at $\beta = 100$? Allow β to decrease by 30%. What is the new V_{EC}? What is the percent change from V_{EC} at $\beta = 100$?

5. Compare the results of Problems 2 and 4. What can be said about the effect of using a higher power supply voltage V_{CC}, on the dc stability of an unstabilized common emitter stage?

6. Design an unstabilized common emitter stage with the following criteria: $V_{CC} = 10$ V, $I_C = 2$ mA, $V_{EB} = 0.6$ V, $\beta = 100$, $V_{RC} = V_{EC}$.

Table 4.7
Bipolar Transistor Amplifier DC Design Formulas

Configuration		Assumptions	R_C	R_E	$R_1\ (=R_B)$	R_2
COMMON BASE	Unstabilized	$V_{RC} = V_{BC}$	$\dfrac{V_{CC}}{2I_{CQ}}$	$\dfrac{\beta+1}{\beta}\left(\dfrac{V_{EE}-V_{EB}}{I_{CQ}}\right)$	—	—
COMMON EMITTER:	Unstabilized	$V_{RC} = V_{EC}$	$\dfrac{V_{CC}}{2I_{CQ}}$	0	$\dfrac{\beta(V_{CC}-V_{EB})}{I_{CQ}}$	∞
	Emitter Stabilized	$V_{RC} = V_{EC}$ $\gamma = \dfrac{V_{RE}}{V_{CC}}$	$\dfrac{V_{CC}(1-\gamma)}{2I_{CQ}}$	$\dfrac{\beta\gamma V_{CC}}{(\beta+1)I_{CQ}}$	$\dfrac{\beta[V_{CC}(1-\gamma)-V_{EB}]}{I_{CQ}}$	∞
	Universal Stabilized	$V_{RC} = V_{EC}$ $\gamma = \dfrac{V_{RE}}{V_{CC}}$ $\delta = I_1/I_B$	$V_{CC}\dfrac{(1-\gamma)}{2I_{CQ}}$	$\dfrac{\beta\gamma V_{CC}}{(\beta+1)I_{CQ}}$	$\dfrac{\beta[V_{CC}(1-\gamma)-V_{EB}]}{\delta I_{CQ}}$	$\dfrac{\beta(\gamma V_{CC}+V_{EB})}{I_{CQ}(\delta-1)}$
	Collector Feedback	$V_{RC} = V_{EC}$ $\gamma = V_{RE}/V_{CC}$ $\delta = I_1/I_B$	$\dfrac{\beta V_{CC}(1-\gamma)}{2I_{CQ}(\beta+\delta)}$	$\dfrac{\beta\gamma V_{CC}}{(\beta+1)I_{CQ}}$	R_F $=\dfrac{\beta[V_{CC}(1-\gamma)-2V_{EB}]}{2\delta I_{CQ}}$	$\dfrac{\beta(\gamma V_{CC}+V_{EB})}{I_{CQ}(\delta-1)}$
COMMON COLLECTOR (EMITTER FOLLOWER)		$V_{RE} = V_{EC}$ $\delta = \dfrac{I_1}{I_B}$	0	$\dfrac{\beta V_{CC}}{(\beta+1)2I_{CQ}}$	$\dfrac{\beta[V_{CC}-2V_{EB}]}{2\delta I_{CQ}}$	$\dfrac{\beta(V_{CC}+2V_{EB})}{2I_{CQ}(\delta-1)}$

7. For the design completed in Problem 6, perform an analysis to determine the correctness of your design. Allow the beta to increase and decrease by 30%. What is the V_{EC} for each β value? What is the percent change from V_{EC} at $\beta = 100$ for both new values?

   Compare the results of Problem 7 with the results of Problem 2. What can be stated about the effects of design changes in collector current upon the dc stability of the unstabilized common emitter amplifier?

8. Design a stabilized common emitter stage with the following criteria: $V_{CC} = 10$ V, $I_C = 20$ mA, $V_{EB} = 0.6$ V, $\beta = 100$, $V_{RC} = V_{EC}$ and $V_{RE} = 10\%$ V_{CC}.

9. For the design completed in Problem 8, perform an analysis to determine the correctness of your design. Allow β to increase and decrease by 30%. What are the new V_{EC} values for each β? What are the percentage changes in V_{EC} for the changes in β (as compared with V_{EC} at $\beta = 100$)?

10. Compare the percentage changes in V_{EC} for Problem 9 *versus* the percentage changes in V_{EC} for Problem 2.

11. Design an emitter stabilized common emitter stage with the following criteria: $V_{CC} = 30$ V, $I_C = 20$ mA, $V_{EB} = 0.6$ V, $\beta = 100$, $V_{RC} = V_{EC}$ and $V_{RE} = 10\%$ V_{CC}.

12. For the design completed in Problem 11, perform an analysis to determine the correctness of your design. Allow β to increase and decrease by 30%. What are the new V_{EC} values? What are the percentage changes in V_{EC} for the changes in β (as compared with V_{EC} at $\beta = 100$)?

   Compare the results of Problem 11 with Problem 9. What can be stated about the effects of an increased power supply voltage upon the stability of a stabilized common emitter stage?

13. Design a stabilized common emitter stage with the following criteria: $V_{CC} = 10$ V, $I_C = 2$ mA, $V_{EB} = 0.6$ V, $\beta = 100$, $V_{RC} = V_{EC}$ and $V_{RE} = 10\%$ V_{CC}.

14. For the design completed in Problem 13, perform an analysis to determine the correctness of your design. Allow β to increase and decrease by 30%. What are the new V_{EC} values for each β? What are the percentage changes in V_{EC} for the changes in β (as compared with V_{EC} at $\beta = 100$)? What can be stated about the effects of different design collector currents upon the transistor amplifier stage stability?

15. Design a stabilized common emitter stage with the following criteria: $V_{CC} = 10$ V, $I_C = 20$ mA, $V_{EB} = 0.6$ V, $\beta = 100$, $V_{RC} = V_{EC}$ and $V_{RE} = 20\%$ V_{CC}.

16. For the design completed in Problem 15, perform an analysis to determine the correctness of your design. Allow β to increase and decrease by 30%.

What are the new V_{EC} values for each β? What are the percentage changes in V_{EC} for the changes in β (as compared with V_{EC} at $\beta = 100$)? What can be stated about the effects of different percentages of the power supply voltage appearing across R_E (compare Problems 16 and 12)?

17. Design a universal stabilized common emitter stage with the following criteria: $V_{CC} = 10$ V, $I_C = 20$ mA, $V_{EB} = 0.6$ V, $\beta = 100$, $V_{RC} = V_{EC}$, $V_{RE} = 10\%$ V_{CC}, $I_1 = 4I_B$.

18. For the design completed in Problem 17, perform an analysis to determine the correctness of the design. Allow β to increase and decrease by 30%. What are the new V_{EC} values for each β? What are the percentage changes in V_{EC} for the changes in β (as compared with V_{EC} at $\beta = 100$)?

19. *Compare* the results of Problem 18 with the results of Problem 8. What can be stated about the percentage changes in V_{EC} for the stabilized common emitter *versus* the universal stabilized common emitter stage (with similar operating currents and voltages)?

20. Design a universal stabilized common emitter stage with the following criteria: $V_{CC} = 10$ V, $I_C = 2$ mA, $V_{EB} = 0.6$ V, $\beta = 100$, $V_{RC} = V_{EC}$, $V_{RE} = 20\%$ V_{CC}, $I_1 = 4I_B$.

21. For the design completed in Problem 20, perform an analysis to determine the correctness of the design. Allow the β to increase and decrease by 30%. What are the new V_{EC} values for each β? What are the percentage changes in V_{EC} for the changes in β (as compared with V_{EC} at $\beta = 100$)?

22. *Compare* the results of Problem 21 with the results of Problem 18. What can be stated about the percentage changes in V_{EC} for the universal stabilized common emitter stage with varying amounts of the power supply voltage appearing across the emitter resistor?

Compare the effect on stability that results due to an increase in the percentage of the power supply voltage appearing across the emitter resistor.

23. Design common emitter stage with universal stabilization to meet the following criteria: $V_{CC} = 10$ V, $I_C = 20$ mA, $V_{EB} = 0.6$ V, $\beta = 100$, $V_{RC} = V_{EC}$, $V_{RE} = 10\%$ V_{CC}, $I_1 = 8I_B$.

24. For the design completed in Problem 23, perform an analysis to determine the correctness of the design. Allow β to increase and decrease by 30%. What are the new V_{EC} values for each β? What are the percentage changes in V_{EC} for the changes in β (as compared with V_{EC} at $\beta = 100$)? Compare the results of this problem with those of Problem 18 results.

25. Design an emitter follower circuit such that it is identical to the design presented in section 4.6.3 (Fig. 4.51) with the exception that the base current is one-half of the current flowing through the R_1 resistor. Use the analysis formulas to verify the correctness of your design.

26. Allow β to increase by 30% in the circuit designed in Problem 25 and determine the new V_{EC} and its percentage change. How does this change compare with the analysis of the design presented in section 4.6.3? Allow β to decrease 30% from the nominal value and repeat the above exercise.

27. If an emitter follower is only required to amplify a "positive-going" signal, to increase the output voltage "swing," it is desirable to bias the transistor such that the emitter is near ground potential. Develop the appropriate design and analysis formulas for the situation where $V_{RE} = V_{EC}$. Repeat the design and analysis presented in section 4.6.3, except let $V_{RE} = 1$ V.

Chapter 5

Bipolar Transistor Amplifier Stage
DC Bias
Circuit Stability Analysis

5.1 STABILITY FACTOR [1, 2]

Up to this point, we have discussed the dc stability of a common emitter amplifier with respect to the effect of changes of beta (β) upon the voltage appearing across the collector resistor (V_{RC}). The voltage across the collector load resistor is directly proportional to the current flowing through the resistor. In most cases, the current flowing through the collector load resistor is simply the collector current, the exception being the common emitter amplifier with collector feedback.

There are other variables that will determine changes in the dc operating voltages and currents in a transistor. The three first-order variables of interest are the change in leakage current, I_{CBO}, the change in the forward base-to-emitter voltage drop V_{EB}, and lastly the change in β. There are second-order effects not normally considered in a treatment of stability in a presentation at this level.

The most obvious factors that can affect leakage current, base-to-emitter voltage, and β are temperature and age. Temperature is usually investigated in most textbook discussions of stability, but aging is excluded because of the difficulty of making long-term predictions of transistor characteristic changes. A typical sequence of events, which can affect transistor stability, might take place as follows: (1) the operating environmental temperature could increase; (2) this temperature increase could cause an increase in the leakage current; and, finally, (3) the increased leakage current would result in an increased collector current.

The increase in leakage current was a severe problem that designers were faced with when only germanium transistors were available. Modern silicon, planar transistors are orders of magnitude lower in leakage current.

Returning to our original consideration, we know that this variable will affect the change in collector current. How can we generate an equation that will allow us to predict the total change in the collector current in terms of I_{CBO}, V_{EB}, and β?

A mathematical tool, referred to as *partial differential equations*, enables us to predict the overall change due to changes in numerous other variables. In our case, we would write the partial differential equation as

$$\frac{dI_C}{dT} = \frac{\partial I_C}{\partial I_{CBO}} \cdot \frac{dI_{CBO}}{dT} + \frac{\partial I_C}{\partial V_{EB}} \cdot \frac{dV_{EB}}{dT} + \frac{\partial I_C}{\partial \beta} \cdot \frac{d\beta}{dT} \qquad (5.1)$$

Each of the partial derivative terms:

$$\frac{\partial I_C}{\partial I_{CBO}}, \frac{\partial I_C}{\partial V_{EB}}, \frac{\partial I_C}{\partial \beta}$$

describe independent rates of change of the collector current with respect to each of the variables: the collector-to-base leakage current, the forward emitter-to-base voltage drop, and the dc β.

Each of these terms are evaluated as functions of the particular bias stabilizing network being employed for the common emitter stage. In other words, these variables are evaluated by examining the equations that are employed to predict the collector current for various common-emitter dc bias configuration. Each term is referred to as a *stability factor*.

Ideally, we would like the rate of change of the collector current with respect to the temperature to approach zero. Hence, if each of the stability factors approached zero, we would achieve that goal. It is worth mentioning again that the stability factor, i.e., the partial derivative terms:

$$\frac{\partial I_C}{\partial I_{CBO}}, \frac{\partial I_C}{\partial V_{EB}}, \frac{\partial I_C}{\partial \beta}$$

are functions of the circuit configuration. They express rates of change of collector current as a function of some other variable, such as the transistor β. From our previous investigations, we know that changes in β will affect the collector current, depending upon the stabilization network employed. By proper design, these terms (stability factors) may be minimized, thus decreasing the overall rate of change of the collector current with respect to the change in temperature.

The regular derivative terms:

$$\frac{dI_{CBO}}{dT}, \frac{dV_{EB}}{dT}, \frac{d\beta}{dT}$$

are simply measured quantities. These values can be obtained in the laboratory by the use of voltmeters, ammeters, and temperature measuring devices. These terms can be minimized by appropriate transistor selection. In some instances, the manufacturer's specification data sheets will supply the above rates of change.

For calculation of the stability factors—whether an unstabilized, emitter stabilized, or universal stabilized, common-emitter circuit is employed—the same analysis may be used.

To develop the stability equation it is desirable to simplify the base bias circuit of the universal stabilized circuit (see Figs. 5.1 and 5.2).

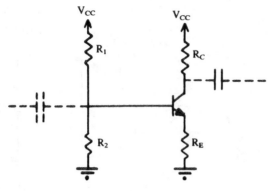

Fig. 5.1

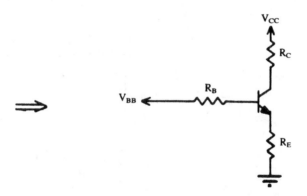

Fig. 5.2

The base bias circuit of a universal stabilized circuit may be thought of as though it were an emitter stabilized circuit. A *Thevenin's theorem* analysis of the two base bias resistors R_1 and R_2 and the V_{CC} power supply may be employed to reduce the configuration to a single resistor R_B and a single power supply V_{BB}. See Figs. 5.3 to 5.5, where

$$R_B = R_1 \mathbin{/\mkern-5mu/} R_2 = \frac{R_1 R_2}{R_1 + R_2} \tag{5.2}$$

180

and

$$V_{BB} = V_{CC}.\frac{R_2}{R_1 + R_2}$$

(5.3)

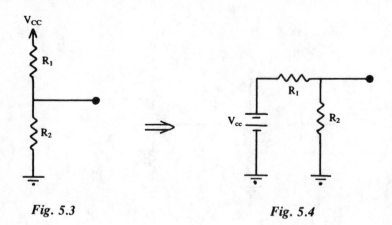

Fig. 5.3   Fig. 5.4

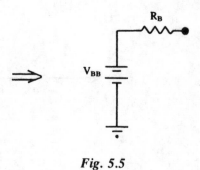

Fig. 5.5

5.1.1 Stability Factor Calculations for Circuit Components, Voltages, and Currents

Up to this point, we have made the assumption that the dc collector current was directly proportional to the dc base current, the constant of proportionality being the dc β of the transistor, i.e.,

$$I_C = \beta I_B = h_{FE} I_B$$

This assumption is not strictly true. The collector current is proportional to the dc β and base current, but it is also composed of leakage current. The leakage current flowing out of the base, I_{CBO}, is amplified by the transistor and forms a larger leakage current in the collector circuit. Mathematically, this current can be expressed as follows:

$$I_C = h_{FE} I_B + (h_{FE} + 1)\, I_{CBO} \tag{5.4}$$
$$\text{(larger leakage current)}$$

In order for us to evaluate each of the stability factors in terms of the specific circuit configuration of interest (e.g., universal stabilization, *et cetera*), we must express the above equation in terms of the circuit components. Referring to the schematic of Fig. 5.6 and employing Kirchoff's voltage law:

$$V_{RE} + V_{EB} + I_B R_B = V_{BB} \tag{5.5}$$

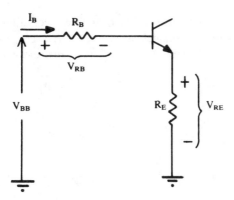

Fig. 5.6

Solving Equation (5.5) with respect to I_B:

$$V_{RE} + V_{EB} + I_B R_B = V_{BB}$$
$$I_E R_E + V_{EB} + I_B R_B = V_{BB}$$
$$(I_C + I_B)\, R_E + V_{EB} + I_B R_B = V_{BB}$$
$$I_C R_E + I_B R_E + V_{EB} + I_B R_B = V_{BB}$$
$$I_B\,(R_E + R_B) = V_{BB} - V_{EB} - I_C R_E$$
$$I_B = \frac{V_{BB} - V_{EB} - I_C R_E}{R_E + R_B} \tag{5.6}$$

Substituting Equation (5.6) into (5.4):

$$I_C = h_{FE}I_B + (h_{FE} + 1)I_{CBO} \qquad (\beta = h_{FE})$$

$$I_C = \beta \left[\frac{V_{BB} - V_{EB} - I_C R_E}{R_E + R_B} \right] + (\beta + 1) I_{CBO}$$

$$I_C = \frac{\beta (V_{BB} - V_{EB})}{R_E + R_B} - \frac{\beta I_C R_E}{R_E + R_B} + (\beta + 1) I_{CBO}$$

$$I_C \left(\frac{R_E + R_B}{R_E + R_B} \right) = \frac{\beta(V_{BB} - V_{EB})}{R_E + R_B} - \frac{\beta I_C R_E}{R_E + R_B}$$
$$+ (\beta + 1) I_{CBO} \left(\frac{R_E + R_B}{R_E + R_B} \right)$$

$$I_C (R_E + R_B) + I_C \beta R_E = \beta (V_{BB} - V_{EB}) + (\beta + 1) I_{CBO} (R_E + R_B)$$

$$I_C [(\beta + 1) R_E + R_B] = \beta (V_{BB} - V_{EB}) + (\beta + 1) I_{CBO} (R_E + R_B)$$

$$I_C = \frac{\beta (V_{BB} - V_{EB}) + (\beta + 1) I_{CBO} (R_E + R_B)}{(\beta + 1) R_E + R_B} \qquad (5.7)$$

Now that we have an expression for I_C with respect to the leakage current, and known or measurable voltages and circuit resistances, we may determine the partial derivative of I_C with respect to the variables of interest, namely I_{CBO}, V_{EB}, and β.

Employing Equation (5.7) by inspection, we have

$$\frac{\partial I_C}{\partial I_{CBO}} = \frac{(\beta + 1) (R_E + R_B)}{R_B + (\beta + 1) R_E} \qquad (5.8)$$

Rewriting Equation (5.7), we have

$$I_C = \frac{\beta (V_{BB} - V_{EB})}{R_B + (\beta + 1) R_E} + \frac{I_{CBO} (\beta + 1)(R_B + R_E)}{R_B + (\beta + 1)R_E}$$

$$I_C = \frac{\beta V_{BB}}{R_B + (\beta + 1) R_E} - \frac{\beta V_{EB}}{R_B + (\beta + 1) R_E}$$
$$+ \frac{I_{CBO} (\beta + 1)(R_B + R_E)}{R_B + (\beta + 1) R_E}$$

$$\frac{\delta I_C}{\delta V_{EB}} = - \frac{\beta}{R_B + (\beta + 1) R_E} \qquad (5.9)$$

or, in another form, we have

$$I_C = \frac{\beta\,(V_{BB} - V_{EB})}{R_B + \beta\,R_E + R_E} + \frac{\beta\,I_{CBO}\,(R_B + R_E)}{R_B + \beta\,R_E + R_E} + \frac{I_{CBO}\,(R_B + R_E)}{R_B + \beta\,R_E + R_E}$$

$$= \frac{\beta\,[V_{BB} - V_{EB} + I_{CBO}\,(R_B + R_E)]}{\beta\,R_E + R_B + R_E} + \frac{I_{CBO}\,(R_B + R_E)}{\beta\,R_E + R_B + R_E}$$

$$= \frac{\beta\,[V_{BB} - V_{EB} + I_{CBO}\,(R_B + R_E)]}{\left(\beta + \dfrac{R_B + R_E}{R_E}\right) R_E} + \frac{I_{CBO}\,(R_B + R_E)}{\left(\beta + \dfrac{R_B + R_E}{R_E}\right) R_E}$$

Let:

$$\gamma = \frac{V_{BB} - V_{EB} + I_{CBO}\,(R_B + R_E)}{R_E}, \quad \Delta = \frac{R_B + R_E}{R_E}$$

substituting into the equation for I_C,

$$I_C = \frac{\beta\,\gamma}{\beta + \Delta} + \frac{I_{CBO}\,\Delta}{\beta + \Delta},$$

thus,

$$\frac{\partial I_C}{\partial \beta} = \gamma \left[\frac{-\beta}{(\beta + \Delta)^2} + \frac{1}{\beta + \Delta}\right] + I_{CBO}\,\Delta \left[\frac{-1}{(\beta + \Delta)^2}\right]$$

$$= \gamma \left[\frac{-\beta}{(\beta + \Delta)^2} + \frac{\beta + \Delta}{(\beta + \Delta)^2}\right] + I_{CBO}\,\Delta \left[\frac{-1}{(\beta + \Delta)^2}\right]$$

$$\frac{\partial I_C}{\partial \beta} = \gamma \left[\frac{\Delta}{(\beta + \Delta)^2}\right] - \frac{I_{CBO}\Delta}{(\beta + \Delta)^2} = \frac{\gamma\,\Delta}{(\beta + \Delta)^2} - \frac{I_{CBO}\,\Delta}{(\beta + \Delta)^2} \tag{5.10}$$

However,

$$I_C = \frac{\beta\,\gamma}{\beta + \Delta} + \frac{I_{CBO}\,\Delta}{\beta + \Delta}$$

$$\frac{I_C}{\beta + \Delta} = \frac{\beta\,\gamma}{(\beta + \Delta)^2} + \frac{I_{CBO}\,\Delta}{(\beta + \Delta)^2}$$

$$\frac{\Delta}{\beta}\left(\frac{I_C}{\beta + \Delta}\right) = \frac{\Delta}{\beta}\frac{\beta\,\gamma}{(\beta + \Delta)^2} + \frac{\Delta}{\beta}\frac{I_{CBO}\,\Delta}{(\beta + \Delta)^2}$$

$$\frac{\Delta I_C}{\beta\,(\beta + \Delta)} - \frac{I_{CBO}\,\Delta^2}{\beta\,(\beta + \Delta)^2} = \frac{\gamma\,\Delta}{(\beta + \Delta)^2} \tag{5.11}$$

184

We have, from (5.10) and (5.11):

$$\frac{\partial I_C}{\partial \beta} = \frac{\Delta I_C}{\beta (\beta + \Delta)} - \frac{I_{CBO} \Delta^2}{\beta (\beta + \Delta)^2} - \frac{I_{CBO} \Delta}{(\beta + \Delta)^2}$$

$$= \frac{\Delta I_C}{\beta (\beta + \Delta)} \left(\frac{\beta + \Delta}{\beta + \Delta}\right) - \frac{I_{CBO} \Delta^2}{\beta (\beta + \Delta)^2} - \frac{I_{CBO} \Delta}{(\beta + \Delta)^2} \left(\frac{\beta}{\beta}\right)$$

$$= \frac{\Delta I_C (\beta + \Delta) - I_{CBO} \Delta^2 - I_{CBO} \Delta \beta}{\beta (\beta + \Delta)^2}$$

$$= \frac{\Delta I_C (\beta + \Delta) - I_{CBO} \Delta (\beta + \Delta)}{\beta (\beta + \Delta)^2}$$

$$= \frac{(I_C - I_{CBO}) \Delta}{\beta (\beta + \Delta)} = \frac{(I_C - I_{CBO}) \left(\dfrac{R_E + R_B}{R_E}\right)}{\beta \left(\beta + \dfrac{R_E + R_B}{R_E}\right)}$$

$$= \frac{(I_C - I_{CBO}) \left(\dfrac{R_E + R_B}{R_E}\right)}{\beta \left(\dfrac{\beta R_E + R_E + R_B}{R_E}\right)}$$

$$\frac{\partial I_C}{\partial \beta} = \frac{(I_C - I_{CBO})(R_E + R_B)}{\beta [(\beta + 1) R_E + R_B]} \tag{5.12}$$

Now that we have derived each of the stability factors in terms of the circuit components, voltages, and currents, we may use these results to examine a physical example.

Example (5.1). Typical data for the plastic encapsulated, *npn* silicon, class C, RF amplifier; 2N3642, may be as shown in Table 5.1.

Row 4 of Table 5.1 was completed by use of the following (these data are sometimes omitted in manufacturers specifications):

$$I_{CEO} = (h_{FE} + 1)I_{CBO}$$

Therefore, at 25°C:

$$I_{CBO} = \frac{I_{CEO}}{h_{FE} + 1} = \frac{0.05(10^{-3}) \text{ mA}}{101} = \frac{0.05(10^{-3}) \text{ mA}}{100} = 5(10^{-7}) \text{ mA}$$

at 65°C

$$I_{CBO} = \frac{I_{CEO}}{h_{FE} + 1} = \frac{1 \text{ mA}}{101} = \frac{10^{-3} \text{ mA}}{101} = \frac{10^{-3} \text{ mA}}{100} = 10^{-5} \text{ mA}$$

Table 5.1

Temperature	25°C	65°C	
β	100	200	$\therefore \dfrac{\Delta\beta}{\Delta T} = \dfrac{100}{40°C} = 2.5/°C \doteq \dfrac{d\beta}{dT}$
I_{CEO}	.05 μA	1 μA	$\therefore \dfrac{\Delta I_{CEO}}{\Delta T} = \dfrac{1 - .05}{40}$ $= .02375 \dfrac{\mu A}{°C} \doteq \dfrac{dI_{CEO}}{dT}$
I_{CBO}	$5(10^{-7})$ mA	(10^{-5}) mA	$\therefore \dfrac{\Delta I_{CBO}}{\Delta T} = \dfrac{(100 - 5)10^{-7}}{40}$ $= 2.375(10^{-7}) \dfrac{mA}{°C} \doteq \dfrac{dI_{CBO}}{dT}$
V_{EB}	0.7 V	0.6 V	$\therefore \dfrac{\Delta V_{EB}}{\Delta T} = -2.5 \dfrac{mV}{°C} \doteq \dfrac{dV_{EB}}{dT}$
			(typical for silicon transistors)

Referring to the universal stabilized common emitter circuit design, $R_1 = 42$ k, $R_2 = 10.67$ k, $V_{CC} = 10$ V, $R_E = 0.19802$ k. Thus,

$$R_B = \frac{R_1 R_2}{R_1 + R_2} = \frac{42\,(10.67)}{42 + 10.67} = 8.5275 \text{ k}$$

$$V_{BB} = V_{CC} \frac{R_2}{R_1 + R_2} = 10 \frac{10.67}{42 + 10.67} = \frac{106.7}{52.67} = 2.03 \text{ V}$$

Substituting into the stability factor equation for leakage current:

$$\frac{\partial I_C}{\partial I_{CBO}} = \frac{(\beta + 1)\,(R_B + R_E)}{R_B + (\beta + 1)R_E} = \frac{101(8.5275 + 0.19802)}{8.5275 + 101\,(0.19802)}$$

$$\frac{\partial I_C}{\partial I_{CBO}} = \frac{881.3}{28.53} = 30.89 \quad \text{(dimensionless)}$$

and

$$\frac{\partial I_C}{\partial V_{EB}} = -\frac{\beta}{R_B + (\beta + 1)R_E} = -\frac{100}{8.5275 + 101(0.19802)}$$

$$\frac{\partial I_C}{\partial V_{EB}} = -\frac{100}{28.5275} = -3.505 \frac{mA}{V}$$

Also,

$$\frac{\partial I_C}{\partial \beta} = \frac{(I_C - I_{CBO})(R_B + R_E)}{\beta[R_B + (\beta + 1)R_E]} = \frac{(5)(8.5275 + 0.19802)}{100[8.5275 + 101(0.19802)]}$$

$$\frac{\partial I_C}{\partial \beta} = \frac{5(8.7255)}{100(28.528)} = \frac{43.63}{2853} = 0.01529 \text{ mA}$$

Restating the equation that expresses collector current change with respect to change in temperature:

$$\frac{dI_C}{dT} = \frac{\partial I_C}{\partial I_{CBO}} \cdot \frac{dI_{CBO}}{dT} + \frac{\partial I_C}{\partial V_{EB}} \cdot \frac{dV_{EB}}{dT} + \frac{\partial I_C}{\partial \beta} \cdot \frac{d\beta}{dT}$$

$$= 30.89 \cdot \left[2.375(10^{-7}) \frac{\text{mA}}{\text{°C}} \right] + \left(-3.505 \frac{\text{mA}}{\text{V}} \right) \left[-2.5(10^{-3}) \frac{\text{V}}{\text{°C}} \right]$$

$$+ 0.01529 \text{ mA} \cdot \left(\frac{2.5}{\text{°C}} \right)$$

$$= 7.337 (10^{-6}) \frac{\text{mA}}{\text{°C}} + 8.763 \frac{\text{mA}}{\text{°C}} + 0.03823 \frac{\text{mA}}{\text{°C}}$$

$$\frac{dI_C}{dT} = 0.047 \frac{\text{mA}}{\text{°C}}$$

We can see that the leakage current for a silicon transistor contributes a negligible amount to the overall transistor collector current change (with respect to temperature).

The change in the collector current for a rise in temperature from room temperature (24°C) to 100°C (a military environment specification temperature) would then be calculated as follows:

$$\frac{\Delta I_C}{\Delta T} \doteq \frac{dI_C}{dT}$$

$$\Delta I_C = \frac{\Delta I_C}{\Delta_T} \cdot \Delta_T = 0.047 \frac{\text{mA}}{\text{°C}} (100°C - 25°C) = 3.525 \text{ mA}$$

Example (5.2). A change in collector current of this magnitude will significantly alter the Q point voltages. If

$$I_{CQ} \text{ (at 25°C)} = 5 \text{ mA}, R_C = 0.9 \text{ k}, V_{RC} = 4.5 \text{ V}$$
$$V_{RE} = 1 \text{ V}, V_{EC} = 4.5 \text{ V}$$

at 100°C, we have

$$I_C \text{ (100°C)} = I_C(25°C) + \Delta I_C = 5 \text{ mA} + 3.525 \text{ mA}$$

I_C (100°C) = 8.525 mA

V_{RC} (100°C) = I_C(100°C)R_C = 8.525 (0.9) = 7.67 V

V_{RE}(100°C) = I_E(100°C)R_E = (I_C + I_B)R_E = $I_C \left(1 + \dfrac{1}{\beta} \right)$

$\qquad = I_C \left(\dfrac{\beta + 1}{\beta} \right) R_E = 8.525 \left(\dfrac{201}{200} \right) (0.19802) = 1.69$ V

$V_{EC} = V_{CC} - V_{RC} - V_{RE} = 10 - 7.67 - 1.69 = 0.63$ V

We can see that voltage V_{EC} has decreased from the original design value of 4.5 V as computed in the previous design example, to a value of 0.63 V. The 0.63 V limits the peak signal excursion to that value (0.63 V) in the output circuit (collector circuit).

Which circuit is more stable with respect to changes in β, the universal stabilized or collector-feedback stabilized common emitter configuration? For a meaningful answer to this question, we must establish that the Q point is to be the same for both configurations under consideration; i.e., I_C, V_{RC}, V_{EC}, β, I_B, and V_{RE}. For the above conditions to be true, R_B for the universal stabilized configuration must be a lesser value than for the collector-feedback stabilized common emitter configuration, since $R_B = R_1 // R_2$ (R_2 does not exist for the emitter stabilized configuration).

Examining the stability factor equation with respect to β:

$$
\begin{aligned}
\frac{\partial I_C}{\partial \beta} &= \left[\frac{I_C - I_{CBO}}{\beta} \right] \left[\frac{R_B + R_E}{R_B + (\beta + 1)R_E} \right] \\[2mm]
&= \left[\frac{I_C - I_{CBO}}{\beta} \right] \left[\frac{R_B + R_E}{R_B + R_E + \beta R_E} \right] \\[2mm]
&= \left[\frac{I_C - I_{CBO}}{\beta} \right] \left[\frac{1}{1 + \dfrac{\beta R_E}{R_B + R_E}} \right]
\end{aligned}
$$

Examining the above reworked equation, as R_B decreases, the denominator

$$\left(1 + \frac{\beta R_E}{R_B + R_E} \right)$$

becomes larger and the entire equation decreases in value. Thus, the stability factor decreases (i.e., more stability).

This completes the analysis of the stability of the common emitter stage with universal stabilization. All of the formulas developed here are valid and usable for the common emitter stage, which can be either with or without stabilization.

Where V_{BB} appears in the equation, it is replaced with V_{CC}, and where R_B appears in the equation, no computations are necessary to determine its value. We merely substitute the given value of R_B into the equation. For the unstabilized common emitter, where R_E appears in the equations, the value zero is substituted in the equations.

5.1.2 Stability of the Common Emitter Stage with Collector Feedback

The equations developed above are not appropriate for analyzing the common emitter stage with collector feedback. Therefore, the following is a development of the appropriate equations and a comparison of the stability of the common emitter stage with universal bias *versus* the common emitter stage with collector feedback. Referring to Fig. 5.7:

$$V_{RF} = V_{RE} + V_{EC} - V_{gnd\text{-}B}$$

$$I_6 = \frac{V_{RF}}{R_F} = \frac{V_{RE} + V_{EC} - V_{gnd\text{-}B}}{R_F} \,,\quad V_{gnd\text{-}B} = V_{RE} + V_{EB}$$

$$\text{Therefore, } I_6 = \frac{V_{RE} + V_{EC} - (V_{RE} + V_{EB})}{R_F} = \frac{V_{EC} - V_{EB}}{R_F} \tag{5.13}$$

$$I_2 = \frac{V_{R2}}{R_2} = \frac{V_{gnd\text{-}B}}{R_2} = \frac{V_{RE} + V_{EB}}{R_2}$$

$$I_B = I_6 - I_2 = \frac{V_{EC} - V_{EB}}{R_F} - \frac{V_{RE} + V_{EB}}{R_2}$$

$$= \frac{V_{EC}}{R_F} - \frac{V_{EB}}{R_F} - \frac{V_{RE}}{R_2} - \frac{V_{EB}}{R_2} \tag{5.14}$$

Also

$$V_{CC} = V_{RE} + V_{EC} + (I_C + I_6)\, R_C$$

Therefore, $V_{EC} = V_{CC} - V_{RE} - (I_C + I_6)\, R_C$

From the preceding equation:

$$V_{EC} = V_{CC} - V_{RE} - I_C R_C - I_6 R_C$$

$$= V_{CC} - V_{RE} - I_C R_C - \left(\frac{V_{EC} - V_{EB}}{R_F}\right) R_C \quad - \left(\frac{V_{EC} - V_{EB}}{R_F}\right) R_C$$

$$\text{Also: } I_C = \beta\, I_B + (\beta + 1)\, I_{CBO} \tag{5.15}$$

$$\text{and } I_E = \frac{I_C - I_{CBO}}{\alpha} = (I_C - I_{CBO}) \left(\frac{\beta + 1}{\beta}\right) \tag{5.16}$$

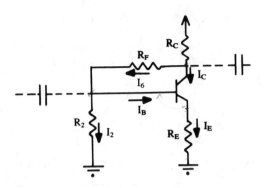

Fig. 5.7

Therefore, $V_{EC} = V_{CC} - I_E R_E - I_C R_C - V_{EC} \dfrac{R_C}{R_F} + V_{EB} \dfrac{R_C}{R_F}$

From (5.16)

$$V_{EC} = V_{CC} - (I_C - I_{CBO})\left(\frac{\beta + 1}{\beta}\right) R_E - I_C R_C - V_{EC}\frac{R_C}{R_F} + V_{EB}\frac{R_C}{R_F}$$

$$= V_{CC} - I_C\left(\frac{\beta + 1}{\beta}\right) R_E + I_{CBO}\left(\frac{\beta + 1}{\beta}\right) R_E$$

$$- I_C R_C - V_{EC}\frac{R_C}{R_F} + V_{EB}\frac{R_C}{R_F}$$

$$V_{EC} + V_{EC}\frac{R_C}{R_F} = V_{CC} + V_{EB}\frac{R_C}{R_F} - I_C\left[\left(\frac{\beta + 1}{\beta}\right) R_E + R_C\right]$$

$$+ I_{CBO}\left(\frac{\beta + 1}{\beta}\right) R_E$$

$$V_{EC} = \frac{V_{CC} + V_{EB}\dfrac{R_C}{R_F} - I_C\left[\left(\dfrac{\beta + 1}{\beta}\right) R_E + R_C\right] + I_{CBO}\left(\dfrac{\beta + 1}{\beta}\right) R_E}{1 + R_C/R_F}$$

$$(5.17)$$

From (5.15) and (5.14)

$$I_C = \beta\left[\frac{V_{EC}}{R_F} - \frac{V_{EB}}{R_F} - \frac{V_{RE}}{R_2} - \frac{V_{EB}}{R_2}\right] + (\beta + 1) I_{CBO}$$

$$I_C = \beta \frac{V_{EC}}{R_F} - \beta \frac{V_{EB}}{R_F} - \beta \frac{V_{RE}}{R_2} - \beta \frac{V_{EB}}{R_2} + (\beta + 1) I_{CBO}$$

$$= \frac{\beta}{R_F} \left\{ \frac{V_{CC} + V_{EB}\dfrac{R_C}{R_F} - I_C \left[\left(\dfrac{\beta + 1}{\beta}\right) R_E + R_C \right] + I_{CBO} \left(\dfrac{\beta + 1}{\beta}\right) R_E}{1 + R_C/R_F} \right\}$$

$$- \frac{\beta}{R_2} I_E R_E - \beta V_{EB} \left(\frac{1}{R_F} + \frac{1}{R_2} \right) + (\beta + 1) I_{CBO}$$

$$= \frac{\beta}{R_F} \left\{ \frac{V_{CC} + V_{EB}\dfrac{R_C}{R_F} - I_C \left[\left(\dfrac{\beta + 1}{\beta}\right) R_E + R_C \right] + I_{CBO} \left(\dfrac{\beta + 1}{\beta}\right) R_E}{\dfrac{R_F + R_C}{R_F}} \right\}$$

$$- \frac{\beta R_E}{R_2} \left[(I_C - I_{CBO}) \left(\frac{\beta + 1}{\beta} \right) \right] - \beta V_{EB} \left(\frac{R_F + R_2}{R_F R_2} \right) + (\beta + 1) I_{CBO}$$

$$= \frac{\beta V_{CC} + \beta V_{EB}\dfrac{R_C}{R_F} - I_C \beta \left[\left(\dfrac{\beta + 1}{\beta}\right) R_E + R_C \right] + \beta I_{CBO} \left(\dfrac{\beta + 1}{\beta}\right) R_E}{R_F + R_C}$$

$$- \frac{\beta R_E}{R_2} \left(\frac{\beta + 1}{\beta} \right) I_C + \frac{\beta R_E}{R_2} \left(\frac{\beta + 1}{\beta} \right) I_{CBO}$$

$$- \beta V_{EB} \left(\frac{R_F + R_2}{R_F R_2} \right) + (\beta + 1) I_{CBO}$$

$$I_C (R_F + R_C) = \beta V_{CC} + \beta V_{EB} \frac{R_C}{R_F} - I_C \beta \left[\left(\frac{\beta + 1}{\beta} \right) R_E + R_C \right]$$

$$+ I_{CBO} (\beta + 1) R_E - \frac{R_E}{R_2} (\beta + 1)(R_F + R_C)$$

$$+ \frac{R_E}{R_2} (\beta + 1) I_{CBO} (R_F + R_C) - \beta V_{EB} \left(\frac{R_F + R_2}{R_F R_2} \right) \cdot (R_F + R_C)$$

$$\curvearrowleft (R_F + R_C) + (\beta + 1) I_{CBO} (R_F + R_C)$$

Collecting similar terms:

$$I_C (R_F + R_C) + I_C \beta \left[\left(\frac{\beta + 1}{\beta} \right) R_E + R_C \right] + I_C \frac{R_E}{R_2} (\beta + 1)(R_F + R_C)$$

$$= \beta V_{CC} + \beta V_{EB} - \beta V_{EB} \left(\frac{R_F + R_2}{R_F R_2} \right) (R_F + R_C) + I_{CBO} (\beta + 1) R_E$$

$$+ I_{CBO} (\beta + 1) \left(\frac{R_E}{R_2} \right) (R_F + R_C) + I_{CBO} (\beta + 1)(R_F + R_C)$$

$$I_C[R_F + R_C + (\beta + 1)R_E + \beta R_C + (\beta + 1)R_E\left(\frac{R_F + R_C}{R_2}\right)$$

$$= \beta\left\{V_{CC} - V_{EB}\left[\left(\frac{R_F + R_2}{R_F R_2}\right)(R_F + R_C) - \frac{R_C}{R_F}\right]\right\} + I_{CBO}$$

$$\times (\beta + 1)[R_E + \frac{R_E}{R_2}(R_C + R_F) + R_F + R_C]$$

$$I_C = \frac{\beta\left\{V_{CC} - V_{EB}\left[\left(\frac{R_F + R_2}{R_F R_2}\right)(R_F + R_C) - \frac{R_C}{R_F}\cdot\frac{R_2}{R_2}\right]\right\}}{R_F + R_C(\beta + 1) + R_E(\beta + 1)\left(1 + \frac{R_F + R_C}{R_2}\right)}$$

$$+ \frac{I_{CBO}(\beta + 1)\left\{R_F + R_C + R_E\left[1 + \left(\frac{R_F + R_C}{R_2}\right)\right]\right\}}{R_F + R_C(\beta + 1) + R_E(\beta + 1)\left(1 + \frac{R_F + R_C}{R_2}\right)}$$

$$= \beta[V_{CC} - V_{EB}\left(\frac{R_F{}^2 + R_F R_C + R_2 R_F + R_C R_2 - R_C R_2}{R_F R_2}\right)$$

$$+ \frac{I_{CBO}(\beta + 1)\left[R_F + R_C + R_E\left(1 + \frac{R_F + R_C}{R_2}\right)\right]}{R_F + (\beta + 1)\left[R_C + R_E\left(1 + \frac{R_F + R_C}{R_2}\right)\right]}$$

$$I_C = \frac{\beta\left[V_{CC} - V_{EB}\left(\frac{R_F}{R_2} + \frac{R_C}{R_2} + 1\right)\right] + I_{CBO}(\beta + 1)}{R_F + (\beta + 1)\left[R_C + R_E\left(1 + \frac{R_F + R_C}{R_2}\right)\right]}$$

$$\times \frac{\left[R_F + R_C + R_E\left(1 + \frac{R_F + R_C}{R_2}\right)\right]}{R_F + (\beta + 1)\left[R_C + R_E\left(1 + \frac{R_F + R_C}{R_2}\right)\right]}$$

$$I_C = \frac{\beta\left[V_{CC} - V_{EB}\left(1 + \frac{R_F + R_C}{R_2}\right)\right] + I_{CBO}(\beta + 1)}{R_F + (\beta + 1)\left[R_C + R_E\left(1 + \frac{R_F + R_C}{R_2}\right)\right]}$$

$$
\times \frac{\left[R_F + R_C + R_E \left(1 + \dfrac{R_F + R_C}{R_2} \right) \right]}{R_F + (\beta + 1)\left[R_C + R_E \left(1 + \dfrac{R_F + R_C}{R_2} \right) \right]} \tag{5.18}
$$

If $R_2 = \infty$, $R_E = 0$:

$$
I_C = \frac{\beta\,(V_{CC} - V_{EB}) + I_{CBO}\,(\beta + 1)\,(R_F + R_C)}{R_F + (\beta + 1)R_C}
$$

which checks against our previous results.

Now that we have an expression for I_C with respect to the leakage current, and known or determined measurable voltages and circuit resistances, we may determine the partial derivatives of I_C with respect to the variables of interest, namely, I_{CBO}, V_{EB} and β.

Employing Equation (5.18), by inspection, we have

$$
\frac{\partial I_C}{\partial I_{CBO}} = \frac{(\beta + 1)\left[R_F + R_C + R_E \left(1 + \dfrac{R_F + R_C}{R_2} \right) \right]}{R_F + (\beta + 1)\left[R_C + R_E \left(1 + \dfrac{R_F + R_C}{R_2} \right) \right]} \tag{5.19}
$$

Rewriting Equation (5.18), we have:

$$
I_C = \frac{\beta\, V_{CC} - \beta V_{EB} \left(1 + \dfrac{R_F + R_C}{R_2} \right)}{R_F + (\beta + 1)\left[R_C + R_E \left(1 + \dfrac{R_F + R_C}{R_2} \right) \right]}
$$

$$
+ \frac{I_{CBO}\,(\beta + 1)\left[R_F + R_C + R_E \left(1 + \dfrac{R_F + R_C}{R_2} \right) \right]}{R_F + (\beta + 1)\left[R_C + R_E \left(1 + \dfrac{R_F + R_C}{R_2} \right) \right]}
$$

Then, by inspection:

$$
\frac{\partial I_C}{\partial V_{EB}} = \frac{-\,\beta \left(1 + \dfrac{R_F + R_C}{R_2} \right)}{R_F + (\beta + 1)\left[R_C + R_E \left(1 + \dfrac{R_F + R_C}{R_2} \right) \right]} \tag{5.20}
$$

Restating Equation (5.18), we have

$$I_C = \frac{\beta \left[V_{CC} - V_{EB} \left(1 + \frac{R_F + R_C}{R_2} \right) \right]}{R_F + (\beta + 1) \left[R_C + R_E \left(1 + \frac{R_F + R_C}{R_2} \right) \right]}$$

$$\frac{+ I_{CBO} (\beta + 1) \left[R_F + R_C + R_E \left(1 + \frac{R_F + R_C}{R_2} \right) \right]}{R_F + (\beta + 1) \left[R_C + R_E \left(1 + \frac{R_F + R_C}{R_2} \right) \right]}$$

Let: $\theta = V_{CC} - V_{EB} \left(1 + \frac{R_F + R_C}{R_2} \right)$,

and $\phi = R_C + R_E \left(1 + \frac{R_F + R_C}{R_2} \right)$ (5.23)

Thus, $I_C = \dfrac{\beta \theta + I_{CBO} (\beta + 1) (R_F + \phi)}{R_F + (\beta + 1) \phi}$ (5.24)

Expanding Equation (5.24):

$$I_C = \frac{\beta\theta + \beta I_{CBO} (R_F + \phi) + I_{CBO} (R_F + \phi)}{R_F + \beta\phi + \phi}$$

$$= \frac{\beta \left[\theta + I_{CBO} (R_F + \phi) \right] + I_{CBO} (R_F + \phi)}{\beta\phi + R_F + \phi}$$

$$= \left[\theta + I_{CBO} (R_F + \phi) \right] \cdot \frac{\beta}{\beta\phi + R_F + \phi}$$

$$+ I_{CBO} (R_F + \phi) \cdot \frac{1}{\beta\phi + R_F + \phi}$$

$$= \frac{\theta + I_{CBO} (R_F + \phi)}{\phi} \cdot \frac{\beta}{\beta + \dfrac{R_F + \phi}{\phi}}$$

$$+ \frac{I_{CBO} (R_F + \phi)}{\phi} \cdot \frac{1}{\beta + \dfrac{R_F + \phi}{\phi}}$$

$$\frac{\partial I_C}{\partial \beta} = \frac{\theta + I_{CBO} (R_F + \phi)}{\phi} \left[\beta \cdot \frac{-1}{\left(\beta + \dfrac{R_F + \phi}{\phi} \right)^2} + \frac{1}{\beta + \dfrac{R_F + \phi}{\phi}} \right]$$

$$+ \frac{I_{CBO} (R_F + \phi)}{\phi} \frac{-1}{\left(\beta + \dfrac{R_F + \phi}{\phi}\right)^2}$$

$$= \frac{\theta + I_{CBO} (R_F + \phi)}{\phi} \left[\frac{-\beta}{\left(\beta + \dfrac{R_F + \phi}{\phi}\right)^2} + \frac{\beta + \dfrac{R_F + \phi}{\phi}}{\left(\beta + \dfrac{R_F + \phi}{\phi}\right)^2} \right]$$

$$- \frac{I_{CBO} (R_F + \phi)}{\phi} \frac{1}{\left(\beta + \dfrac{R_F + \phi}{\phi}\right)^2} \frac{\phi}{\phi}$$

$$= \frac{[\theta + I_{CBO} (R_F + \phi)] (R_F + \phi)}{\phi^2 \left(\beta + \dfrac{R_F + \phi}{\phi}\right)^2}$$

$$- \frac{I_{CBO} (R_F + \phi) \phi}{\phi^2} \frac{1}{\left(\beta + \dfrac{R_F + \phi}{\phi}\right)^2}$$

$$= \frac{\theta (R_F + \phi) + I_{CBO} (R_F + \phi)^2 - I_{CBO} (R_F + \phi) \phi}{\phi^2 \left(\beta + \dfrac{R_F + \phi}{\phi}\right)^2}$$

$$= \frac{\theta (R_F + \phi) + I_{CBO} (R_F + \phi) (R_F + \phi - \phi)}{(\beta \phi + R_F + \phi)^2}$$

$$= \frac{\theta (R_F + \phi) + I_{CBO} (R_F + \phi) R_F}{[R_F + (\beta + 1) \phi]^2} = \frac{(\theta + I_{CBO} R_F) (R_F + \phi)}{[R_F + (\beta + 1) \phi]^2}$$

$$\frac{\partial I_C}{\partial \beta} = \frac{\theta + I_{CBO} R_F}{R_F + (\beta + 1)\phi} \cdot \frac{R_F + \phi}{R_F + (\beta + 1) \phi} \tag{5.25}$$

However, from (5.24):

$$I_C = \frac{\beta\theta + I_{CBO} (\beta + 1) (R_F + \phi)}{R_F + (\beta + 1) \phi}$$

$$I_C - I_{CBO} = \frac{\beta\theta + I_{CBO} (\beta + 1) (R_F + \phi)}{R_F + (\beta + 1)\phi} - I_{CBO} \left[\frac{R_F + (\beta + 1) \phi}{R_F + (\beta + 1) \phi} \right]$$

$$= \frac{\beta\theta + I_{CBO} (\beta + 1)(R_F + \phi) - I_{CBO} [R_F + (\beta + 1) \phi]}{R_F + . (\beta + 1) \phi}$$

$$= \frac{\beta\theta + I_{CBO} [(\beta + 1) R_F + (\beta + 1) \phi - R_F - (\beta + 1) \phi]}{R_F + (\beta + 1) \phi}$$

$$I_C - I_{CBO} = \frac{\beta\theta + \beta I_{CBO}R_F}{R_F + (\beta + 1)\,\phi} = \frac{\beta\,(\theta + I_{CBO}\,R_F)}{R_F + (\beta + 1)\,\phi}$$

$$\frac{I_C - I_{CBO}}{\beta} = \frac{\theta + I_{CBO}\,R_F}{R_F + (\beta + 1)\,\phi} \qquad\qquad (5.26)$$

and from (5.25) and (5.26):

$$\frac{\partial I_C}{\partial \beta} = \frac{I_C - I_{CBO}}{\beta} \cdot \frac{R_F + \phi}{R_F + (\beta + 1)\,\phi}$$

$$\frac{\partial I_C}{\partial \beta} = \frac{I_C - I_{CBO}}{\beta}\left\{ \frac{R_F + R_C + R_E\left(1 + \dfrac{R_F + R_C}{R_2}\right)}{R_F + (\beta + 1)\left[R_C + R_E\left(1 + \dfrac{R_F + R_C}{R_2}\right)\right]} \right\}$$

$$(5.27)$$

5.2   STABILITY CALCULATIONS FOR THE UNIVERSAL STABILIZED COMMON EMITTER CIRCUIT

Now that we have developed the stability formulas for a common emitter stage with collector feedback, we can compare these formulas with those for the universal stabilized common-emitter stage. It can be seen that, except for the resistor subscripts, there is considerable similarity between the equations with the exception that R_C appears in the collector feedback equations.

A computational example is appropriate for comparison purposes with the universal stabilized circuit. In the universal stabilized circuit that was examined for stability, $I_1 = 4I_B$, $V_{RE} = 1$ V, $V_{CC} = 10$ V, $I_{CQ} = 5$ mA, $V_{EB} = 0.6$ V, $V_{RC} = V_{EC} = 4.5$ V.

Example (5.3).   Given: $R_C = .86538$ k, $R_E = .19802$ k, $R_F = 19.5$ k, $R_2 = 10.67$ k. In a manner similar to the steps followed in computing the stability of the universal stabilized common emitter stage:

$$\frac{d\beta}{dT} = \frac{\Delta\beta}{\Delta T} = \frac{2.5}{°C}$$

$$\frac{dI_{CBO}}{dT} \doteq \frac{\Delta I_{CBO}}{\Delta T} = 2.375\,(10^{-7})\ \text{mA/°C}$$

$$\frac{dV_{EB}}{dT} \doteq \frac{\Delta V_{EB}}{\Delta T} = -2.5\ \frac{\text{mV}}{°C}$$

From (5.19):

$$\frac{\partial I_C}{\partial I_{CBO}} = \frac{(\beta + 1)\left[R_F + R_C + R_E\left(1 + \dfrac{R_F + R_C}{R_2}\right)\right]}{R_F + (\beta + 1)\left[R_C + R_E\left(1 + \dfrac{R_F + R_C}{R_2}\right)\right]}$$

$$= \frac{101\left[19.5 + 0.86538 + 0.19802\left(1 + \dfrac{19.5 + 0.86538}{10.67}\right)\right]}{19.5 + 101\left[0.86538 + 0.19802\left(1 + \dfrac{19.5 + 0.86538}{10.67}\right)\right]}$$

$$= \frac{101\,[19.5 + 0.86538 + 0.19802\,(2.90925)]}{19.5 + 101\,[0.86538 + 0.19802\,(2.90925)]}$$

$$= \frac{101(19.5 + 0.86538 + 0.57609)}{19.5 + 101\,(0.86538 + 0.57609)}$$

$$\frac{\partial I_C}{\partial I_{CBO}} = \frac{101(19.5 + 1.4415)}{19.5 + 101\,(1.4415)} = \frac{101\,(20.94)}{19.5 + 145.58} = \frac{2115.1}{165.09} = 12.812$$

(dimensionless)

From (5.20):

$$\frac{\partial I_C}{\partial V_{EB}} = \frac{-\beta\left(1 + \dfrac{R_F + R_C}{R_2}\right)}{R_F + (\beta + 1)\left[R_C + R_E\left(1 + \dfrac{R_F + R_C}{R_2}\right)\right]}$$

$$= \frac{-100(2.909)}{165.09} = -1.762\,\frac{\text{mA}}{\text{V}}$$

From (5.27):

$$\frac{\partial I_C}{\partial \beta} = \frac{I_C - I_{CBO}}{\beta}\left\{\frac{R_F + R_C + R_E\left(1 + \dfrac{R_F + R_C}{R_2}\right)}{R_F + (\beta + 1)\left[R_C + R_E\left(1 + \dfrac{R_F + R_C}{R_2}\right)\right]}\right\}$$

$$\frac{\partial I_C}{\partial \beta} = \frac{(5 - 0)}{100}\left(\frac{20.94}{165.09}\right) = 0.006342\text{ mA}$$

The overall rate of change of collector current with respect to changes in temperature is given by the Equation (5.1):

$$\frac{\partial I_C}{\partial T} = \frac{\partial I_C}{\partial I_{CBO}} \cdot \frac{dI_{CBO}}{dT} + \frac{\partial I_C}{\partial V_{EB}} \cdot \frac{dV_{EB}}{dT} + \frac{\partial I_C}{\partial \beta} + \frac{d\beta}{dT}$$

$$I_C + I_6 = I_C + \delta I_B = I_C + \delta \left[\frac{I_C - (\beta + 1) I_{CBO}}{\beta} \right]$$

$$= I_C + \frac{\delta I_C}{\beta} - \left(\frac{\beta + 1}{\beta} \right) \delta I_{CBO}$$

$$= I_C \left(1 + \frac{\delta}{\beta} \right) - \left(\frac{\beta + 1}{\beta} \right) \delta I_{CBO}$$

Example (5.4). Neglecting the second term, we have, at 100°C:

$$I_C + I_6 = 6.52 \left(1 + \frac{4}{100} \right) = 6.52 \, (1.04) = 6.781 \text{ mA}$$

However, β changes at a rate of 2.5/°C. Therefore,

$$\beta(100°C) = \beta(25°C) + \frac{\Delta \beta}{\Delta T} \times \Delta T = 100 + (100°C - 25°C) \frac{2.5}{°C}$$
$$\beta(100°C) = 100 + 75(2.5) = 100 + 187.5 = 287.5$$

Thus,

$$V_{RC} (100°C) = (I_C + I_6) \text{ (at } 100°C) \cdot R_C = 6.781 \, (0.86538) = 5.868 \text{ V}$$
$$V_{RE} (100°C) = I_E (100°C) \cdot R_E = (I_C (100°C) - I_{CBO} (100°C))$$
$$\times \left(\frac{\beta(100°C) + 1}{\beta(100°C)} \right) R_E$$
$$= 6.52 \left(\frac{288.5}{287.5} \right) (0.19802)$$
$$= 1.296 \text{ V}$$

Substituting the measured values for the regular derivatives and calculated values for the partial derivatives we have

$$\frac{dI_C}{dT} = 12.812 \left[2.375 \, (10^{-7}) \frac{\text{mA}}{°C} \right] + \left(-1.762 \frac{\text{mA}}{\text{V}} \right) \left[\frac{-2.5(10^{-3})\text{V}}{°C} \right]$$
$$+ 0.006342 \text{ mA} \left(\frac{2.5}{°C} \right)$$

$$\frac{dI_C}{dT} = 0.02026 \ \frac{mA}{°C}$$

The change in the collector current for a rise in temperature from room temperature (25°C) to 100°C, would then be calculated as follows:

$$\frac{\Delta I_C}{\Delta T} \doteq \frac{dI_C}{dT} = 0.02026 \ \frac{mA}{°C}$$

$$\Delta I_C = \frac{\Delta I_C}{\Delta T} \cdot \Delta T = 0.02026 \ \frac{mA}{°C} \cdot 75°C = 1.52 \ mA$$

at 100°C

$$I_C \ (100°C) = I_C \ (25°C) + \Delta I_C = 5 \ mA + 1.52 \ mA = 6.52 \ mA$$
$$V_{RC} = (I_C + I_6) \ R_C$$

However,

$$I_C = \beta I_B + (\beta + 1) I_{CBO}$$

Therefore, $\beta I_B = I_C - (\beta + 1) I_{CBO}$

$$I_B = \frac{I_C - (\beta + 1) I_{CBO}}{\beta}$$

where

$$V_{EC} \ (100°C) = V_{CC} \ (100°C) - V_{RC} \ (100°C) - V_{RE} \ (100°C)$$
$$= 10 - 5.868 - 1.296$$
$$V_{EC} \ (100°C) = 2.836 \ V$$

We can see that this is somewhat of an improvement as compared to the universal stabilized common emitter stage where V_{EC} (100°C) = .6309 V.

Percent change in V_{EC}:

Universal Stabilized
$$\left[\frac{4.5 - .6309}{4.5} \right] \times 100 = -86\%$$

Collector Feedback
$$\left[\frac{4.5 - 2.836}{4.5} \right] \times 100 = -37\%$$

Table 5.2

	Universal Stabilized	*Collector Feedback*
I_C	5.0 mA	5.0 mA
V_{RC}	4.5 V	4.5 V
V_{RE}	1.0 V	1.0 V
V_{EC}	4.5 V	4.5 V
$\delta \left(= \dfrac{I_1}{I_B} \right)$	4	$\delta \left(= \dfrac{I_G}{I_B} \right)$ 4
β	100	100
$\dfrac{\partial I_C}{\partial I_{CBO}}$	30.89	12.82
$\dfrac{\partial I_C}{\partial V_{EB}}$	$-3.505 \dfrac{\text{mA}}{\text{V}}$	$-1.762 \dfrac{\text{mA}}{\text{V}}$
$\dfrac{\partial I_C}{\partial \beta}$	.01529 mA	.006342 mA

In each case, the circuit parameters of Table 5.2 were given at 25°C.

In the chapter dealing with small-signal analysis of bipolar transistor amplifiers, we will see how the voltage amplification of each of these stabilized common emitter circuits compares with the others. The chapter dealing with small signal analysis of bipolar transistor amplifiers.

PROBLEMS

1. Predict the new collector current and collector-to-emitter voltage if the operating temperature changes from 25°C to 100°C for the same transistor employed in the universal stabilized circuit and if the transistor is now used in a unstabilized common emitter circuit. Given: $I_C = 5$ mA, $V_{EC} = V_{RC} = 5$ V, $V_{CC} = 10$ V, (25°C) = 100. *Compare* with the results shown in the text.

2. Predict the new collector current and collector-to-emitter voltage if the operating temperature changes from 25°C to 100°C for the same transistor employed in the universal stabilized circuit and if the transistor is now used in a stabilized common emitter circuit. Given: $I_C = 5$ mA, $V_{EC} = V_{RC} = 4.5$ V, $V_{RE} = 1$ V, $V_{CC} = 10$ V, (25°C) = 100. *Compare* with the results shown in the text.

3. Predict the new collector current and collector to emitter voltage if the operating temperature changes from 25°C to 125°C for the same transistor employed in the collector feedback example in Example (5.3). Will the transistor become saturated?

REFERENCES

1. Watson, J., Semiconductor Circuit Design (New York: Halsted Press, 1966).
2. Pascoe, R., Fundamentals of Solid-State Electronics (New York: John Wiley and Sons, 1976).

Chapter 6

Field-Effect Transistor Amplifier Stage
DC and AC
Graphical Design and Analysis

6.1 FIELD EFFECT TRANSISTOR AMPLIFIER GRAPHICAL DESIGN AND ANALYSIS

In previous chapters we discussed the general operation and biasing of field-effect transistors (FET). We noted that a field-effect transistor could be thought of as a voltage-controlled resistor, where voltage applied between the source-to-gate terminal pair controlled the drain current. We also determined the required component values to yield desired dc bias conditions.

At this point, we are ready to employ the FET as an amplifier of an alternating current signal. If we think of the FET as a variable resistor, then it should be desirable to have one-half of the power supply voltage across the transistor and one-half the power supply voltage across the drain resistor (see Fig. 6.1 (a,b)).

This will enable maximum signal swing.

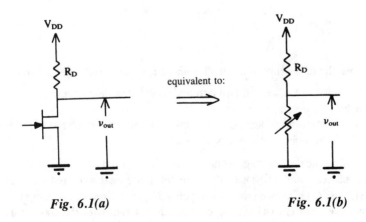

equivalent to:

Fig. 6.1(a)   Fig. 6.1(b)

If we refer to a family of output curves for a JFET transistor (see Fig. 6.2), we see that we can choose an operating point at many locations along the curves. The crosshatched areas are not acceptable for quiescent operating points (Q points).

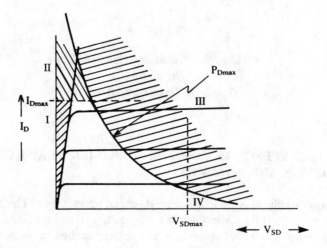

Fig. 6.2

In region I, drain current cannot be made to flow. Operation in region II would exceed the manufacturer's rated maximum drain current. Operation in region III would exceed the transistor's maximum power dissipation. Region IV operation would cause voltage breakdown between the source and drain junctions.

Let us choose our operating point along the line:

$$V_{SDQ} = \frac{V_{DD}}{2}$$

We choose the drain current at the Q point at a value, which meets two criteria:

1. We must not exceed the maximum power dissipation rating of the transistor;
2. We must allow at least one V_{SG} curve to be present above the curve on which the operating point is chosen.

If we now locate a voltage value, which is representative of the desired or required power supply voltage, V_{DD}, on the horizontal axis and connect that point with the Q point, we have developed a dc "load line." Assuming no other resistances are present (which are connected in series with the transistor), the dc and ac load lines are identical.

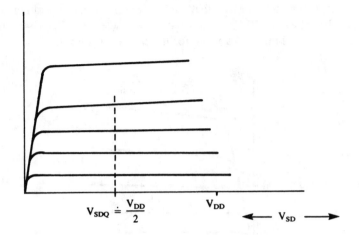

Fig. 6.3

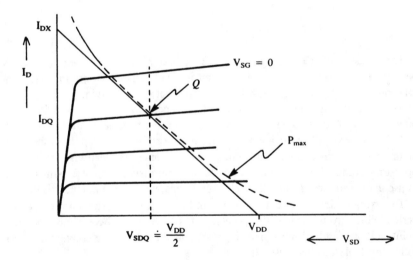

Fig. 6.4

The ac load line is the total load to which the transistor amplifier must deliver alternating (or signal) current, as it performs the function of an amplifier.

To determine the resistance value of R_D, we apply Ohm's law as follows:

$$R_D = \frac{V_{DD}}{I_X}$$

where I_X is the point of intersection between the load line and vertical axis.

Referring to Fig. 6.5, we see that for a MPF-102 JFET, operating with a power supply voltage of 20 V, we have arbitrarily selected a $V_{SGQ} = -0.8$ V resulting in $I_{DQ} = 2$ mA and a value for R_D can be calculated.

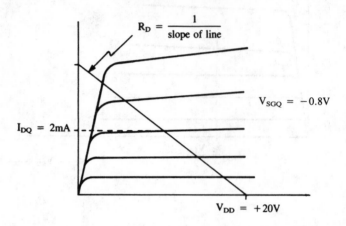

$$R_D = \frac{1}{\text{slope of line}}$$

$V_{SGQ} = -0.8V$

$I_{DQ} = 2mA$

$V_{DD} = +20V$

Fig. 6.5

Let us assume an input signal voltage excursion of a 0.2 V peak (= 0.4 V peak-to-peak). Because the transistor and its load resister R_D are in series, the total power supply voltage must be "shared" between the two.

We note in Fig. 6.6 that two dots have been drawn where the load line intersects two curves, adjacent to the V_{SGQ} curve. If we label these points (1) and (2), and extend two straight lines from the origin to these points, the slope of these lines represents the instantaneous resistance of the transistor at these operating points. Because the resistance of the transistor at these points is not of use in determining the amplifier characteristics, we will not pursue this further.

Referring to the output curves shown in Fig. 6.7, if we extend horizontal and vertical lines from each of the points, (1), (2), and Q, the various operating point voltages and currents may be read and recorded. For convenience, the following table has been established.

Table 6.1

Point	V_{SG}	I_D	V_{SD}
1	−0.2 Vdc	4.40 mA	5 Vdc
2	−1 Vdc	1.56 mA	14.6 Vdc
Q	−0.6 Vdc	3 mA	10 Vdc
Δ	1 − 0.2 = −0.8 V p.p.	4.40 − 1.56 = 2.84 mA p.p.	14.6 − 5 = 9.6 V p.p.

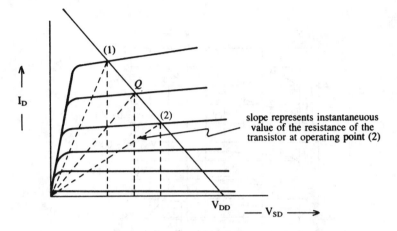

Fig. 6.6

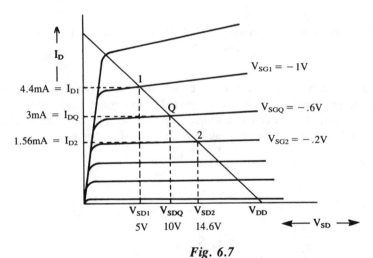

Fig. 6.7

The voltage gain of an amplifier is usually defined as the output voltage change divided by the input voltage change. Therefore, we have

$$A_V = \frac{\Delta V_{out}}{\Delta V_{in}} = \frac{V_{SD2} - V_{SD1}}{V_{SG2} - V_{SG1}} = \frac{14.6\ V - 5\ V}{-1\ V - (-0.2)\ V} = -12\ V \qquad (6.1)$$

It is important to emphasize that the minus sign preceding the 12 does not imply a loss in voltage gain, but merely a phase reversal between the output and input voltages (see Fig. 6.8).

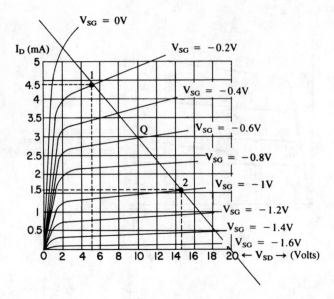

Fig. 6.8 Output curves for an MPF-102 JFET (from a Tektronix type 575 curve tracer)

Example (6.1). If we were to apply a one millivolt positive-going sine wave (peak, peak-to-peak, or RMS) between the ground-to-gate points, and if we view the output waveform for our common source amplifier (see Fig. 6.9), we would notice a 12 mV sine wave (peak, peak-to-peak, or RMS) that was negative-going.

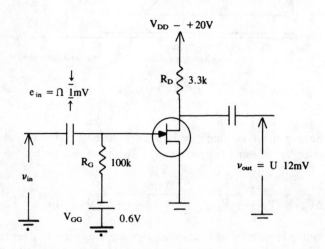

Fig. 6.9

$$R_D = \frac{\Delta E}{\Delta I} = \frac{V_{DD} - V_{SDQ}}{I_{DQ}}$$

$$= \frac{20 - 10}{3}$$

$$= 3.33 \text{ k}$$

The above design could function correctly, but would most likely not be employed in a practical application because of the need for two power supplies (V_{GG} and V_{DD}).

Example (6.2). Let us redesign the common source amplifier and employ self-bias. If we retain the same operating Q point, as in the above example, then (referring to Fig. 6.10) let $V_{SDQ} = 10$ V, $V_{SGQ} = -0.6$ V, and $I_{DQ} = 3$ mA. These data automatically fix the source resistor value:

$$R_S = \frac{V_{SGQ}}{I_{DQ}} = \frac{0.6 \text{ Vdc}}{3 \text{ mA dc}} = 0.2 \text{ k}$$

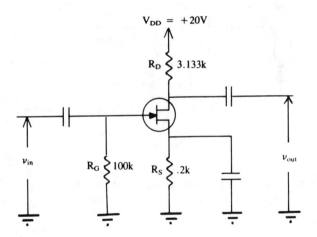

Fig. 6.10

Because the Q point has not been changed, the slope of the line connecting V_{DD} to the Q point is the same. The slope of this line represents the total dc resistance of all resistors in series with the transistor, $R_S + R_D$. Therefore, $R_S + R_D = 3.33$k, and then we may compute R_D as follows:

$$R_D = (R_S + R_D) - R_S = 3.33 \text{ k} - 0.2 \text{ k}$$
$$R_D = 3.133 \text{ k}$$

If the source resistance is bypassed with a capacitor whose reactance is a negligible value compared to the source resistance ($X_{CS} = 0.1R_S$), then alternating or signal currents will only develop an ac signal across R_D (not R_S). In this instance, R_D is referred to as the ac load and a load line with a slope of $1/R_D$ is referred to as the ac load line. To construct an ac load line, we proceed as follows (see Fig. 6.11).

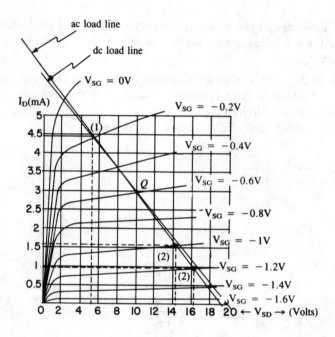

Fig. 6.11

We assume a ΔI_D. Let us choose $\Delta I_D = 3$ mA (from $I_{DQ} = 3$ mA to $I_D = 0$ mA). For this change in I_D we expect a change in V_{SD} computed as follows:

$$\Delta V_{SD} = \Delta I_D Z_L = \Delta I_D R_D = 3 \text{ mA } (3.133 \text{ k}) = 9.399 \text{ V} = 9.4 \text{ V}$$

If we start at the Q point and "travel" down 3 mA, we reach the horizontal axis. We must then travel 9.4 V to the right and we arrive at 19.4 V (10 V + 9.4 V). By connecting $V_{SD} = 19.4$ V with the Q point, we have the ac load line.

If we again construct the data in tabular form (Table 6.2), and computing the voltage gain, we have

$$A_V = \frac{\Delta V_{out}}{\Delta V_{in}} = \frac{V_{SD2} - V_{SD1}}{V_{SG2} - V_{SG1}} = \frac{16.3 - 5.4}{-1 - (-0.2)} = -11.1$$

Table 6.2

Point	V_{SG}	I_D	V_{SD}
1	−0.2 Vdc	4.45 mA dc	5.4 Vdc
2	−1 Vdc	1.56 mA dc	14.3 Vdc
Q	−0.6 Vdc	3 mA dc	10 Vdc
Δ	0.8 V p.p.	2.89 mA p.p.	8.9 V p.p.

We note that there is a slight loss in voltage gain between the fixed and self-biased circuits. This may be partially explained by the fact that as the slope of the ac load line steepens, there is simply less of a voltage swing for a ΔV_{SD} along the horizontal axis.

Example (6.3). Suppose that we have the circuit shown in Fig. 6.12.

The total ac load through which the signal current from the transistor must flow to generate an ac voltage is R_D in parallel with R_L. (Do not forget that V_{DD} is at ac ground potential because of the large electrolytic capacitors connected across the output terminal of our power supply.) Therefore, the "top" of R_D is at ac ground potential.

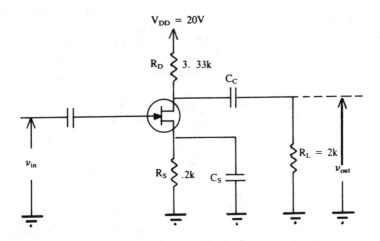

Fig. 6.12

The ac load line should then have a slope that is representative of:

$$Z_L = R_D /\!/ R_L = 3.133 \text{ k} /\!/ 2 \text{ k} = 1.22 \text{ k}$$

In order to determine the voltage gain for the case where $R_L = 2k$, we must plot the ac load line as done in the previous example.

Example (6.4). Let us take a change in $I_D = 3$ mA. Then

$$\Delta V = \Delta I \, Z_L = 3 \text{ mA } (1.22 \text{ k}) = 3.66 \text{ V}$$

referring to Fig. 6.13,

ac load line:

$$\text{slope} = \frac{1}{R_D /\!/ R_L}$$

$$= \frac{1}{3.133 \text{ k} /\!/ 2 \text{ k}}$$

dc load line:

$$\text{slope} = \frac{1}{R_S + R_D}$$

$$= \frac{1}{0.2 \text{ k} + 3.133 \text{ k}}$$

Traveling from the Q point down 3 mA to the horizontal axis and then 3.66 V to the right, we arrive at 13.66 V ($V_{SDQ} + \Delta V_{SD} = 10 \text{ V} + 3.66 \text{ V}$). If we place a dot on the horizontal axis at 13.66 V and connect this with the Q point, we have plotted the ac load line. See Fig. 6.14. We again construct a table of input and output voltages (Table 6.3) and computing the voltage gain, we obtain

$$A_V = \frac{\Delta V_{out}}{\Delta V_{in}} = \frac{V_{SD2} - V_{SD1}}{V_{SG2} - V_{SG1}} = \frac{11.6 \text{ V} - 8 \text{ V}}{-1 \text{ V} - (-0.2 \text{ V})} = -4.5$$

Table 6.3

Point	V_{SG}	I_D	V_{SD}
1	-0.2 Vdc	4.70 mA dc	8 Vdc
2	-1 Vdc	1.54 mA dc	11.6 Vdc
Q	-0.6 Vdc	3 mA dc	10 Vdc
$\mid \Delta \mid$	0.8 V p.p.	3.16 mA p.p.	3.6 V dc

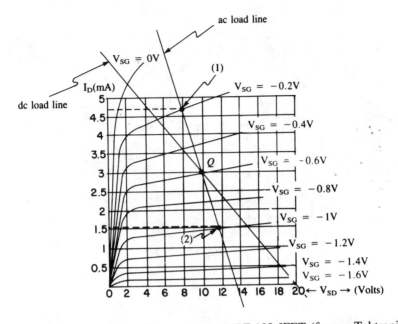

Fig. 6.13 Input and Output Curves for an MPF-102 JFET (from a Tektronix type 575 curve tracer)

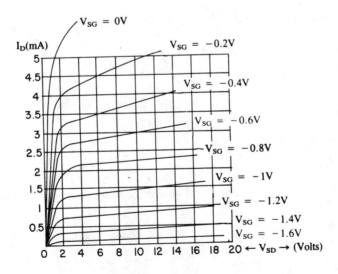

Fig. 6.14(a)

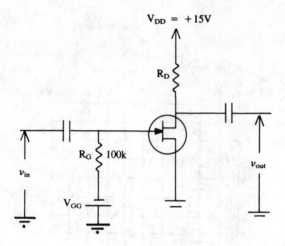

Fig. 6.14(b)

We notice that there is a considerable loss in gain with the addition of a 2 k load resistor connected to our original self-biased amplifier. Thus, we may conclude that as the ac load impedance is decreased in value, a corresponding decrease in voltage gain will result. Let us check two ratios, namely the ratio of the load resistance for self-biased circuits with

$$Z_{load1} = R_D = 3.133 \text{ k}$$

and

$$Z_{load2} = R_D /\!/ R_L = 1.22 \text{ k}$$

Therefore,

$$\frac{Z_{L1}}{Z_{L2}} = \frac{3.133}{1.22} = 2.56$$

and the ratio of their computed voltage gains by graphical techniques:

$$\frac{A_{V1}}{A_{V2}} = \frac{11.1}{4.5} = 2.46; \qquad 2.46 \doteq 2.56 \text{ (a 4\% difference)}$$

So, we see that the voltage gains are (almost exactly) directly proportional to the load resistance. If active amplifier devices were truly linear devices, then the voltage gains of transistor amplifiers would be exactly proportional to the load impedances. In a subsequent chapter, we will investigate transistor amplifiers that are assumed to be perfectly linear.

PROBLEMS

1. For the JFET output curves and schematic shown in Fig. 6.14, design a fixed bias circuit such that

   $$V_{DD} = +15 \text{ V}, V_{SG} = -1 \text{ V}, I_{DQ} = 1.5 \text{ mA}$$

   Predict the voltage gain if an input voltage of 0.2 V peak (0.4 V p.p.) is applied.

2. For the JFET output curves shown in Fig. 6.15, design a self-biased circuit such that

   $$V_{DD} = +15 \text{ V}, V_{SGQ} = -1 \text{ V}, I_{DQ} = 1.5 \text{ mA}$$

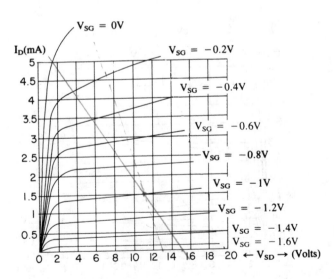

Fig. 6.15(a)

214

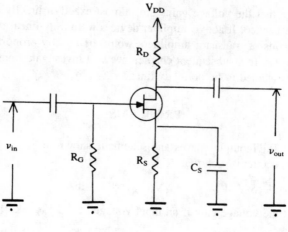

Fig. 6.15(b)

Predict the voltage gain if an input voltage of 0.2 V peak (0.4 V p.p.) is applied.

3. Repeat Problem 2, if a load resistance of 1.8 k is connected to the output of the JFET amplifier. See Fig. 6.16.

$V_{DD} = +15$ V, $V_{SGQ} = -1$ V, $I_{DQ} = 1.5$ mA

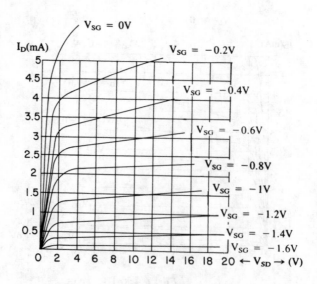

Fig. 6.16(a)

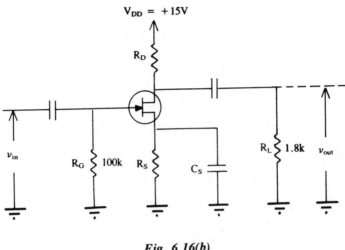

Fig. 6.16(b)

Predict the voltage gain if an input voltage of 0.2 V (0.4 V p.p.) is applied.

Chapter 7

Bipolar Transistor Amplifier Stage
DC and AC
Graphical Design and Analysis

7.1 INTRODUCTION

In a previous chapter, we examined how to bias, or "turn on," a bipolar transistor. If all that we could do with a transistor was dissipate dc power, it would have little value. Therefore, we would like to examine how to use a transistor to amplify a weak alternating current signal in order to make the signal stronger. For example, the weak ac electrical signal from the terminals of a phonograph cartridge could not be directly connected to the terminals of a loudspeaker and expected to produce audible sounds.

Although a small signal graphical design or analysis is rarely considered in an industrial setting, this approach helps to bridge the gap between the concepts of design and analysis of the transistor dc bias theory, and the more abstract concepts of the small-signal equivalent circuit theory of the transistor.

In Chapter 1, the reader may recall that we used a mechanical analogy to describe the bias voltage across the series connection of the collector load resistor and the collector-to-emitter junction. This analogy consisted of a weight suspended by a spring from the ceiling. We noted that for the spring to have maximum upward and downward travel, the weight should be suspended midway between floor and ceiling.

We also noted that variations in the level of base current affected the collector current, which in turn affected the emitter-to-collector voltage. If we had a transistor that was properly biased, such that $V_{RC} = V_{EC}$, and then had some method of adding a small amount of current to the quiescent base current, all of the other Q point voltages and currents would be altered.

Let us look at the base circuit for a common emitter connection (see Fig. 7.1). If we connect a capacitor to the base and apply a signal to the capacitor, an incremental current ($+\Delta I$) would be added to the quiescent base current (if the signal voltage source was instantaneously positive-going).

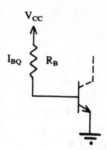

Fig. 7.1

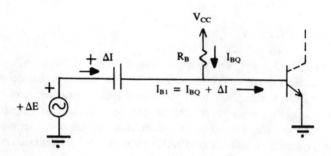

Fig. 7.2

Note: For an ac signal, at any instant, we can think of the capacitor as being a short.

The increase in the base current from I_{BQ} to I_{B1} results in a corresponding increase in the collector current from I_{CQ} to I_{C1}. This, in turn, results in an increase in the voltage drop across the collector load resistor.

We note that the output voltage as measured from ground to collector would be a decreasing voltage.

Correspondingly, if we examine the base circuit with a negative-going signal connected through a capacitor to the base, we note that a small amount of the quiescent base current is diverted away from the base and into the signal generator (see Fig. 7.4).

Thus, the resulting base current I_{B2} is slightly less than the quiescent value. This is also true of the resulting collector current I_{C2} and voltage drop across the collector load resistor, V_{EC2}.

The decrease in the voltage drop across the collector load resistor will result in a corresponding increase in the voltage from emitter to collector.

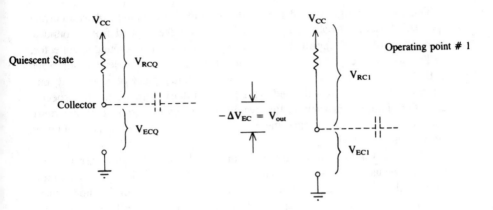

Fig. 7.3

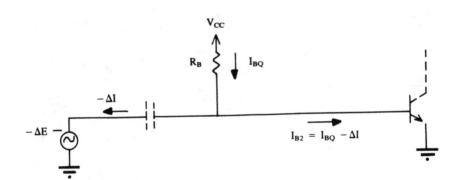

Fig. 7.4

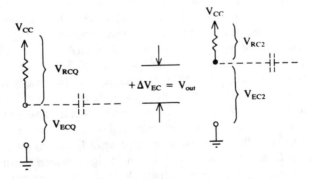

Fig. 7.5

Thus, even though a bipolar transistor is commonly thought of as a current operated device, a small change in the signal voltage applied from ground to base will result in an amplified signal voltage appearing from ground to collector.

The above description may be sufficient for a qualitative understanding, such as that required for a technician to service an installed piece of equipment, but, for design purposes, we will need to predict quantitative parameters for a specific amplifier. Parameters such as the overall current gain (current amplification), voltage gain, input impedance, the dc bias conditions, *et cetera* can be obtained via graphical analysis.

Algebraic solutions of circuits containing a diode and resistor are not convenient because the diode is a nonlinear device. If we use a graphical solution, we would plot the line representing the resistor in series with the diode, backwards, starting from the power supply voltage. (See Fig. 7.6(b).)

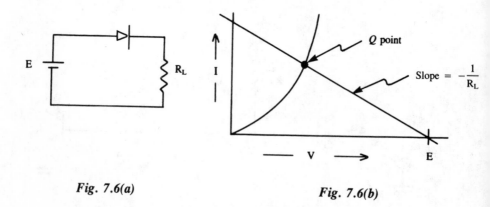

Fig. 7.6(a)   Fig. 7.6(b)

If the diode curve was nonlinear with a shape as shown in Fig. 7.7, the same graphical technique could be used to determine the Q point curve or voltage. We note in Fig. 7.7 that the shape of the nonlinear curve is the same as for a bipolar transistor if we plot the collector current I_C *versus* the emitter-to-collector voltage V_{EC} at a constant base current. Note what happens to the operating point should we increase the base current from $I_{BQ} = 10$ µA to $I_{B1} = 20$ µA.

Thus, the operating point has "slid" up the load line from the quiescent point, or "rest" position, to point (1). This results in an increase in the collector current from 4 mA to 6 mA and a decrease in the emitter-to-collector voltage from 5 V to 2 V. A decrease in the base current would have resulted in the operating point "sliding" down (and to the right) on the load line from the Q point to point (2), with a corresponding decrease in I_C and an increase in V_{EC}. It is important to note that as we vary the input signal source (feeding into the base), we are varying the output voltage (V_{EC}) dc level. We usually are *not* interested in having the dc voltage appear at the input of the next amplifier.

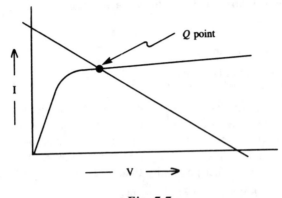

Fig. 7.7

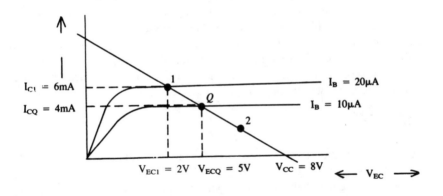

Fig. 7.8

The reason for the undesirability of having the dc level being coupled from one stage to the next is that each stage has its own required dc bias conditions. In Chapter 6, we saw that the dc bias conditions are established by the computed resistor values.

Thus, we are interested in having only the *change* in dc level (which represents an ac signal) coupled from one stage to the next. A capacitor connected from the output of one stage to the input of the next is referred to as a coupling capacitor. This capacitor serves a dual function: it *blocks the dc voltage* from one stage that would affect the dc bias voltage on an adjacent stage and it *couples the signal* from the output of one stage to a subsequent stage. In effect, a coupling capacitor "skims off" the variation in the dc level at the output of an amplifier stage and passes it on to a succeeding stage.

Let us proceed through a complete design and analysis example for a common base transistor connection using a graphical approach.

7.1.1 Graphical Analysis Procedures for a Common Base Transitor Amplifier

Two possibilities exist:

1. The load line is given; or
2. The operating point is given.

This analysis assumes that the collector-to-base dc voltage is given or can be estimated. It also assumes that the base-to-emitter power supply value is given or can be estimated. Another value, which must be given or estimated, is the emitter input current swing. (See Fig. 7.11 for a circuit schematic.)

The analysis for the first possibility (load line) is accomplished as follows:

Example (7.1). Referring to Fig. 7.9, make a dot on the horizontal axis at a point equal to the collector-to-base power supply voltage. The curves shown in the figure are referred to as the "common base characteristic curves" (see Ch. 1). Draw a line from this dot, diagonally upward and to the left, with the appropriate slope. This slope can be calculated by starting at V_{CC} and assuming any desired voltage. For convenience, it may be desirable to count off the full value of V_{CC} to the left of the dot. This will enable us to plot the load line and to plot the second point (establishing the load line) on the vertical axis. The slope can be calculated by the following equation:

$$\Delta I_C = \frac{\Delta V_C}{R_L} \tag{7.1}$$

The value of I should be marked off on the vertical axis, e.g., assume $V_{CC} = 12$, $\Delta V_C = 12$. If R_L is given 2 k, $\Delta I_C = 6$ mA.

Now, a convenient value for the Q point must be chosen. The value for I_{EQ} is usually chosen so that the input or output signal can "swing" equally in either direction before reaching "saturation" or "cut-off." For example, in Fig. 7.9, I_{EQ} is chosen as 3 mA. Next, we select a value for the input emitter "swing." The peak value of emitter current chosen is added to I_{EQ} to establish one of two input current swing limits. In order to establish the other signal point, the chosen peak current value is subtracted from I_{EQ}. The points are plotted on the common base characteristic curves (Fig. 7.9) at the intersection of the load line and the appropriate I_E lines.

First, the value of the output voltage signal (peak-to-peak) is determined from Fig. 7.9. This is accomplished by dropping vertical lines from points (1) and (2) to the horizontal axis on Fig. 7.9. ΔV_{out} may then be calculated:

$$\Delta V_{out} = V_{CB2} - V_{CB1} = 10 - 2.2 = 7.8 \text{ Vp.p.} \tag{7.2}$$

Second, the value of the output current signal (peak-to-peak) is determined from Fig. 7.9. This is accomplished by extending horizontal lines from points

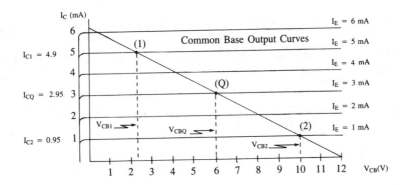

Fig. 7.9

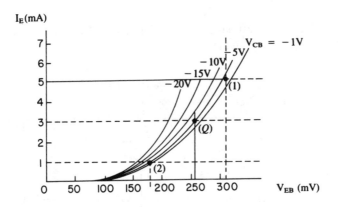

Fig. 7.10

(1) and (2) to the vertical axis. ΔI_{out} may then be calculated:

$$\Delta I_{out} = I_{C1} - I_{C2} = 4.9 - 0.95 = 3.95 \text{ mA} \tag{7.3}$$

Third, the value of the input voltage swing must be determined. Refer to Fig. 7.10.

At point (1), $I_{E1} = 5$ mA and $V_{CB1} = 2.2$ V. To plot point (1) locate $I_{E1} = 5$ mA on the vertical axis. Draw a horizontal line from left to right across the graph. Between the $V_{CB} = -1$ and $V_{CB} = -5$ curves, locate a point on the horizontal line. Label this point (1). Repeat this procedure for the quiescent point (Q) and point (2). Fourth, drop vertical lines from points (1), (2), and (Q) down to the vertical axis. Read and record the corresponding values of V_{BE} for each point. In this example: $V_{BE1} = 315$ mV, $V_{BEQ} = 258$ mV, $V_{BE2} = 177$ mV. Fifth, calculate the value of ΔV_{in}:

$$\Delta V_{in} = V_{BE1} - V_{BE2} = 315 - 177 = 138 \text{ mV} \qquad (7.4)$$

With the above information, the voltage gain A_V, current gain A_1, power gain A_P, and ac input resistance can be calculated:

$$A_V = \frac{\Delta V_{out}}{\Delta V_{in}} = \frac{7.8}{0.138} = 56.5, \qquad \text{from equations (7.2) and (7.4)}$$

$$A_I = \frac{\Delta I_{out}}{\Delta I_{in}} = \frac{3.95}{4} = 0.987, \qquad \text{from equations (7.3) and (7.1)}$$

$$A_P = A_V A_I = 56.5(0.987) = 55.8$$

$$r_{ac} = \frac{\Delta V_{in}}{\Delta I_{in}} = \frac{0.138}{4} = 0.0345 \text{ k} = 34.5 \ \Omega$$

Sixth, the values of the various resistors must be established for proper dc and ac operation (refer to Fig. 7.11).

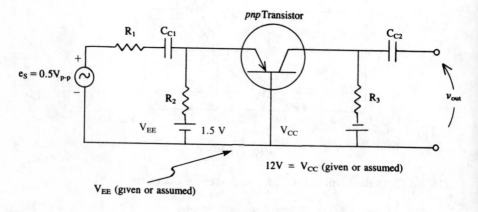

Fig. 7.11

R_2 must limit the forward dc bias current into the base-to-emitter junction. I_{EdcQ} was chosen to be 3 mA. The base-to-emitter drop was found to be 258 mV at $I_{Edc} = 3$ mA. Therefore, the total voltage across R_2 is

$$V_{R2} = 1.5 - 0.258 = 1.242 \text{ V}$$

The dc current in the loop is 3 mA. Thus,

$$R_2 = \frac{1.242}{3} = 0.414 \text{ k} = 414 \ \Omega$$

Finally, the value of R_1 must be determined (see Fig. 7.12).

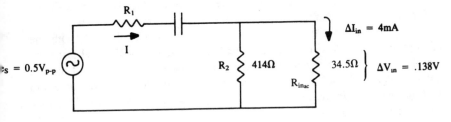

Fig. 7.12

To calculate the value of R_1, it may be assumed for this example, that the ac signal current through R_1 is equal to the signal current into the base-to-emitter junction. This is a reasonable assumption because the ratio of R_2 to the transistor input resistance is more than ten to one. Therefore, the ac voltage across R_1 is equal to the source voltage minus the base-to-emitter drop. Therefore,

$$e_{R1} = e_S - \Delta V_{in} = 0.5\ V_{p\text{-}p} - 0.138\ V_{p\text{-}p} = 0.362\ V_{p\text{-}p}$$

Thus,

$$R_1 = \frac{e_{R1}}{\Delta I_{in}} = \frac{0.362\ V}{4\ mA_{p\text{-}p}} = 0.0905\ V = 90.5\ \Omega$$

In the beginning of this analysis, two possible combinations of given data were presented. The second set of data included an operating point rather than a load resistance value (load line). If we are concerned with the operating point, then the load line is constructed from the V_{CC} power supply voltage on the horizontal axis through the Q point and the remainder of the analysis is as presented here.

7.1.2 Graphical Analysis Procedures for a Common Emitter Transistor Amplifier

Two possibilities exist:

1. The load line is given; or
2. The operating point is given.

This analysis assumes that the collector-to-emitter dc voltage is given or can be estimated. It also assumes that the emitter-to-base power supply value is given or can be estimated. Another value, which must be given or estimated, is the base input current swing.

The analysis for the first possibility is accomplished as follows:

Example (7.2). Referring to Fig. 7.13, make a dot on the horizontal axis at a point equal to the collector-to-emitter power supply voltage. The curves shown in Fig. 7.13 are referred to as the "common emitter characteristic curves" (Ch. 1). Draw a line from this dot, diagonally upward and to the left, with the appropriate slope. This slope can be calculated by starting at V_{CC} and assuming any desired voltage. For convenience, it may be desirable to count off the full value of V_{CC} to the left of the dot. This enables us to plot the load line and to plot the second point (establishing the load line) on the vertical axis. The slope can be calculated by the following equation:

$$\Delta I_C = \frac{V_C}{R_L}$$

This value of I should be marked off on the vertical axis.

Example (7.3). Assume $V_{CC} = 12$, $\Delta V_{CE} = 12$ Vp.p., if R_L is given as 2 k, $\Delta I_C = 6$ mAp.p.

First a convenient value for the Q point must be chosen. The value for I_{Bdc} is usually chosen so that the input or output signal can swing an equal amount in either direction before reaching "saturation" or "cut-off." In Fig. 7.13, I_{Bdc} is chosen as 20 μA. The next step is to consider the input current base swing. The peak base current is added to the base current quiescent value (chosen earlier) to establish one of the two input signal current swing limits. To establish the other signal point, the peak current is subtracted from I_B quiescent. These points are plotted on the common emitter curves at the intersections of the load line and the appropriate I_B lines. For this example:

$$I_Q = 20 \ \mu A \qquad I_B \text{ peak } = 10 \ \mu A$$

Therefore, at point 1: $I_{B1} = 20 \ \mu A + 10 \ \mu A = 30 \ \mu A$
at point 2: $I_{B2} = 20 \ \mu A - 10 \ \mu A = 10 \ \mu A$
Therefore, $\Delta I_{in} = I_{B1} - I_{B2} = 30 \ \mu A - 10 \ \mu A = 20 \ \mu Ap.p.$

Second, the value of the output voltage signal (peak-to-peak) is determined from Fig. 7.13. This is accomplished by dropping vertical lines from points (1) and (2) to the horizontal axis. ΔV_{out} may then be calculated:

$$\Delta V_{out} = V_{CE2} - V_{CE1} = 8 \text{ V} - 4 \text{ V} = 4 \text{ Vp.p.}$$

Third, the value of the output current signal (peak-to-peak) is determined from Fig. 7.13. This is accomplished by extending horizontal lines from points (1) and (2) to the vertical axis. ΔI_{out} may then be calculated:

$$\Delta I_{out} = I_{C1} - I_{C2} = 4 \text{ mA} - 2 \text{ mA} = 2 \text{ mAp.p.}$$

Fourth, the value of the input voltage swing must be determined. Refer to Fig. 7.14. At point (1), $I_{B1} = 30\ \mu A$ and $V_{BE} = 0.6$ V. To plot point (1) locate $I_{B1} = 30\ \mu A$ on the horizontal axis. Draw a vertical line from point (1) to the horizontal axis. Then draw a horizontal line from $V_{BE} = 0.6$ V across the graph. The intersection of these two lines is point (1). Repeat this procedure for the Q point and for point (2) as well.

Fifth, draw in the appropriate lines to find the value of V_{BE} for all three points. Read and record the corresponding values of V_{BE} for each point.

Example (7.4). $V_{BE1} = 0.6$ V, $V_{BEQ} = 0.4$ V, $V_{BE2} = 0.2$ V. First, calculate the value of ΔV_{in}:

$$\Delta V_{in} = V_{BE1} - V_{BE2} = 0.6 - 0.2 = 0.4\ \text{Vp.p.}$$

With the above information, voltage gain A_V, current gain A_I, power gain A_P, and ac input resistance can be calculated:

$$A_V = \frac{\Delta V_{out}}{\Delta V_{in}} = \frac{4\ \text{V}}{0.4\ \text{V}} = 10$$

$$A_I = \frac{\Delta I_{out}}{\Delta I_{in}} = \frac{2\ \text{mA}}{20\ \mu A} = 100$$

$$A_P = A_V \cdot A_I = (10)(100) = 1000$$

$$r_{ac\ in} = \frac{\Delta V_{in}}{\Delta I_{in}} = \frac{0.4}{20\ \mu A} = 20\ \text{k}\Omega\ \text{(not typical)}$$

Second, the values of the various resistors must be established for proper dc and ac operation. Referring to Fig. 7.15, R_2 must limit the forward dc bias current in the emitter-to-base junction. I_{Bdc} was chosen to be 20 μA. The emitter-to-base drop was found to be 0.4 V (see Fig. 7.14) at $I_{Bdc} = 20\ \mu A$. Therefore, the total voltage across R_2 is

$$V_{R2} = 8\ \text{V} - 0.4\ \text{V} = 7.6\ \text{V}$$

The dc current in the loop is 20 μA.

Therefore, $R_2 = \dfrac{7.6\text{V}}{20\ \mu A} = 380\ \text{k}\Omega$

Note: For a common emitter circuit, V_{BB} and V_{CC} are quite often chosen to be the same power supply (to reduce circuit costs).

Third, the value of R_1 must be determined.

To calculate the value of R_1, it may be assumed for this example, that the ac signal current through R_1 is equal to the signal current into the emitter-to-base junction. This is a reasonable assumption because the ratio of R_2 to the transistor input resistance is more than ten to one. Therefore, the ac voltage across R_1 is equal to the source voltage minus the emitter-to-base drop.

$$e_{R1} = e_S - \Delta V_{in} = 1\ V_{p.p.} - 0.4\ V_{p.p.} = 0.6\ V_{p.p.}$$

Therefore, $R_1 = \dfrac{e_{R1}}{\Delta I_{in}} = \dfrac{0.6\ V}{20\ \mu A} = 30\ k\Omega$

In the beginning of this analysis, two possible combinations of given data were presented. The second set of data included an operating point rather than a load resistance value (load line). If we are concerned with the operating point the load line is constructed from the V_{CC} power supply voltage on the horizontal axis through the Q point and the remainder of the analysis is as presented here.

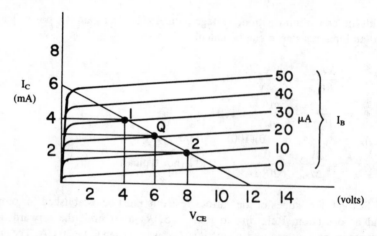

Fig. 7.13

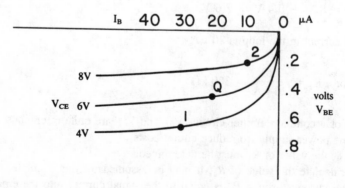

Fig. 7.14

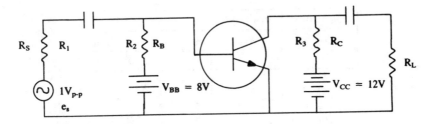

Fig. 7.15

PROBLEMS

1. By using the *npn* common-base input and output transistor characteristic curves shown in Fig. 7.16 (a,b), design a common base amplifier stage with the following parameters:

$$V_{CC} = 10 \text{ V}, \; V_{EQ} = 4 \text{ mA}, \; \Delta I_E = 2 \text{ mA(p.p.)}, \; V_{CBQ} = 5 \text{ V}$$

Determine the value of R_c, R_1, R_E, A_V, A_I, $R_{in\ ac}$.
For the 2N2477 transistor:

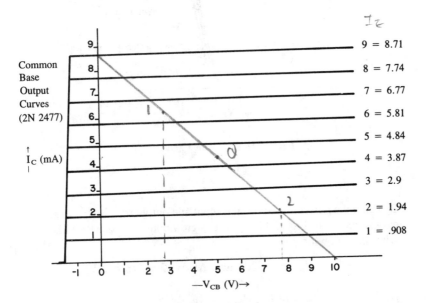

Fig. 7.16(a) Output Curves

VCB = 10 V	VCB = 5 V	VCB = 1 V			
VBE	VBE	IE	VBE	VBE	IE
		0	0	0	0
.64	.65	1	.65	.1	0
.67	.67	2	.675	.2	0
.68	.69	3	.69	.3	0
.695	.7	4	.71	.4	0
.71	.715	5	.72	.5	.2
.72	.73	6	.735	.65	.7
.735	.74	7	.74	.7	3.0
.74	.75	8	.755	.73	5.0
.75	.76	9	.77	.8	7.5

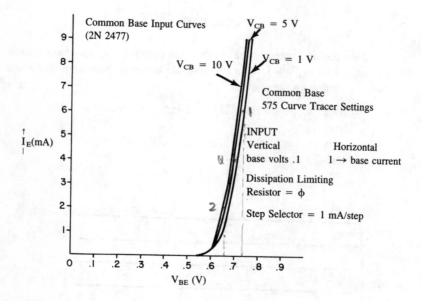

Fig. 7.16(b) Input Curves

2. By using the *pnp* common-emitter input and output transistor characteristic curves shown in Fig. 7.17 (a,b), design a common base amplifier stage with the following parameters:

$$V_{CC} = -12 \text{ V}, \quad I_{BQ} = -200 \text{ μA}, \quad \Delta I_C = 2 \text{ mA(p.p.)}, \quad V_{CBQ} = -6 \text{ V}$$

Determine the value of R_C, R_1, R_E, A_V, A_I, $R_{in \text{ ac}}$.

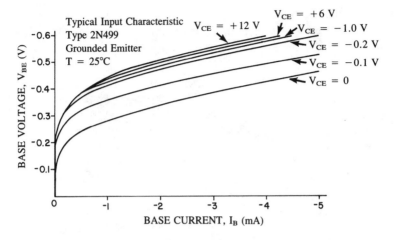

Fig. 7.17(a) Output Curves

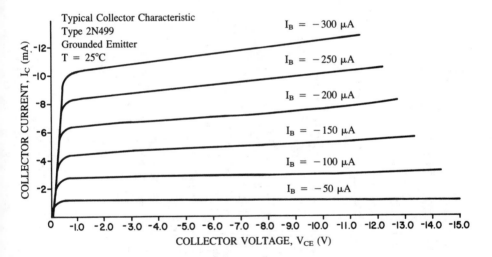

Fig. 7.17(b) Input Curves

3. By using the *pnp* common-emitter input and output transistor characteristic curves shown in Fig. 7.18(a,b), design a common emitter amplifier stage with the following parameters:

$$V_{CC} = -16 \text{ V}, I_{BQ} = .3 \text{ mA}, \Delta I_B = 400 \text{ }\mu\text{A(p.p.)}, V_{EC} = -8 \text{ V}$$

Determine the value of R_C, R_B, A_V, A_I, $R_{in\ ac}$.
Use a single V_{CC} power supply for both the base and collector circuits
($V_{CC} = -16$ V)

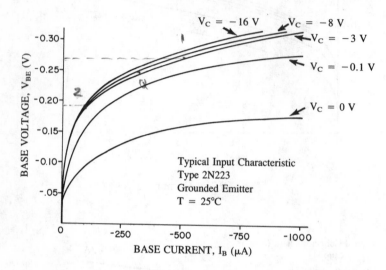

Fig. 7.18(a)

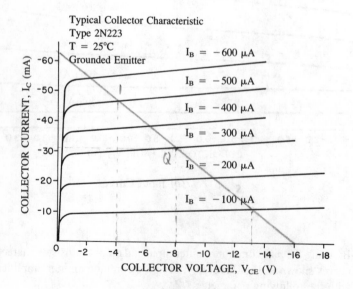

Fig. 7.18(b)

8.1 FIELD-EFFECT TRANSISTOR SMALL-SIGNAL AMPLIFIER ANALYSIS

To this point, we have discussed the general characteristics of FETs, their biasing, and lastly, the graphical design and analysis of FET amplifiers. We noted that the graphical design approach required a considerable investment of time, and also required that we had access to the output and transfer characteristic curves for the particular transistor which we were considering employing. If simple algebraic techniques could be developed for predicting the performance of a FET amplifier, with respect to its ability to increase the voltage level of a varying alternating current signal (amplification), and also predicting the amplifier input and output impedance, this would be a considerable improvement over graphical techniques. A technique developed for this purpose is called *small-signal analysis*.

In performing small-signal analysis, we assume that the signal levels which we are applying to the transistor amplifier stage are very small. Thus, with a small signal applied to the amplifier, the response of the amplifier can be considered linear. In simple terms, this implies that if we apply a one millivolt sine wave to the amplifier, and we note two mV at the output, then, if we apply 10 mV, we would expect to "see" 20 mV at the output.

If the applied signal is too large the amplifier no longer acts as a linear device. So, then, we may ask, at what voltage level does the amplifier become nonlinear? The answer is that there is no point at which the amplifier can be said to be linear or nonlinear. The decision that the designer must address concerns the level of nonlinearity that can be tolerated in a particular design. If an 8 V peak-to-peak signal is noted at the output of an amplifier stage operating with a 10 V supply voltage, chances are that the amplifier is not operating very linearly.

234

8.1.1 Small-Signal Analysis Techniques

At this point, it is appropriate to talk about small-signal analysis techniques. In performing a small-signal analysis, we "make-believe" that the signal which we "see" at the output terminal-to-ground is created by a small signal generator located inside of the active device (bipolar transistor, FET, or electron tube).

Whether we assume that there is a current generator or a voltage generator inside the active device is not critical, with the exception that certain active devices, (i.e., bipolar and field effect transistors), perform more like current sources than voltage sources.

A typical small-signal equivalent circuit for a transistor might appear as shown in Fig. 8.1. We note that in the input circuit there is a voltage generator in series with a resistor, and in the output circuit there is a current generator in parallel with a resistance or conductance.

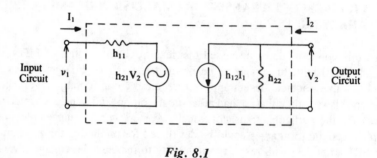

Fig. 8.1

The lower-case h is used to represent the word "hybrid." The h-parameters are used to denote a mixture of current sources and voltage sources and resistances and conductances.

For a field-effect transistor, the source-to-gate junction is a back-biased diode (except for enhancement-mode MOSFETs), and none of the output voltage reflects back to the input circuit. Therefore, the voltage generator is not included in the equivalent circuit. Because a back-biased diode is for all practical purposes an open circuit, i.e., infinite resistance, the input circuit for a FET is as shown in Fig. 8.2.

We note that nothing is attached to the gate terminal (inside of the transistor), but, for the output circuit, we include a current generator and a shunt resistor (see Fig. 8.3). The arrow indicating current direction is pointing downwards. The reason for this is that if an incremental positive voltage is applied to the gate, the drain-to-source channel "opens further" and allows more current to flow from the drain to source. Therefore, the transistor experiences an incremental increase in drain current downward through the transistor. However, the problem remains of how much drain current change is developed for a change in source-to-gate voltage. First, since the device is assumed to be linear, we may write that the drain current is proportional to the source-to-gate voltage:

$$\Delta I_D \sim \Delta V_{SG}$$

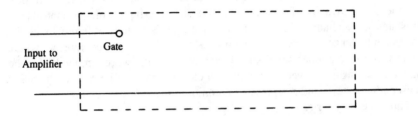

Fig. 8.2 FET Input Equivalent Circuit

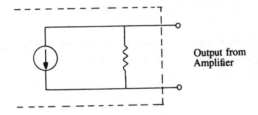

Fig. 8.3 FET Output Equivalent Circuit

We also learned that proportionality signs can be removed and replaced with equality signs if we include a proportionality constant (k). Therefore,

$$\Delta I_D = k\Delta V_{SG}$$

or, rewriting,

$$k = \frac{\Delta I_D}{\Delta V_{SG}}$$

Because we are interested in evaluating the proportionality constant while allowing only the drain current and source-to-gate voltage to vary, all other parameters should be held constant. The only other variable in a common source connected FET is the source-to-drain voltage V_{SD}, which must be held constant (but not zero). Thus, we can write the following:

$$k = \left. \frac{\Delta I_D}{\Delta V_{SG}} \right|_{V_{SD}} = \text{constant}$$

We note that the ratio has dimensions of amps divided by volts. Therefore, k is measured in units of conductance. Because the input voltage controls, or translates into, changes in the output current, k is usually referred to as the "transconductance of the field-effect transistor." The symbol is g_m or g_{fs}, where the lower-case g implies small signal, m represents the word "mutual" for the mutual relationship between the input and output circuits, f represents "forward transconductance," and s represents common "source" connection.

Thus, we may write

$$g_m = \lim_{\Delta V_{SG} \to 0} \left. \frac{\Delta I_D}{\Delta V_{SG}} \right|_{V_{SD}} = \text{constant} \tag{8.1}$$

We must approach the limit if we expect g_m to represent a truly linear relationship. Thus, mathematically, we can express (8.1) as

$$g_m = \frac{\partial I_D}{\partial V_{SG}} \tag{8.2}$$

For practical purposes, we can express the transconductance equation as follows:

$$g_m = \left. \frac{\Delta I_D}{\Delta V_{SG}} \right|_{V_{SD}} = \text{constant} \tag{8.3}$$

Example (8.1). If we examine the output curves for a FET, as shown in Fig. 8.4, the transconductance can be determined by drawing a vertical line through the Q point. We note that all points along the vertical line represent constant values of the source-to-drain voltage. If we place a dot at one curve above and below the V_{SGQ} curve and project horizontal lines to the vertical axis, we can now determine g_m:

$$g_m = \left. \frac{\Delta I_D}{\Delta V_{SG}} \right|_{V_{SD}} = \text{constant}$$

$$= \left. \frac{I_{D2} - I_{D1}}{V_{SG2} - V_{SG1}} \right|_{V_{SDQ}}$$

$$= \left. \frac{3.8 - 2.25 \text{ mA}}{-0.4 - (-0.8) \text{ V}} \right|_{V_{SD} \, = \, +10 \text{ V}}$$

$$= 3.875 \text{ mmho}$$

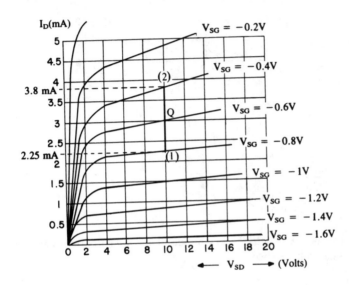

Fig. 8.4

Another method for determining the transconductance relies on the fact that the transfer characteristic curves are plots of drain current *versus* source-to-gate voltage. If we locate the Q point on the transfer characteristic curve and draw a tangent line through the Q point, the slope of the line is the transconductance at the operating point.

Example (8.2). Determine the transconductance at $V_{SG} = -0.6$ V. (See Fig. 8.5.)

$$g_m = \left. \frac{\Delta I_D}{\Delta V_{SG}} \right|_{V_{SD} \, = \, \text{constant}}$$

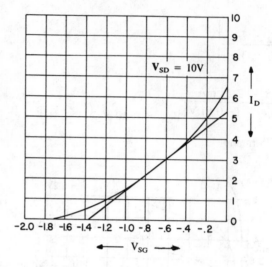

Fig. 8.5

$$= \frac{5.3 \text{ mA}}{1.33 \text{ V}}$$

$$= 3.98 \text{ mmho}$$

We can now evaluate the magnitude of the signal current developed by an input signal voltage, because

$$\Delta I_D = g_m \Delta V_{SG}$$

To complete our small signal equivalent circuit, we must evaluate the output resistance r_d, shown in the FET equivalent circuit of Fig. 8.6. Since the resistance shown is that which the ac signal current generator must supply, we use a lower-case r to represent ac resistance. The value of the resistance could *not* be measured by an ohmmeter. The lower-case d subscript indicates that the resistance is in the drain-to-source channel.

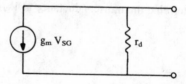

Fig. 8.6

From our previous dc circuit analysis work, we learned that resistance values can be interpreted as the slope of a line drawn on a current *versus* voltage plot.

Because r_d is the output resistance, if we refer to the output curves, we should be able to determine its value. First, let us formally define r_d. Resistance is voltage divided by current, and since r_d is an ac resistance, we must use changes in current and voltage. Therefore,

$$r_d = \frac{\Delta V_{out}}{\Delta I_{out}} = \frac{\Delta V_{SD}}{\Delta I_D} \tag{8.4}$$

However, for this ac resistance, we must allow only two variables to change. Therefore,

$$r_d = \left. \frac{\Delta V_{SD}}{\Delta I_D} \right|_{V_{SG} = \text{constant}} \tag{8.5}$$

or more precisely,

$$r_d = \lim_{\Delta I_{out} \to 0} \left. \frac{\Delta V_{SD}}{\Delta I_D} \right|_{V_{SG} = \text{constant}}$$

or

$$r_d = \frac{\partial V_{SD}}{\partial I_D} \tag{8.6}$$

For the purpose of evaluating r_d, we will use the definition:

$$r_d = \left. \frac{\Delta V_{SD}}{\Delta I_D} \right|_{V_{SG} = \text{constant}}$$

Referring to the output curves shown in Fig. 8.7, if we draw a line through the Q point and tangent to the $V_{SG} = 6$ V line, we see that the line may be connected to a horizontal line to form a right triangle. The size of the right triangle is not critical, but, the larger the triangle, the more accurate our results will be. Computing:

$$r_d = \left. \frac{\Delta V_{SD}}{\Delta I_D} \right|_{V_{SGQ}} = \frac{20 - 0 \text{ V}}{3.4 - 2.65 \text{ mA}} = 26.7 \text{ k}\Omega$$

240

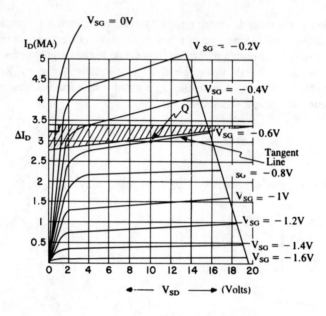

Fig. 8.7

Now, let us define voltage gain, or voltage amplification, as the change in output voltage divided by the change in input voltage. Thus,

$$A_V = \frac{\Delta V_{out}}{\Delta V_{in}} \tag{8.7}$$

and for a common-source connected FET,

$$A_V = \frac{\Delta V_{SD}}{\Delta V_{SG}} \tag{8.8}$$

Referring to the schematic (Fig. 8.8(a)) and equivalent circuit (Fig. 8.8(b)) shown below, we note that the current $g_m\Delta V_{SG}$ flows through r_d and R_D and generates the output voltage, $-\Delta V_{SD}$. Thus, according to Ohm's law:

$$-\Delta V_{SD} = g_m\Delta V_{SG} \ (r_d \ /\!/ \ R_D)$$

or

$$\frac{\Delta V_{SD}}{\Delta V_{SG}} = A_V = -g_m \frac{r_dR_D}{r_d + R_D} \tag{8.9}$$

Dividing the numerator and denominator by r_d, we obtain

$$A_V = -g_m \frac{R_D}{1 + R_D/r_d}$$

and if, $r_d \gg R_D$, which is quite often the situation ($26.7\ \text{k} \gg 3.133\ \text{k}$), then

$$A_V \doteq -g_m R_D \qquad (8.10)$$

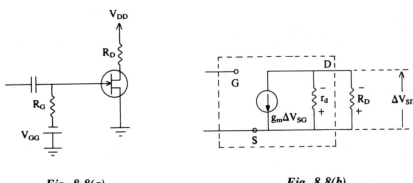

Fig. 8.8(a)   Fig. 8.8(b)

If the amplifier is connected to another stage, or some other load, then we have the schematic (Fig. 8.9(a)) and equivalent circuit (Fig. 8.9(b)) shown below where Z_L is defined as all *external* resistances connected from the drain to ground. Thus, $Z_L = R_D \parallel R_L$.

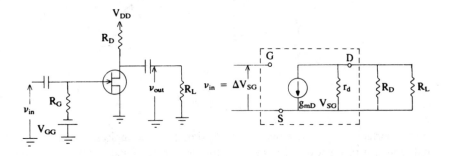

Fig. 8.9(a)   Fig. 8.9(b)

$$A_V = -g_m(r_d \parallel Z_L) \qquad (8.11)$$

or

$$A_V \doteq -g_m Z_L \qquad (8.12)$$

Example (8.3). Compute the voltage gain of the circuit shown in Fig. 8.10, where

$$r_d \parallel Z_L = r_d \parallel R_D \parallel R_L$$
$$= 26.7 \text{ k} \parallel 3 \text{ k} \parallel 6 \text{ k}$$
$$= 1.86 \text{ k}$$

and

$$A_V = -g_m(r_d \parallel Z_L)$$
$$= -2.2 \text{ mmho } (1.86 \text{ k})$$
$$= -4.09$$

given

$$g_m = 2.2 \text{ mmho}$$
$$r_d = 26.7 \text{ k}\Omega$$

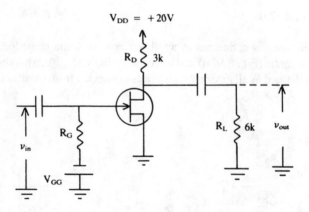

$V_{DD} = +20V$

R_D　3k

R_L　6k　　v_{out}

R_G

v_{in}

V_{GG}

Fig. 8.10

It is important to emphasize that the minus sign does not imply a loss in voltage, but merely a phase reversal between output and input signals. Referring to the input circuit of Fig. 8.11, we see that input impedance is only the value of R_G. At low frequencies, capacitance does not affect the input impedance.

Let us determine the output impedance of a FET common-source connected amplifier. In order to do this, we should first define what we mean by the output impedance. The output impedance of a FET common-source amplifier stage is the Norton equivalent resistance. Referring to the circuit shown in Fig. 8.12, we "stand" at the output lead of the stage and "look" to the left, determining all impedances to ground, while ignoring all impedances "behind us." Referring to the equivalent circuit of Fig. 8.13, we open all current sources and "see" $r_d \parallel R_D$. Thus,

$$Z_0 = \frac{r_d R_D}{r_d + R_D} \qquad (8.13)$$

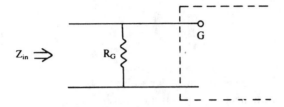

Fig. 8.11

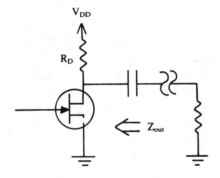

Fig. 8.12

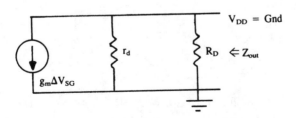

Fig. 8.13

Note: Remember that V_{DD} is at ac ground potential.
If we divide the numerator and denominator of (8.13) by r_d, we have

$$Z_0 = \frac{R_D}{1 + \dfrac{R_D}{r_d}}$$

and, if $r_d \gg R_D$, then

$$Z_0 \doteq R_D \tag{8.14}$$

Example (8.4).   For the circuit shown in Fig. 8.14, determine the values of the input and output impedances, given g_m = 2.2 mmho, r_d = 26.7 k.

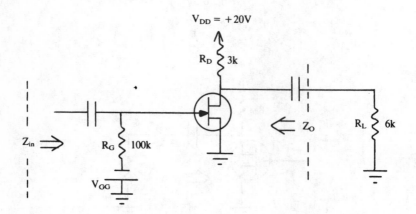

Fig. 8.14

Note: The negative terminal of V_{GG} is at ac ground potential and all capacitors are "shorts."

Hence, we derive

$$Z_{in} = R_G = 100 \text{ k}$$

for the input impedance, and

$$Z_{out} = r_d \mathbin{/\mkern-5mu/} R_D = 26.7 \text{ k} \mathbin{/\mkern-5mu/} 3 \text{ k} = 2.697 \text{ k}$$

for the output impedance.

8.1.2   The Common-Source Connected FET Amplifier with an Unbypassed Source Resistor

We recall from Chapter 3, on FET biasing, that the use of self-biasing eliminated the need for a second power supply, V_{GG}. Let us see what effect the source resistor has upon the amplifier characteristics.

Referring to the schematic in Fig. 8.15, we can draw the equivalent circuit. Note that the source terminal is no longer at ac or dc ground potential.

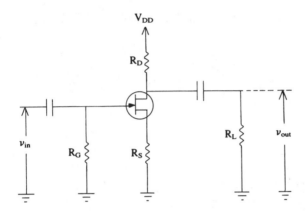

Fig. 8.15

Voltage Gain for FET (R_S unbypassed)

Referring to Fig. 8.16(a), let us derive the voltage gain (A_V) for a FET with an unbypassed source resistor (R_s).

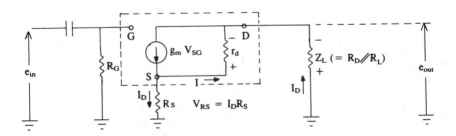

Fig. 8.16(a)

$$V_{SG} = -I_D R_S + e_{in}, \; e_{out} = I_D R_S + V_{SD} \Rightarrow V_{SD} = e_{out} - I_D R_S$$

also,

$$e_{out} = -I_D Z_L \Rightarrow I_D = -\frac{e_{out}}{Z_L}$$

Employing Kirchoff's current law at S:

$$\Sigma I_{out} = \Sigma I_{in} \Rightarrow I_D + I = g_m V_{SG} \Rightarrow I_D - \frac{V_{SD}}{r_d} = g_m V_{SG}$$

Since

$$I = -\frac{V_{SD}}{r_d}$$

Thus,

$$I_D = g_m V_{SG} + V_{SD}/r_d$$

$$I_D = g_m V_{SG} + \frac{e_{out} - I_D R_S}{r_d} = g_m V_{SG} + \frac{e_{out}}{r_d} - \frac{I_D R_S}{r_d}$$

$$I_D + \frac{I_D R_S}{r_d} - \frac{e_{out}}{r_d} = g_m (-I_D R_S + e_{in})$$

$$I_D + \frac{I_D R_S}{r_d} - \frac{e_{out}}{r_d} = -g_m I_D R_S + g_m e_{in}$$

$$I_D + \frac{I_D R_S}{r_d} - \frac{e_{out}}{r_d} + g_m I_D R_S = g_m e_{in}$$

$$I_D \left(1 + \frac{R_S}{r_d} + g_m R_S\right) - \frac{e_{out}}{r_d} = g_m e_{in}$$

$$-\frac{e_{out}}{Z_L} \left(1 + \frac{R_S}{r_d} + g_m R_S\right) - \frac{e_{out}}{r_d} = g_m e_{in}$$

$$-e_{out} \left(\frac{1 + \dfrac{R_S}{r_d} + g_m R_S}{Z_L} + \frac{1}{r_d}\right) = g_m e_{in}$$

$$-e_{out} \left(\frac{r_d + R_S + g_m R_S r_d + Z_L}{Z_L r_d}\right) = g_m e_{in}$$

$$\frac{e_{out}}{e_{in}} = \frac{-g_m Z_L r_d}{r_d + R_S (1 + g_m r_d) + Z_L}$$

$$\frac{e_{out}}{e_{in}} = \frac{-g_m Z_L}{1 + \frac{R_S}{r_d}(1 + g_m r_d) + \frac{Z_L}{r_d}}$$

$$\frac{e_{out}}{e_{in}} = A_V = \frac{-g_m Z_L}{1 + R_S\left(g_m + \frac{1}{r_d}\right) + \frac{Z_L}{r_d}} \tag{8.15}$$

Thus, if $r_d \gg Z_L$ and $r_d \gg 1$, then

$$A_V \doteq \frac{-g_m Z_L}{1 + g_m R_S} \tag{8.16}$$

Also, $Z_L = R_D$ if there is no external load.

Voltage Gain for FET (with source resistor and with bypass capacitor)

If the source resistor is shunted with a capacitor, then the voltage gain equation will revert to that for a common-source FET amplifier with fixed bias, i.e.,

$$A_V = -g_m (Z_L /\!/ r_d)$$

Example (8.5). Compute the voltage gain for the circuit shown in Fig. 8.16(b), with and without the source resistor bypassed with a capacitor, given $r_d = 26.7$ k, $g_m = 2.2$ mmho.

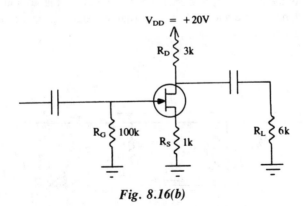

Fig. 8.16(b)

With source resistor:

$$\text{exact: } A_V = \frac{-g_m Z_L}{1 + R_S\left(g_m + \frac{1}{r_d}\right) + \frac{Z_L}{r_d}}$$

$$= \frac{-2.2\,(2)}{1 + 1\left(2.2 + \dfrac{1}{26.7}\right) + \dfrac{2}{26.7}} = -1.33$$

approximate: $A_V = -\dfrac{g_m\,Z_L}{1 + g_m\,R_S} = -\dfrac{2.2\,(2)}{1 + 2.2\,(1)} = -1.375$

% diff = 3.38%

With bypassed resistor:

exact: $A_V = -g_m\,(Z_L \mathbin{/\!/} r_d) = -2.2\left[\dfrac{2(26.7)}{2 + 26.7}\right] = -4.09$

approximate: $A_V = -g_m\,Z_L = -2.2\,(2) = -4.4$

% diff = 7.5%

We can see that the voltage gain is approximately three times greater when the source resistor is bypassed. The input impedance computation remains the same as for a common-source FET amplifier without a source resistor, i.e.,

$$Z_{in} = R_G \tag{8.17}$$

Let us see what happens to the output impedance when an unbypassed source resistor is present. The following section considers an equivalent circuit and the derivation of the output impedance. Note from Fig. 8.17 that we must disconnect the input signal source and apply an external test signal generator to the output terminals, where $Z_{out} = e_{test}/I_{test}$.

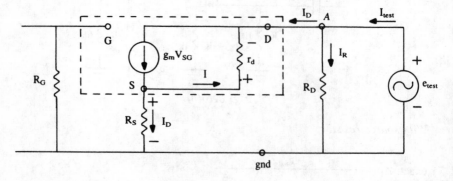

Fig. 8.17 Analysis of Z_{out}

First, employing Kirchoff's current law at S:

$$\Sigma\, I_{out} = \Sigma\, I_{in}$$

also, $V_{SG} = -I_D\, R_S$

$$I_D + I = g_m\, V_{SG}$$

Therefore, $I_D + I = g_m\, (-I_D\, R_S) \Rightarrow I = -I_D - I_D\, g_m\, R_S$

$$I = -I_D(1 + g_m\, R_S) \tag{8.18}$$

Second, employing Kirchoff's current law at A:

$$\Sigma\, I_{out} = \Sigma\, I_{in}$$
$$I_D + I_R = I_{test}$$

where $I_R = \dfrac{e_{test}}{R_D}$

Therefore, $I_D + \dfrac{e_{test}}{R_D} = I_{test}$

Third, employing Kirchoff's voltage law by summing voltages around the outside loop, starting at the ground terminal counter clockwise:

$$e_{test} + I\, r_d - I_D\, R_S = 0$$

and, from (8.18), we have

$$e_{test} - I_D\, (1 + g_m\, R_S)\, r_d - I_D\, R_S = 0$$
$$e_{test} = I_D\, (1 + g_m\, R_S)\, r_d + I_D\, R_S = I_D\, [(1 + g_m\, R_S)\, r_d + R_S]$$

but $I_D + \dfrac{e_{test}}{R_D} = I_{test}$

Therefore, $I_D = I_{test} - \dfrac{e_{test}}{R_D}$

or, $e_{test} = \left(I_{test} - \dfrac{e_{test}}{R_D} \right) [(1 + g_m\, R_S)\, r_d + R_S]$

$$= I_{test}\, [(1 + g_m\, R_S)\, r_d + R_S] - \dfrac{e_{test}}{R_D}\, [(1 + g_m\, R_S)\, r_d + R_S]$$

$$e_{test} + \dfrac{e_{test}}{R_D}\, [(1 + g_m\, R_S)\, r_d + R_S] = I_{test}\, [(1 + g_m\, R_S)\, r_d + R_S]$$

$$e_{test} \left[1 + \dfrac{(1 + g_m\, R_S)\, r_d}{R_D} + \dfrac{R_S}{R_D} \right] = I_{test}\, [(1 + g_m\, R_S)\, r_d + R_S]$$

$$\frac{e_{test}}{I_{test}} = Z_0 = \frac{(1 + g_m R_S) r_d + R_S}{1 + \dfrac{(1 + g_m R_S) r_d}{R_D} + \dfrac{R_S}{R_D}}$$

$$= \frac{R_D [(1 + g_m R_S) r_d + R_S]}{R_D + (1 + g_m R_S) r_d + R_S}$$

Therefore,

$$Z_0 = \frac{R_D}{1 + \dfrac{R_D}{r_d (1 + g_m R_S) + R_S}} \tag{8.19}$$

Hence, if $r_d \gg R_D$, we have

$$Z_0 \doteq R_D \tag{8.20}$$

Example (8.6). Compute the output impedance of the circuit shown in Fig. 8.18, given $g_m = 2.2$ mmho, $r_d = 26.7$ k.

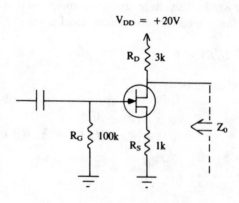

Fig. 8.18

$$Z_0 = \frac{R_D}{1 + \dfrac{R_D}{r_d (1 + g_m R_S) + R_S}} = \frac{3 \text{ k}}{1 + \dfrac{3}{26.7 [1 + 2.2(1)] + 1}}$$

$$Z_0 = 2.9 \text{ k}$$

According to the approximation formula:

$$Z_0 \doteq R_D = 3 \text{ k}$$

If the source resistor is bypassed:

$$Z_0 = R_D \mathbin{/\!/} r_d = 3 \text{ k} \mathbin{/\!/} 26.7 \text{ k} = 2.7 \text{ k}$$

8.2   THE SOURCE FOLLOWER (COMMON DRAIN CIRCUIT)

An additional FET amplifier configuration of value to circuit designers is the *source follower*. The output signal is taken from the source of the FET amplifier, while the input is applied to the gate (as is the common source circuit). We note from the schematic of Fig. 8.19 that there is no R_D in the circuit. The features of the source follower of value to the circuit designer are high input impedance, low output impedance, and its capability of delivering more signal current to a load than a common source circuit. In order to obtain these features, we must sacrifice voltage gain, which is slightly less than unity. This occurs because the signal voltage at the source is approximately equal to the signal voltage at the gate (for load resistances that do not approach a short). Why is the signal voltage at the source approximately equal to that at the gate?

We recall that as the gate becomes incrementally positive, an incremental increase in the drain current occurs. The incremental increase in the drain current causes an incremental increase in voltage across the source resistance. Thus, the voltage across the source "follows" the ground-to-gate voltage (see Figs. 8.19 and 8.20).

Let us derive the equation for voltage gain of a source follower. Noting that

$$A_V = \frac{e_{out}}{e_{in}}$$

we see that by Kirchoff's voltage law and Ohm's law:

$$V_{SG} = -e_{out} + e_{in}$$
and

$$e_{out} = I_D Z_S$$

therefore,

$$I_D = \frac{e_{out}}{Z_S}$$

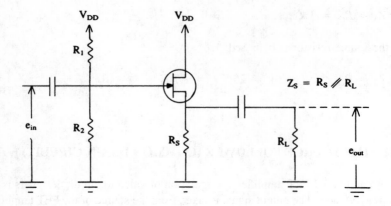

Fig. 8.19 Source Follower

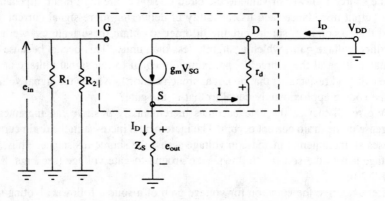

Fig. 8.20 Equivalent Circuit

By Kirchoff's current law at node S:

$$\Sigma I_{in} = \Sigma I_{out}$$

$$g_m V_{SG} = I_D + I = \frac{e_{out}}{Z_S} + \frac{e_{out}}{r_d}$$

where $I = \dfrac{e_{out}}{r_d}$

Therefore,

$$g_m \left(-e_{out} + e_{in} \right) = \frac{e_{out}}{Z_S} + \frac{e_{out}}{r_d}$$

$$-g_m e_{out} + g_m e_{in} = e_{out}\left(\frac{1}{Z_S} + \frac{1}{r_d}\right)$$

$$e_{out}\left(\frac{1}{Z_S} + \frac{1}{r_d}\right) + g_m e_{out} = g_m e_{in}$$

$$e_{out}\left(\frac{Z_S + r_d}{r_d Z_S} + g_m\right) = g_m e_{in}$$

$$\frac{e_{out}}{e_{in}} = \frac{g_m}{\dfrac{Z_S + r_d}{Z_S r_d} + g_m} = \frac{1}{1 + \dfrac{r_d + Z_S}{g_m r_d Z_S}} = \frac{1}{1 + \dfrac{1 + Z_S/r_d}{g_m Z_S}}$$

$$\frac{e_{out}}{e_{in}} = A_V = \frac{1}{1 + \dfrac{1 + Z_S/r_d}{g_m Z_S}} \tag{8.21}$$

Hence, if $r_d \gg Z_S$, then

$$A_V \doteq \frac{1}{1 + \dfrac{1}{g_m Z_S}}$$

$$A_V \doteq \frac{g_m Z_S}{1 + g_m Z_S} \tag{8.22}$$

Example (8.7). Compute the voltage gain for the source-follower stage shown in Fig. 8.21, given

$$g_m = 2.2 \text{ mmho}$$

and

$$r_d = 26.7 \text{ k}\Omega$$

and

$$Z_S = R_S \mathbin{/\mkern-5mu/} R_L = 0.5 \text{ k}$$

$$A_V = \frac{1}{1 + \dfrac{1 + Z_S/r_d}{g_m Z_S}} = \frac{1}{1 + \dfrac{1 + 0.5/26.7}{2.2\,(0.5)}} = 0.519$$

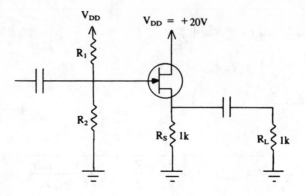

Fig. 8.21

Example (8.8).   Compute the voltage gain for the source-follower stage shown in Fig. 8.22, given

$$g_m = 2.2 \text{ mmho}$$
$$r_d = 26.7 \text{ k}$$

and

$$Z_S = R_S \mathbin{/\mkern-5mu/} R_L$$
$$= 1 \text{ k} \mathbin{/\mkern-5mu/} 0.05 \text{ k}$$
$$= 0.0476 \text{ k}$$

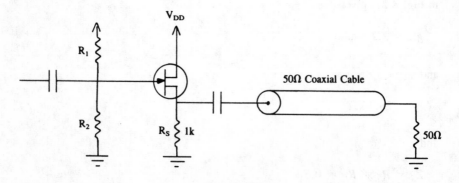

Fig. 8.22

$$A_V = \cfrac{1}{1 + \cfrac{1 + Z_S/r_d}{g_m Z_S}} = \cfrac{1}{1 + \cfrac{1 + 0.0476/26.7}{2.2(0.0476)}} = 0.0947$$

The *input impedance* is simply all resistances at the amplifier terminals connected to ac ground. Therefore,

$$Z_{in} = R_1 \ /\!/ \ R_2 \qquad (8.23)$$

The *output impedance* can be determined by "forcing" ac current into the output terminals, via the voltage developed by the test-generator connected to the output of the amplifier stage.

Examining the circuit shown in Figs. 8.23 and 8.24, we note that $V_{SG} = -e_{test}$ (since the gate is at ac ground potential). Using Kirchoff's current law at node S, we have

$$\Sigma I_{in} = \Sigma I_{out}$$
$$gm \ V_{SG} + I_{test} = I_R + I$$

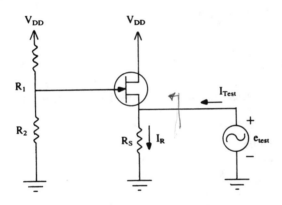

Fig. 8.23

Therefore, $g_m \ (-e_{test}) + I_{test} = \dfrac{e_{test}}{R_S} + \dfrac{e_{test}}{r_d}$

$$e_{test} \left(g_m + \frac{1}{R_S} + \frac{1}{r_d} \right) = I_{test}$$

$$\frac{e_{test}}{I_{test}} = \frac{1}{g_m + \dfrac{1}{R_S} + \dfrac{1}{r_d}} = \frac{R_S}{g_m R_S + 1 + \dfrac{R_S}{r_d}}$$

$$\frac{e_{test}}{I_{test}} = Z_0 = \frac{R_S}{g_m R_S + 1 + \dfrac{R_S}{r_d}} \qquad (8.24)$$

256

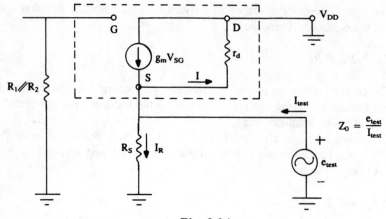

Fig. 8.24

Hence, if $r_d \gg R_S$, we have

$$Z_0 = \frac{R_S}{1 + g_m R_S} \tag{8.25}$$

Example (8.9). Determine the output impedance of the source follower shown in Fig. 8.25, given

$g_m = 2.2$ mmho
$r_d = 26.7$ kΩ

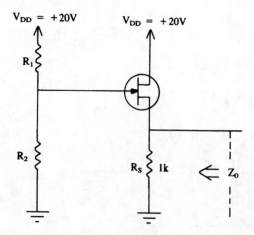

Fig. 8.25

$$Z_0 = \frac{R_S}{g_m R_S + 1 + \dfrac{R_S}{r_d}} = \frac{1 \text{ k}}{2.2(1) + 1 + \dfrac{1}{26.7}} = 0.309 \text{ k}$$

$$Z_{out} = 309 \ \Omega$$

We see that the source follower has a relatively low output impedance value.

8.3 THE COMMON GATE CIRCUIT

The common gate circuit is the least prevalent of the three configurations (common source, drain [source follower], and gate). The common gate circuit exhibits a voltage gain greater than unity, a low input impedance, and an output impedance approximately equal to the drain resistor value. The output impedance is defined to include all the bias resistors.

Let us derive the formula for the voltage gain by drawing the schematic (Fig. 8.26(a)) and its equivalent circuit (Fig. 8.26(b)).

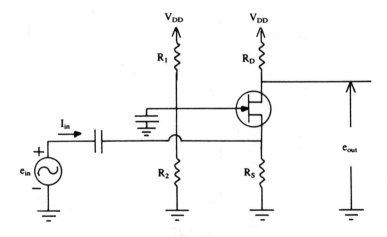

Fig. 8.26(a)

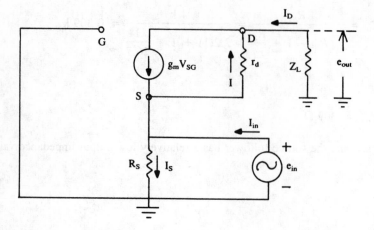

Fig. 8.26(b)

Note: The gate is at ac ground potential.

In some high frequency applications, the source resistor may be connected in series with the signal source as shown in Fig. 8.27. However, for our application, we will analyze Fig. 8.26, where we note that

$$V_{SG} = -e_{in} \quad \text{and} \quad I_D = -\frac{e_{out}}{Z_L}$$

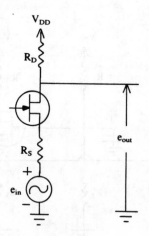

Fig. 8.27

By Kirchoff's voltage law:

$$e_{in} - I\,r_d - e_{out} = 0$$

or $I = \dfrac{e_{in} - e_{out}}{r_d}$

Writing Kirchoff's current law at node D:

$$\Sigma\,I_{in} = \Sigma\,I_{out}$$
$$I_D + I = g_m\,V_{SG}$$
$$-\frac{e_{out}}{Z_L} + \frac{e_{in}}{r_d} - \frac{e_{out}}{r_d} = g_m\,(-e_{in})$$
$$e_{in}\left(\frac{1}{r_d} + g_m\right) = e_{out}\left(\frac{1}{Z_L} + \frac{1}{r_d}\right)$$

$$\frac{e_{out}}{e_{in}} = A_V = \frac{g_m + \dfrac{1}{r_d}}{\dfrac{1}{r_d} + \dfrac{1}{Z_L}} = \frac{g_m\,Z_L + \dfrac{Z_L}{r_d}}{1 + \dfrac{Z_L}{r_d}} \tag{8.26}$$

Hence, if $r_d >> Z_L$, we have

$$A_V \doteq g_m\,Z_L \tag{8.27}$$

We note that the expression for A_V is a positive quantity denoting the fact that there is no phase inversion between output and input.

Example (8.10). Calculate the voltage gain for the common gate amplifier shown in Fig. 8.28, given,

$$g_m = 2.2 \text{ mmho}$$
$$r_d = 26.7 \text{ k}$$
$$Z_L = R_D \,/\!/\, R_L = 6\text{ k} \,/\!/\, 12\text{ k} = 4\text{ k}$$

Note: gate bias resistors are not shown.

$$A_V = \frac{g_m\,Z_L + \dfrac{Z_L}{r_d}}{1 + Z_L/r_d} = \frac{2.2\,(4) + 4/26.7}{1 + 4/26.7} = 7.78$$

To determine the *input impedance* of the common gate amplifier, we again refer to Fig. 8.28 and Fig. 8.26(b). At node S:

260

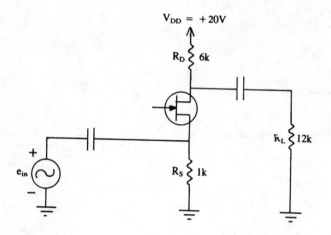

Fig. 8.28

$$\Sigma I_{in} = \Sigma I_{out}$$
$$I_{in} + g_m V_{SG} = I + I_S$$

$$I_{in} + g_m(-e_{in}) = \frac{e_{in} - e_{out}}{r_d} + \frac{e_{in}}{R_S}$$

$$I_{in} - g_m e_{in} = \frac{e_{in}}{r_d} - \frac{e_{out}}{r_d} + \frac{e_{in}}{R_S}$$

but $A_V = \dfrac{e_{out}}{e_{in}}$

$$I_{in} - g_m e_{in} = \frac{e_{in}}{r_d} - \frac{e_{in} A_V}{r_d} + \frac{e_{in}}{R_S}$$

or $e_{out} = e_{in} A_V$

$$I_{in} - g_m e_{in} = \frac{e_{in}}{r_d} - \frac{1}{r_d}\left[\frac{e_{in}\left(g_m + \frac{1}{r_d}\right)}{\frac{1}{r_d} + \frac{1}{Z_L}}\right] + \frac{e_{in}}{R_S}$$

$$I_{in} = g_m e_{in} + \frac{e_{in}}{r_d} - e_{in}\left(\frac{g_m + \frac{1}{r_d}}{1 + \frac{r_d}{Z_L}}\right) + \frac{e_{in}}{R_S}$$

$$= e_{in}\left(g_m + \frac{1}{r_d} - \frac{g_m + \frac{1}{r_d}}{1 + \frac{r_d}{Z_L}} + \frac{1}{R_S}\right)$$

$$\frac{e_{in}}{I_{in}} = \cfrac{1}{g_m + \cfrac{1}{r_d} - \left(\cfrac{g_m + \cfrac{1}{r_d}}{1 + \cfrac{r_d}{Z_L}}\right) + \cfrac{1}{R_S}}$$

$$\frac{e_{in}}{I_{in}} = Z_{in} = \cfrac{1}{\left(g_m + \cfrac{1}{r_d}\right)\left(1 - \cfrac{1}{1 + \cfrac{r_d}{Z_L}}\right) + \cfrac{1}{R_S}} \tag{8.28}$$

Hence, if $r_d \gg Z_L$ and $r_d \gg 1$, we have

$$Z_{in} \doteq \cfrac{1}{g_m + \cfrac{1}{R_S}} \tag{8.29}$$

Example (8.11). Determine the input impedance for the circuit shown in Fig. 8.29.

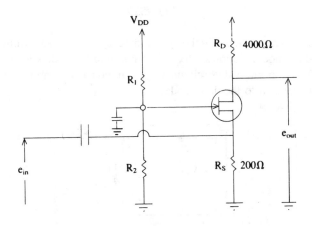

Fig. 8.29

If we had selected a FET with a $g_m = 8$ mmho, $r_d = 20$ k, and a source resistor of 200 Ω, the input impedance would be computed to be 85 Ω. A low input impedance of 85 Ω would very nearly match a 75 Ω coaxial cable. With a 4000 Ω load, the voltage gain would be

$$A_V \doteq g_m Z_L = 8(4) = 32$$

A video signal transmitted via a coaxial cable would be a compatible input to a common gate amplifier with the above characteristics.

Let us derive the equation for the output impedance of a common gate amplifier stage. Once again it is appropriate to emphasize the meaning of the phrase *output impedance*. To determine the output impedance, we must stand at the output terminal, disregard all circuitry behind us, and "look into" the amplifier stage.

Because we are examining the entire amplifier stage, and not just the FET, the drain bias resistor must also be included, as indicated in Fig. 8.30.

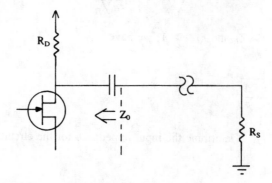

Fig. 8.30

Because we are interested in the output impedance as "seen" while "looking into" the drain node, we must "force" current into the node with a test signal. Figure 8.31 is an equivalent circuit of a common gate amplifier with a test generator connected to the drain node.

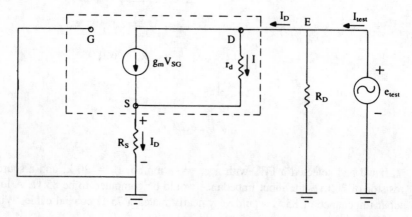

Fig. 8.31

We note that: $V_{SG} = -I_D R_S$ and $Z_0 = e_{test}/I_{test}$

at Node E:

$$\Sigma I_{out} = \Sigma I_{in}$$

$$I_D + \frac{e_{test}}{R_D} = I_{test}$$

Therefore, $I_D = I_{test} - \dfrac{e_{test}}{R_D}$

By Kirchoff's Voltage Law:

$$I_D R_S + I r_d = e_{test}$$

Therefore, $I = \dfrac{e_{test} - I_D R_S}{r_d}$

at node D:

$$\Sigma I_{out} = \Sigma I_{in}$$

$$g_m V_{SG} + I = I_D$$

$$g_m(-I_D R_S) + \frac{e_{test} - I_D R_S}{r_d} = I_D$$

$$-g_m I_D R_S + \frac{e_{test}}{r_d} - \frac{I_D R_S}{r_d} = I_D$$

$$\frac{e_{test}}{r_d} = I_D \left(1 + g_m R_S + \frac{R_S}{r_d}\right)$$

$$\frac{e_{test}}{r_d} = \left(I_{test} - \frac{e_{test}}{R_D}\right)\left(1 + g_m R_S + \frac{R_S}{r_d}\right)$$

$$\frac{e_{test}}{r_d} = I_{test}\left(1 + g_m R_S + \frac{R_S}{r_d}\right) - \frac{e_{test}}{R_D}\left(1 + g_m R_S + \frac{R_S}{r_d}\right)$$

$$e_{test}\left(\frac{1}{r_d} + \frac{1 + g_m R_S + \dfrac{R_S}{r_d}}{R_D}\right) = I_{test}\left(1 + g_m R_S + \frac{R_S}{r_d}\right)$$

$$\frac{e_{test}}{I_{test}} = Z_{out} = \frac{1 + g_m R_S + \dfrac{R_S}{r_d}}{\dfrac{\dfrac{R_D}{r_d} + 1 + g_m R_S + \dfrac{R_S}{r_d}}{R_D}}$$

$$Z_{out} = \frac{R_D}{1 + \dfrac{R_D}{r_d\left[1 + \left(g_m + \dfrac{1}{r_d}\right)R_S\right]}} \tag{8.30}$$

We note that if $r_d \gg R_D$, then

$$Z_0 \doteq R_D \tag{8.31}$$

If we wish to consider the output impedance of only the FET, from (8.30):

$$\frac{e_{test}}{I_{test}} = Z_0 = \frac{1 + g_m R_S + \dfrac{R_S}{r_d}}{\dfrac{\dfrac{1}{r_d} + 1 + g_m R_S + \dfrac{R_S}{r_d}}{R_D}}$$

If $R_D \to \infty$

$$Z'_0 = r_d (1 + g_m R_S) + R_S$$

Because r_d is usually a very large number, we can see that the output impedance of the transistor, without the drain bias resistor, is a large number. Let us determine the output impedance for the common gate circuit shown in Fig. 8.32, given $g_m = 2.5$ mmho, $r_d = 100$ k.

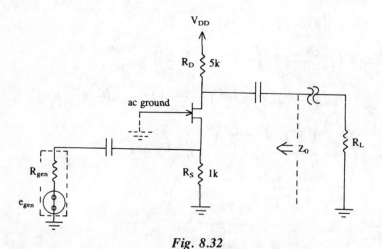

Fig. 8.32

It is important to note that where R_S appears in Equation (8.30), a more general expression would have R_S replaced with Z_S, where $Z_S = R_S \mathbin{/\mkern-5mu/} R_{gen}$ for a shunt-fed source (as above), or $Z_S = R_S + R_{gen}$ for a series-fed source. Without considering R_{gen} for the above,

$$Z_0 = \frac{R_D}{1 + \dfrac{R_D}{r_d\left[1 + \left(g_m + \dfrac{1}{r_d}\right)Z_S\right]}} \qquad (8.32)$$

Therefore,

$$Z_0 = \frac{5\text{ k}}{1 + \dfrac{5\text{ k}}{100\text{ k}\left[1 + \left(2.5 + \dfrac{1}{100}\right)1\text{ k}\right]}} = 4.93\text{ k}$$

Because use of the analysis formulas requires that we know the value of g_m, we need to have a convenient method for determining its value. We could refer to the manufacturer's specifications, or use a curve tracer to obtain a photograph of the output curves, and thereby determine g_m. The additional complication that we encounter is that we must also know the dc Q point, since g_m is not independent of dc operating conditions.

If we know the circuit component values and the power supply voltage then we can use the relationship:

$$g_m = g_{mo}\left(1 - \frac{V_{SG}}{V_P}\right)$$

where

$$g_{mo} = \frac{2\,I_{DSS}}{V_P}$$

Determining the values of both I_{DSS} and V_P can be accomplished with nothing more than a dc voltmeter and a dc power supply. You will note that the following derivation generates a formula which enables the user to determine the numerical value of V_{SG} in terms of the numerical values for V_P, I_{DSS}, and V_{gG}.

8.3.1 Derivation of the Formula for the DC Source-to-Gate Voltage of a Self-Biased FET Common Source Amplifier

Refer to Fig. 8.33. Thus,

$$V_{SG} = -I_D R_S + V_{gG}$$

Therefore, $I_D R_S = V_{gG} - V_{SG}$

266

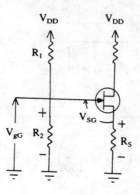

Fig. 8.33

or $I_D = \dfrac{V_{gG} - V_{SG}}{R_S}$

$I_D = I_{DSS} \left(1 - \dfrac{V_{SG}}{V_P}\right)^2$

Substituting:

$\dfrac{V_{gG} - V_{SG}}{R_S} = I_{DSS} \left(1 - \dfrac{V_{SG}}{V_P}\right)^2$

$\dfrac{V_{gG} - V_{SG}}{I_{DSS}\,R_S} = \left(1 - \dfrac{V_{SG}}{V_P}\right)^2 = 1 - \dfrac{2V_{SG}}{V_P} + \dfrac{V_{SG}{}^2}{V_P{}^2}$ $\left(\begin{array}{l}\text{where:}\\ V_{gG}\text{ is determined}\\ \text{by the voltage}\\ \text{divider rule.}\end{array}\right)$

$\dfrac{V_{gG}}{I_{DSS}\,R_S} - \dfrac{V_{SG}}{I_{DSS}\,R_S} = 1 - \dfrac{2V_{SG}}{V_P} + \dfrac{V_{SG}{}^2}{V_P{}^2}$

$\dfrac{V_{SG}{}^2}{V_P{}^2} - 2\,\dfrac{V_{SG}}{V_P} + \dfrac{V_{SG}}{I_{DSS}\,R_S} + 1 - \dfrac{V_{gG}}{I_{DSS}\,R_S} = 0$

$V_{SG}{}^2 - 2\,V_{SG}\,V_P + \dfrac{V_{SG}\,V_P{}^2}{I_{DSS}\,R_S} + V_P{}^2 - \dfrac{V_{gG}\,V_P{}^2}{I_{DSS}\,R_S} = 0$

$V_{SG}{}^2 + V_{SG}\left(\dfrac{V_P{}^2}{I_{DSS}\,R_S} - 2\,V_P\right) + V_P{}^2\left(1 - \dfrac{V_{gG}}{I_{DSS}\,R_S}\right) = 0$

Employing the quadratic formula:

$$V_{SG} = \frac{2V_P - \dfrac{V_P{}^2}{I_{DSS} R_S}}{2}$$

$$\pm \frac{\sqrt{\dfrac{V_P{}^4}{I_{DSS}{}^2 R_S{}^2} - \dfrac{4 V_P{}^3}{I_{DSS} R_S} + 4 V_P{}^2 - 4 V_P{}^2\left(1 - \dfrac{V_{gG}}{I_{DSS} R_S}\right)}}{2}$$

$$= V_P - \frac{1}{2}\frac{V_P{}^2}{I_{DSS} R_S} \pm \frac{1}{2}\sqrt{\frac{V_P{}^4}{I_{DSS}{}^2 R_S{}^2} - \frac{4 V_P{}^3}{I_{DSS} R_S} + \frac{4 V_P{}^2 V_{gG}}{I_{DSS} R_S}}$$

$$= V_P\left(1 - \frac{1}{2}\frac{V_P}{I_{DSS} R_S}\right)$$
$$\pm \frac{V_P}{2}\sqrt{\left(\frac{V_P}{I_{DSS} R_S}\right)^2 - 4\left(\frac{V_P}{I_{DSS} R_S}\right) + 4\left(\frac{V_{gG}}{I_{DSS} R_S}\right)}$$

$$= V_P\left(1 - \frac{1}{2}\frac{V_P}{I_{DSS} R_S}\right)$$
$$\pm \frac{V_P}{2}\sqrt{\left(\frac{V_P}{I_{DSS} R_S}\right)^2 + 4\left(\frac{V_{gG} V_P}{V_P I_{DSS} R_S}\right) - 4\frac{V_P}{I_{DSS} R_S}}$$

$$= V_P\left(1 - \frac{1}{2}\frac{V_P}{I_{DSS} R_S}\right)$$
$$\pm \frac{V_P}{2}\sqrt{\left(\frac{V_P}{I_{DSS} R_S}\right)^2 + 4\left(\frac{V_{gG}}{V_P}\frac{V_P}{I_{DSS} R_S} - \frac{V_P}{I_{DSS} R_S}\right)}$$

$$V_{SG} = V_P\left(1 - \frac{1}{2}\frac{V_P}{I_{DSS} R_S}\right)$$
$$\pm \frac{V_P}{2}\sqrt{\left(\frac{V_P}{I_{DSS} R_S}\right)^2 + 4\left(\frac{V_P}{I_{DSS} R_S}\right)\left(\frac{V_{gG}}{V_P} - 1\right)}$$

$$(8.34)$$

The reader should note that the expression for V_{SG} is well suited for evaluation by use of a programmable calculator, or the BASIC language on a microcomputer or mainframe computer.

Rewriting (8.34), we have

$$V_{SG} = V_P\left(1 - \frac{1}{2}\frac{V_P}{I_{DSS}R_S}\right) \pm \frac{V_P}{2}\left[\left(\frac{V_P}{I_{DSS}R_S}\right)^2 + 4\left(\frac{V_P}{I_{DSS}R_S}\right)\left(\frac{V_{gG}}{V_P} - 1\right)\right]^{1/2}$$

Table 8.1
Field-Effect Transistor Amplifiere—Small-Signal Analysis Formulas

Configuration	A_V	Z_{in}	Z_{out}
Common Source (bypassed source resistor or no source resistor)	$A_V = -g_m(Z_L \parallel r_D)$ $A_V \doteq g_m Z_L$	$Z_{in} = R_1 \parallel R_2$	$Z_0 = r_D \parallel R_D$ $Z_0 \doteq R_D$
Common Source (unbypassed source resistor)	$A_V = \dfrac{-g_m Z_L}{1 + R_S\left(g_m + 1/r_D\right) + \dfrac{Z_L}{r_D}}$	$Z_{in} = R_1 \parallel R_2$	$Z_0 = \dfrac{R_D}{1 + \dfrac{R_D}{r_D + R_S(1 + g_m r_D)}}$ $Z_0 \doteq R_D$
Common Drain (source follower)	$A_V = \dfrac{1}{1 + \dfrac{1 + Z_S/r_D}{g_m Z_S}}$ $A_V \doteq \dfrac{g_m Z_S}{1 + g_m Z_S}$	$Z_{in} = R_1 \parallel R_2$	$Z_0 = \dfrac{R_S}{1 + g_m R_S + \dfrac{R_S}{r_D}}$ $Z_0 \doteq \dfrac{R_S}{1 + g_m R_S}$
Common Gate	$A_V = \dfrac{g_m Z_L + Z_L/r_D}{1 + Z_L/r_D}$ $A_V \doteq g_m Z_L$	$Z_{in} = \dfrac{1}{(g_m + 1/r_D)\left(1 - \dfrac{1}{1 + r_D/Z_L}\right) + \dfrac{1}{R_S}}$ $Z_{in} \doteq \dfrac{1}{g_m + 1/R_S}$	$Z_0 = \dfrac{R_D}{1 + \dfrac{R_D}{r_D[1 + (g_m + 1/r_D)Z_S]}}$ $Z_0 \doteq R_D$

Substituting,

$K_1 = V_P$ (include negative sign)

$K_2 = \dfrac{V_P}{I_{DSS} R_S}$ (include negative sign for V_P)

$K_3 = \dfrac{V_{gG}}{V_P} - 1$ (include negative sign for V_P)

Thus,

$$V_{SG} = K_1 \left(1 - \frac{1}{2}K_2\right) \pm \frac{K_1}{2}\sqrt{K_2^2 + 4K_2 K_3}$$

For a Sharp 5813 programmable calculator, the following keystrokes are necessary:

| 2NDF | LRN | K2 | ÷ | 2 | +/− | + | 1 | = | K1 | STO | 2NDF |

| K4 | 4 | K2 | K3 | + | K2 | X^2 | = | 2NDF | $\sqrt{X}$ | K1 | ÷ | 2 |

| = | +/− | + | 2NDF | K4 | = | 2NDF | LRN |

PROBLEMS

1. For the set of FET output curves shown in Fig. 8.34, determine the transconductance and small signal drain resistance at a Q point of $V_{SDQ} = +10$ V, $V_{SGQ} = -1$ V.

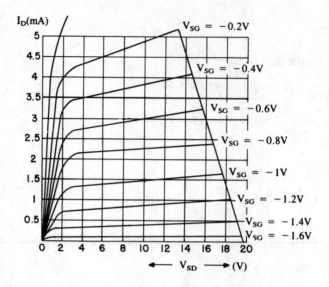

Fig. 8.34

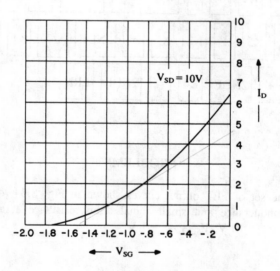

Fig. 8.35

2. By use of the transfer characteristic curves shown in Fig. 8.35, determine the transconductance at the Q point of $V_{SDQ} = +10$ V, $V_{SGQ} = -1$ V.

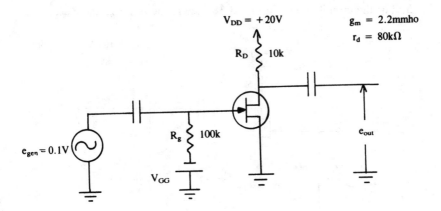

Fig. 8.36

3. For the circuit shown in Fig. 8.36, determine the voltage gain. For an input signal of 0.1 V, what would the output voltage be?

4. For the circuit shown in Fig. 8.37, determine the voltage gain. What is the output impedance?

   What is the input impedance?

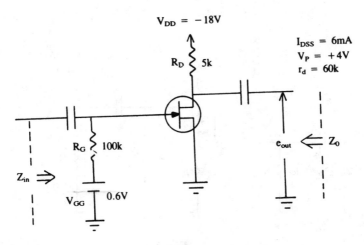

Fig. 8.37

272

5. For the circuit shown in Fig. 8.38, determine A_V, Z_{in}, and Z_{out}.

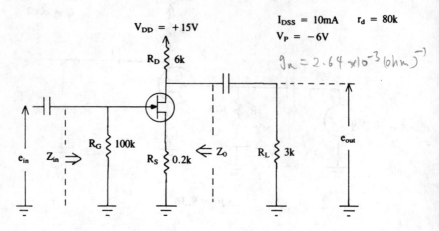

$$g_n = 2.64 \times 10^{-3} (\text{ohm})^{-1}$$

Fig. 8.38

6. For the circuit shown in Fig. 8.39, determine A_V, Z_{in}, and Z_{out}.

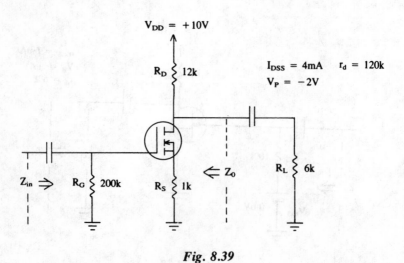

Fig. 8.39

7. For the circuit shown in Fig. 8.40, determine A_V, Z_{in}, and Z_{out}.

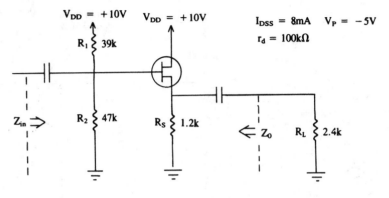

Fig. 8.40

8. For the circuit shown in Fig. 8.41, determine A_V, Z_{in}, and Z_{out}.

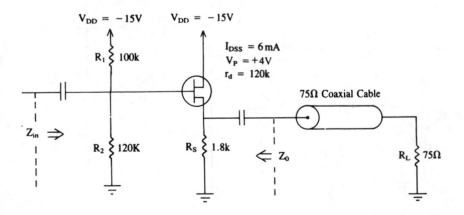

Fig. 8.41

9. For the circuit shown in Fig. 8.24, determine A_V, Z_{in}, and Z_{out}.

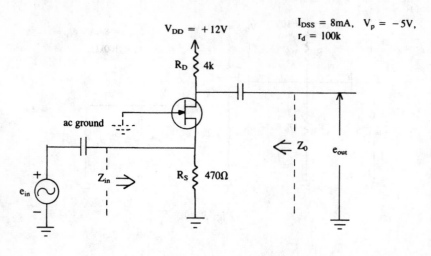

Fig. 8.42

Note: gate dc bias resistors are not shown.

10. For the circuit shown in Fig. 8.43, determine A_V, Z_{in}, and Z_{out}.

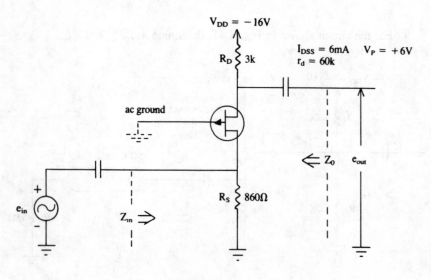

Fig. 8.43

Chapter 9

Bipolar Transistor Amplifier Stage—Small-Signal Analysis

9.1 SMALL-SIGNAL ANALYSIS

We noted in a previous chapter that the values of voltage gain, current gain, power gain, average input resistance, output resistance, and the resistors, which are necessary for dc bias of a bipolar transistor, were determined graphically. Two disadvantages of the graphical method are that it is time consuming and the input characteristic curves are not readily available. Also, due to the close spacing of the input curves, inaccuracies will occur.

If a method of predicting the ac characteristics of an active device (bipolar and field-effect transistors, electron tubes, *et cetera*) could be developed, then we would not need to employ a graphical approach. Various methods have been developed for approximating the ac characteristics of an active device. The accuracy that these methods yields for predicting the voltage gain of a transistor, for example, is a function of several considerations. Two of these considerations are:

1. If a very small signal is used, the percentage difference between the predicted *versus* measured values of voltage gain, *et cetera*, improves;
2. The accuracy of the values computed by measurement techniques of certain, as yet unspecified, characteristics of an active device, referred to as the small signal parameters, is related to the signal level (i.e., the more signal, the less validity these measured values have).

Although the bipolar transistor is a nonlinear device, certain linear equivalent circuits may be utilized for analysis purposes.

By the use of linear equivalent circuits, the accuracy of predicting transistor amplifier characteristics will be limited by the above mentioned conditions.

A common equivalent circuit employed for low frequency circuit analysis, is shown below in Fig. 9.1. We can see that the input circuit consists of a voltage generator in series with a resistor and the output circuit consists of a current generator in parallel with a shunt resistor.

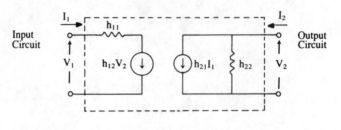

Fig. 9.1

What we have here is a circuit with both voltage and current sources and series and parallel resistors.

First, something should be said about the unusual nomenclature for each component in the circuit. We note that each is labeled, h_{xx}. In fact, all four of the circuit components are labeled h with double subscripts. The letter h is representative of the word "hybrid." Hybrid is appropriate in that we have combined both a series-connected voltage source and resistor as well as a parallel-connected current source and conductance in our circuit, which represents a transistor (or, for that matter, any active device, e.g., field-effect transistor, electron tubes, *et cetera*). You may wonder why we say that "one circuit" is used, when the schematic shows two distinct circuits. The reason for alluding to a "single" circuit is that any voltage appearing in the output circuit develops a corresponding voltage within the input circuit and, conversely, any current that is forced to flow in the input circuit (due to a signal source, e.g., a phonograph cartridge, *et cetera*) will cause a corresponding current to flow in the output circuit. The fact that signals in the input affect the output and *vice versa* leads to the designation of the bipolar transistor as a *bilateral* device.

Let us examine the subscripts on each of the h terms in the circuit shown. First, h_{11} represents a resistance. (The quantity h_{11} is measured in ohms.) The first 1 in the subscript refers to the first circuit (the input circuit). The second 1 refers to the effect of the first circuit on the resistance value. The value h_{12} is a dimensionless quantity. It is a ratio, which represents the small percentage of output voltage that appears in series with the other input components and voltage sources. The effect of the voltage source $h_{12}V_2$, is to increase slightly the total voltage that appears across h_{11}. The increase in the voltage across h_{11}, increases the amount of input current that will flow (for the same signal source voltage). Thus, the input resistance "appears" to the signal source to be slightly less than if h_{12} were not present in the equivalent circuit. For a typical common-emitter connected circuit, $h_{12}V_2$ will usually introduce less than one percent of the output voltage into the input circuit.

In the output circuit, the parameter h_{22} typically is a large shunt resistance (low conductance) that appears across the current generator, measured in mhos.

The first subscript 2 refers to the second circuit and the second subscript 2 refers to the effect that the second circuit has on the conductance value.

The ac resistance value of h_{11} and h_{22} will depend upon the geometry (size and shape of the internal construction) of the active device as well as the configuration (common emitter, common base, *et cetera*).

The last parameter to be discussed is h_{21}. The first subscript 2 is an indicator that h_{21} is a parameter appearing in the second (or output) circuit and the second subscript 1 is an indicator that the parameter is a measure of the effect of the input circuit on the output circuit. The value h_{21} is a dimensionless quantity, which may be thought of as a ratio of output current to input current. It may have a value from 20 to 800 (for a bipolar transistor).

9.1.1 Analysis Formulas

Since we dealt with bipolar transistors in the previous chapter and we are interested in developing formulas which will enable us to predict the performance of a bipolar transistor in an amplifier circuit, it is useful to dispense with the general numerical subscripts for the h parameters and use a more specific notation. The h parameter subscripts in common use for a bipolar transistor are as shown in Table 9.1.

We note in the table that the first subscript denotes the electrical characteristics for the parameter (e.g., o = output; thus, h_{oe} = output conductance). The second subscript denotes the type of connection (e.g., e = common emitter; thus, h_{oe} = common emitter output conductance).

It is important to note at this point that the series resistance, h_{ie}, is not the value of the input resistance that is "seen" by a previous stage (signal source).

Table 9.1

CONNECTION or CONFIGURATION	h_{11} Series Resistance	h_{22} Output Conductance	h_{21} Forward Current Transfer Ratio	h_{12} Reverse Voltage Transfer Ratio
Common Emitter	h_{ie}	h_{oe}	h_{fe}	h_{re}
Common Base	h_{ib}	h_{ob}	h_{fb}	h_{rb}
Common Collector	h_{ic}	h_{oc}	h_{fc}	h_{rc}

278

The input resistance, as we will see shortly, is a function of the base bias resistors as well as the input resistance of the transistor. The input resistance of the transistor, exclusive of the base bias resistors, is slightly less than h_{ie} (for a common emitter connection).

Further, we should also note that the term "output resistance" does *not* mean the impedance that the transistor "sees" looking outward toward the next stage. The opposite, in fact, is the case. To determine the output impedance we must "stand" at the terminals of the succeeding stage and consider the Thevenin equivalent resistance of the transistor from its collector to ground in parallel with the collector resistor, R_C (see Fig. 9.2). The output resistance of the transistor connected as an amplifier, is not equal to $1/h_{oe}$, as might be anticipated. The reason for this apparent discrepancy is that all of the h parameters enter into the determination of the output resistance (of the transistor) as well as the impedance of the signal source. With all these factors accounted for, the output impedance would still be a considerable value. The collector resistor R_C, which is required for proper dc bias, is essentially in parallel with the output resistance of the transistor. The resistance of R_C is usually considerably less than the output resistance of the transistor (without bias components). Therefore, since R_C is in parallel with the transistor, *the output resistance* of the *transistor* connected as an amplifier, is essentially the value of R_C (see Fig. 9.3).

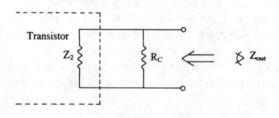

Fig. 9.2

To enable the reader to gain a "feel" for typical value of the input and output resistance and voltage and current gains of an amplifier, Table 9.2 contains typical values for the transistor, both with and without bias resistors.

We have subjectively discussed the small signal characteristic of the bipolar transistor. Now it is appropriate to develop the formulas which will enable us to predict the numerical values of the voltage and current gain and input and output resistance. Redrawing our small signal equivalent circuit, we have the schematic of Fig. 9.4. The student may wonder why the current generator arrow points downward for $h_{21}I_1$. We recall that a positive-going signal voltage (for example, in a common-emitter connected amplifier) results in an increase in I_b which results in an increase in I_c (or more current flowing downward through the transistor).

Table 9.2
Typical Characteristics for a Single-Stage Bipolar
Transistor Amplifier—Common Emitter Configuration

CONFIGURA-TION	Input Resistance	Voltage Gain	Current Gain	Output Resistance [1]	Output Resistance [2]
Common Emitter	500— 4 kΩ	20— 100 Ω	20— 100 Ω	30— 100 kΩ	5 kΩ
Common Base	15— 50 Ω	20— 100 Ω	1 MΩ	1 MΩ	5 kΩ

(1) Without Collector Load Resistor
(2) With Collector Load Resistor (R_C = 5 kΩ)

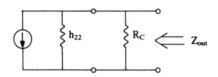

Fig. 9.3

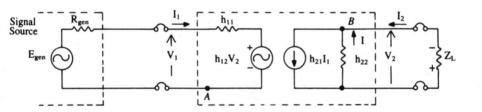

Fig. 9.4

Let us define the following:

Current gain = $A_I = I_2/I_1$ (9.1)

Voltage gain = $A_v = V_2/V_1$ (9.2)

Input Resistance = $Z_1 = V_1/I_1$ (9.3)

Output Resistance = $Z_2 = V_2/I_2$ (9.4)

Development of the Formula for the Voltage Gain

Applying Kirchoff's voltage law around the input loop starting at A, (c.c.w.) we have

$$h_{12} V_2 + h_{11} I_1 - V_1 = 0 \tag{9.5}$$

Using Kirchoff's current law at node B, we have

$$I + I_2 = h_{21} I_1 \tag{9.6}$$

but

$$I = -V_2 h_{22}$$

and

$$I_2 = -V_2/Z_L$$

Substituting, we have

$$-V_2 h_{22} - V_2/Z_L = h_{21} I_1$$

leading to

$$-V_2 \left(h_{22} + \frac{1}{Z_L} \right) = h_{21} I_1$$

and solving for I_1,

$$I_1 = -\frac{V_2}{h_{21}} \left(h_{22} + \frac{1}{Z_L} \right)$$

Therefore, we may replace I_1 in (9.5):

$$h_{12} V_2 + h_{11} \left[-\frac{V_2}{h_{21}} \left(h_{22} + \frac{1}{Z_L} \right) \right] - V_1 = 0$$

$$h_{12} V_2 - V_2 \left(\frac{h_{11} h_{22}}{h_{21}} + \frac{h_{11}}{h_{21} Z_L} \right) - V_1 = 0$$

$$V_2 \left(h_{12} - \frac{h_{11} h_{22}}{h_{21}} + \frac{h_{11}}{h_{21} Z_L} \right) = V_1$$

OR

$$A_V = \frac{V_2}{V_1} = \frac{1}{h_{12} - \dfrac{h_{11}\,h_{22}}{h_{21}} - \dfrac{h_{11}}{h_{21}\,Z_L}} = \frac{1}{h_{12}\,\dfrac{h_{21}\,Z_L}{h_{21}\,Z_L} - \dfrac{h_{11}\,h_{22}\,Z_L}{h_{21}\,Z_L} - \dfrac{h_{11}}{h_{21}\,Z_L}}$$

$$A_V = \frac{h_{21}\,Z_L}{-h_{11} + (h_{12}\,h_{21} - h_{11}\,h_{22})\,Z_L} = \frac{-h_{21}\,Z_L}{h_{11} + (h_{11}\,h_{22} - h_{12}\,h_{21})\,Z_L}$$

$$(9.7)$$

Development of the Formula for the Current Gain

Let us develop the formula for the current gain, A_I.

Restating (9.6), we have

$$I + I_2 = h_{21}\,I_1 \qquad \text{but } I = -V_2\,h_{22}$$

$$-V_2\,h_{22} + I_2 = h_{21}\,I_1 \quad \text{but } I_2 = -\frac{V_2}{Z_L} \text{ or } V_2 = -I_2\,Z_L$$

Therefore, $-(-I_2\,Z_L)\,h_{22} + I_2 = h_{21}\,I_1$

$$I_2\,(Z_L\,h_{22} + 1) = h_{21}\,I_1$$

Thus,

$$\frac{I_2}{I_1} = A_I = \frac{h_{21}}{1 + h_{22}\,Z_L} \qquad\qquad\qquad (9.8)$$

Development of the Formula for the Input Resistance

Let us develop the formulas for the input resistance Z_1.

Restating (9.5):

$$h_{12}\,V_2 + h_{11}\,I_1 - V_1 = 0 \qquad \text{but } I_2 = -V_2/Z_L \qquad \text{or } V_2 = -I_2\,Z_L$$

Therefore, $h_{12}\,(-I_2\,Z_L) + h_{11}\,I_1 = V_1 \qquad \text{or } I_2\,(-h_{12}\,Z_L) + h_{11}\,I_1 = V_1$

Restating (9.8): $\dfrac{I_2}{I_1} = \dfrac{h_{21}}{1 + h_{22}\,Z_L}$

Therefore, $I_2 = \dfrac{h_{21}\,I_1}{1 + h_{22}\,Z_L}$

Hence,

$$\frac{h_{21}\,I_1}{1 + h_{22}\,Z_L}\,(-h_{21}\,Z_L) + h_{11}\,I_1 = V_1$$

$$I_1\left(\frac{-h_{12}\,h_{21}\,Z_L}{1 + h_{22}\,Z_L} + h_{11}\right) = V_1$$

Thus,

$$\frac{V_1}{I_1} = Z_1 = h_{11} - \frac{h_{12} \, h_{21} \, Z_L}{1 + h_{22} \, Z_L} \tag{9.9}$$

It is of interest to note that the input resistance Z_1 is a function of the output load connected to the amplifier.

Determination of the Output Impedance

Let us determine the output impedance, Z_2.

Recalling from dc circuit analysis, in order to determine a Thevenin equivalent resistance of a complex network containing voltage and current sources in addition to various resistance values, it is necessary to short all voltage sources and open all current sources. Thus, we have the schematic of Fig. 9.5.

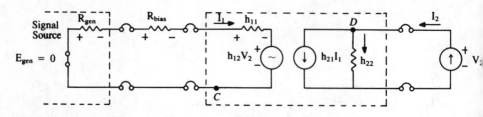

Fig. 9.5

Summing the voltages around the input loop ($R_{source} = R_{gen} + R_{bias}$) or ($R_{source} = R_{gen}/\!/R_{bias}$) at C (c.c.w.):

$$h_{12} \, V_2 + h_{11} \, I_1 + I_1 \, R_{source} = 0$$

$$h_{12} \, V_2 + I_1 \, (h_{11} + R_S) = 0 \qquad \text{Therefore, } I_1 = \frac{-h_{12} \, V_2}{h_{11} + R_S}$$

Summing the currents into node D:

$$I_2 = V_2 \, h_{22} + h_{21} \, I_1 = V_2 \, h_{22} + h_{21} \left(\frac{-h_{12} \, V_2}{h_{11} + R_S} \right)$$

$$I_2 = V_2 \left(h_{22} - \frac{h_{12} \, h_{21}}{h_{11} + R_S} \right)$$

$$\frac{V_2}{I_2} = Z_2 = \frac{1}{h_{22} - \dfrac{h_{12} \, h_{21}}{h_{11} + R_S}} \tag{9.10}$$

It is worth pointing out that the output impedance, as can be seen by examining formula (7.6), is a function of the generator (or source) resistance. It may seem somewhat difficult to accept the idea that the load connected to an amplifier will affect its input resistance (impedance) and, conversely, the generator (or source) resistance will affect its output resistance (impedance). Remember, however, that the transistor is a bilateral device, or, as was pointed out in Chapter 1, a transistor may be thought of as three blocks (n-p-n) of silicon material that are touching each other. Therefore, there is an isolation deficiency between the input and output circuits.

Earlier we stated that numerical subscripts for the h parameters could be replaced by letter subscripts representative of the h parameter value as a function of a particular configuration. Let us examine the amplifier characteristics for some arbitrary but typical h parameter values:

Example (9.1). Choose $h_{fe} = 100$, $h_{re} = .0001$, $h_{oe} = 2(10^{-5})$, $h_{ie} = 3000$.

If we had a dc design as shown in Fig. 9.6, then we would compute the following:

Voltage Gain Computations

$$A_V = \frac{-h_{21}\,Z_L}{h_{11} + (h_{11}\,h_{22} - h_{12}\,h_{21})\,Z_L} \text{ or } A_V = \frac{-h_{fe}\,Z_L}{h_{ie} + (h_{ie}h_{oe} - h_{re}h_{fe})\,Z_L}$$

$$= \frac{-100\,(5000)}{3000 + [3(10^3)\,2(10^{-5}) - 10^{-4}\,(10^2)]\,5(10^3)}$$

$$= -\frac{5000\,(100)}{3250} = -153.8 \tag{9.11}$$

$$\text{(dimensionless)}$$

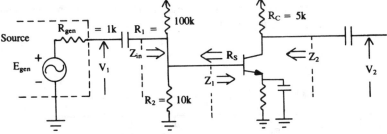

Fig. 9.6

It is important to point out that the negative sign, associated with the resulting voltage gain computation, does *not* imply a loss in voltage, but merely a phase reversal of the signal between the input and output terminals. Hence, if a sine wave voltage is applied to the base of a common emitter stage and at one instant in time the input sine wave is positive-going, the output will be negative-going at the same instant (and usually amplified considerably).

Current Gain Computation

$$A_I = \frac{h_{21}}{1 + h_{22} Z_L} \quad \text{or} \quad A_I = \frac{h_{fe}}{1 + h_{oe} Z_L} \tag{9.12}$$

$$= \frac{100}{1 + 2(10^{-5}) \, 5 \, (10^3)} = \frac{100}{1.1} = 90.9 \text{ (dimensionless)}$$

Input Impedance Computation

$$Z_1 = h_{11} - \frac{h_{12} \, h_{21} \, Z_L}{1 + h_{22} \, Z_L} \quad \text{OR} \quad Z_1 = h_{ie} - \frac{h_{re} \, h_{fe} \, Z_L}{1 + h_{oe} \, Z_L} \tag{9.13}$$

$$= 3000 - \frac{10^{-4} \, (10^2) \, 5(10^3)}{1 + 2(10^{-5}) \, 5(10^3)} = 3000 - \frac{50}{1.1}$$

$$Z_1 = 2954 \ \Omega$$

We note that the input impedance Z_{11} is approximately equal to h_{ie}.

Output Impedance Computation

$$Z_2 = \frac{1}{h_{22} - \dfrac{h_{12} \, h_{21}}{h_{11} + R_S}} \quad \text{OR} \quad Z_2 = \frac{1}{h_{oe} - \dfrac{h_{re} \, h_{fe}}{h_{ie} + R_S}} \tag{9.14}$$

R_S is all the resistances that we "see" by standing at the transistor base and "looking" away from the transistor. We must also be sure to ignore all resistances that are behind us. (See Fig. 9.7).

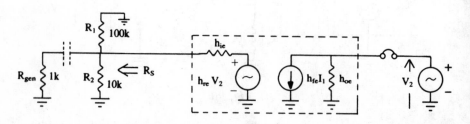

Fig. 9.7

It is important to note the following three considerations in determining R_S:

1. The generator voltage is set to zero.
2. The capacitor reactance is considered to be negligible and, therefore, does not disturb the signal (i.e., a short).
3. The V_{CC} supply is essentially at ac ground potential. Remember that our power supply has a very large filter capacitor across its output terminals to short all ripple to ground.

Therefore,

$$R_S = 100 \text{ k} \mathbin{/\mkern-5mu/} 10 \text{ k} \mathbin{/\mkern-5mu/} 1 \text{ k} = \frac{1}{\dfrac{1}{100} + \dfrac{1}{10} + \dfrac{1}{1}} = 0.9 \text{ k}\Omega$$

$$Z_2 = \frac{1}{2(10^{-5}) - \dfrac{(10^{-4})(10^2)}{3000 + 900}} = \frac{1}{2(10^{-5}) - \dfrac{10^{-2}}{3900}}$$

$$= \frac{1}{2(10^{-5}) - \dfrac{10000(10^{-6})}{3900}}$$

$$= \frac{1}{2(10^{-5}) - 2.56(10^{-6})} = \frac{1}{2(10^{-5}) - 0.256(10^{-5})}$$

$$= \frac{10^5}{1.74} = \frac{10(10^4)}{1.74}$$

$$Z_2 = 5.74(10^4) \ \Omega$$

We see that the value of Z_2 approaches $1/h_{oe}$ ($= 50$ k).

Let us examine some limiting factors for Z_1 and Z_2 (common emitter connected).

$$Z_1 = h_{ie} - \frac{h_{re} h_{fe} Z_L}{1 + h_{oe} Z_L}$$

Rewriting (9.13) by dividing the second term by Z_L, we have

$$Z_1 = h_{ie} - \frac{h_{re} h_{fe}}{h_{oe} + \dfrac{1}{Z_L}}$$

As Z_L tends to infinity ($Z_L \to \infty$), we have

$$Z_1 = h_{ie} - \frac{h_{re} h_{fe}}{h_{oe}}$$

and for our example:

$$Z_1 = 3000 - \frac{10^{-4}\,(10^2)}{2(10^{-5})} = 3000 - \frac{10^{-2}}{2(10^{-5})} = 3000 - 0.5(10^3)$$
$$Z_1 = 2500 \; \Omega$$

We had originally computed

$$Z_1 = 2950 \; \Omega$$

As Z_L tends toward zero ($Z_L \to 0$), we have

$$Z_1 = h_{ie} - 0 = h_{ie} = 3000 \; \Omega$$

We can see that the maximum effect that wide variations in Z_L can have upon the input resistance (for this example) is:

$$\% \text{ change } Z_1 = \left| \frac{3000 - 2500}{3000} \right| \times 100 = 17\%$$

of the value when $Z_L = 0$. For Z_2:

$$Z_2 = \frac{1}{h_{oe} - \dfrac{h_{re}\,h_{fe}}{h_{ie} + R_S}}$$

If R_S tends toward zero ($R_S \to 0$), we have

$$Z_2 = \frac{1}{h_{oe} - \dfrac{h_{re}\,h_{fe}}{h_{ie}}} = \frac{1}{2(10^{-5}) - \dfrac{10^4\,(10^2)}{3000}} = \frac{1}{2(10^{-5}) - \dfrac{10^{-2}}{3000}}$$
$$Z_2 = 60 \text{ k}\Omega$$

If R_S tends toward infinity ($R_S \to \infty$), we have

$$Z_2 = \frac{1}{h_{oe}} = \frac{1}{2(10^{-5})} = 50 \text{ k}$$

As R_S varies from 0 to infinity the output impedance will vary:

$$\text{Max } \% \text{ change } Z_2 = \left| \frac{60 - 50}{60} \right| \cdot 100 = 17\%$$

We have used various symbols to denote circuit parameters (other than the h parameters) in developing and using our small-signal circuit analysis formulas. Let us define each circuit parameter in Table 9.3 and define how the value of each is to be determined.

Table 9.3

CIRCUIT PARAMETER	Description	Procedure for determining the numerical nature of the circuit parameter
Z_L	Load Impedance	Stand at the collector of the transistor, look away from the transistor, and ignore all circuitry behind you. Determine the total impedance of all components connected parallel to ground (R_C ∥ Next Stage).
Z_1	Transistor Input Impedance	Stand at the base of the transistor, look into the transistor, and ignore all circuitry behind you. Determine the impedance to ground by use of the h parameter formula.
Z_{in}	Amplifier Stage Input Impedance	Stand at the input terminal to the transistor amplifier, look into the amplifier, and ignore all circuitry behind you. Determine the total impedance of all components in parallel to ground (R_1 ∥ R_2 ∥ Z_1).
Z_2	Transistor Output Impedance	Stand at the collector terminal of the transistor, look into the transistor, and ignore all circuitry behind you. Determine the impedance to ground by use of the h parameter formula.
Z_{out}	Amplifier Stage Output Impedance	Stand at the output terminal of the amplifier stage, look *into* the amplifier stage, and ignore all circuitry behind you. Determine the total impedance of all components in parallel to ground (R_C ∥ Z_2).
R_S	Source Resistance	Stand at the base of the transistor, look away from the transistor and ignore all circuitry behind you. Determine the total impedance of all components in parallel to ground (R_1 ∥ R_2 ∥ R_{gen}).

288

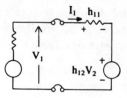

Fig. 9.8

Up to this point, we have used the small-signal transistor h parameters without mentioning how they were mathematically defined, or how they could be determined. Referring to the equivalent circuit of the transistor input in Fig. 9.8, we have

$$h_{12} V_2 + h_{11} I_1 - V_1 = 0$$

Because we are only interested in evaluating h_{11}, if V_2 is set equal to zero,

$$0 + h_{11} I_1 = V_1$$

or

$$h_{11} = \frac{V_1}{I_1} \bigg|_{V_2 = 0} \tag{9.15}$$

The notation, $V_2 = 0$ implies that the ac output voltage is zero. This does *not* mean that the dc bias voltages are zero. To measure h_{11}, we could construct a circuit with the proper dc bias conditions, but short the signal at the collector to ground with a capacitor. See Fig. 9.9 (i.e., $\Delta V_2 = 0$).

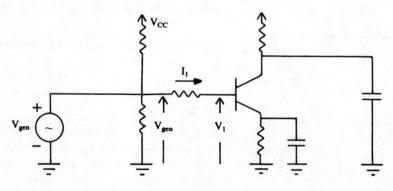

Fig. 9.9

We could then input a signal to the base and measure the input voltage V_1 and input current I_1, and compute

$$h_{11} \left(= \frac{V_1}{I_1} \bigg|_{V_2 = 0} \right)$$

In a similar fashion, if we open the input circuit, $I_1 = 0$, we have

$$h_{12} V_2 + h_{11} I_1 = V_1$$
$$(= 0)$$

for $I_1 = 0$, we have
$$h_{12} V_2 = V_1$$

$$h_{12} = \frac{V_1}{V_2} \bigg|_{I_1 = 0} \tag{9.16}$$

We could devise a circuit shown in Fig. 9.10. Each choke would prevent signal current from flowing in the input circuit.

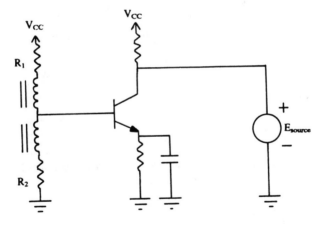

Fig. 9.10

Redrawing the output circuit, we have the schematic of Fig. 9.11.

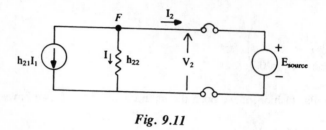

Fig. 9.11

At node F:

$$I_2 = I + h_{21} I_1$$

but

$$I = V_2 h_{22}$$

$$I_2 = h_{22} V_2 + h_{21} I_1$$

If we do not allow any input current to flow, $I_1 = 0$. Then,

$$I_2 = h_{22} V_2 + 0$$

or

$$h_{22} = \left. \frac{I_2}{V_2} \right|_{I_1 = 0} \tag{9.17}$$

The circuit shown in Fig. 9.12 enables us to measure h_{22}. To determine h_{21}, we must set $V_2 = 0$. Thus, we have

$$I_2 = h_{22} V_2 + h_{21} I_1$$

for $V_2 = 0$, we have

$$I_2 = h_{21} I_1 \qquad \text{or} \qquad h_{21} = \left. \frac{I_2}{I_1} \right|_{V_2 = 0} \tag{9.18}$$

To accomplish this measurement, we could use a small sensing resistor in the collector as shown in Fig. 9.13, so that the collector voltage does not vary appreciably, but we may still detect a small voltage across the resistor. By use of Ohm's law we can then calculate the value of I_2.

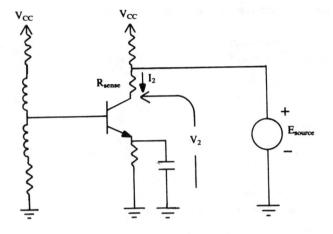

Fig. 9.12

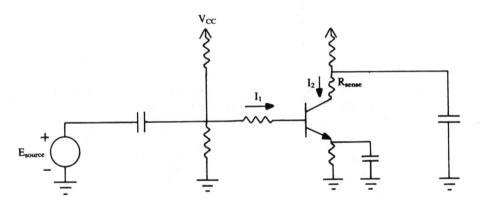

Fig. 9.13

From the student's perspective it would be an inconvenience to construct the appropriate circuits and implement the necessary measurements. If manufacturer's specification sheets containing input and output curves are available, or if photographs of these curves can be taken, then all of the *h* parameters may be determined.

9.1.2 Graphical Determination of the *h* Parameters

Recalling that the voltages and currents in the definition of the *h* parameters are ac signals (not dc bias voltages), we could write each of the parameters as follows:

$$h_{11} = \left.\frac{\Delta V_1}{\Delta I_1}\right|_{V_2 = \text{constant dc level (i.e. } V_2 \text{ ac } = 0)} \tag{9.19}$$

$$h_{12} = \left.\frac{\Delta V_1}{\Delta V_2}\right|_{I_1 = \text{const.}} \tag{9.20}$$

the Δ implies small changes in dc levels (or low level ac signals)

$$h_{21} = \left.\frac{\Delta I_2}{\Delta I_1}\right|_{V_2 = \text{const.}} \tag{9.21}$$

$$h_{22} = \left.\frac{\Delta I_2}{\Delta V_2}\right|_{I_1 = \text{const.}} \tag{9.22}$$

Determination of h_{fe}

Example (9.2). Referring to the common emitter output curves, if we draw a vertical line through the Q point, we have the following:

$$h_{21} = \left.\frac{\Delta I_2}{\Delta I_1}\right|_{V_2 = \text{const.}}$$

Therefore, for common emitter: $h_{fe} = \left.\dfrac{\Delta I_C}{\Delta I_B}\right|_{V_{EC} = \text{const.}}$

$$= \frac{I_{C2} - I_{C1}}{I_{B2} - I_{B1}} = \frac{4 - 2.2}{0.04 - 0.02} = 90$$

at $V_{EC} = 5$ V

Note: If we draw a vertical line through the Q point at every operating point on the line, the voltage from emitter to collector is 5 V!

Determination of h_{oe}

Again we refer to the output curves, as given by Fig. 9.15.

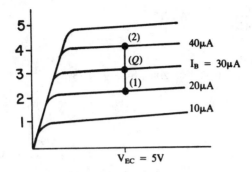

Fig. 9.14

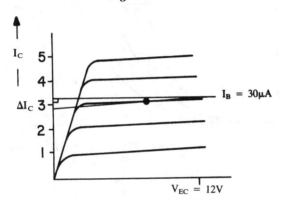

Fig. 9.15

Example (9.3). If we draw a tangent line through the operating point and then complete a right triangle (as large a triangle as possible for maximum accuracy), we can now determine h_{oe}:

$$h_{oe} = \frac{\Delta I_C}{\Delta V_{EC}}\bigg|_{I_B = \text{const.}} = \frac{I_{C2} - I_{C1}}{V_{EC2} - V_{EC1}}\bigg|_{I_B = \text{const.}}$$

$$= \frac{3.2 - 2.8}{12 - 0}\bigg|_{I_B = 30\ \mu A} = \frac{0.4\ mA}{12\ V}$$

$$= 0.0333\ (10^{-3})\ mho = 33\ mmho$$

As a point of interest, let us see what the *approximate* output resistance is by determining the reciprocal of the output conductance:

$$r_{out} \doteq \frac{1}{0.0333(10^{-3})} = 30 \text{ k}\Omega$$

Determination of h_{re}

Refer to the input curves, as given by Fig. 9.15.

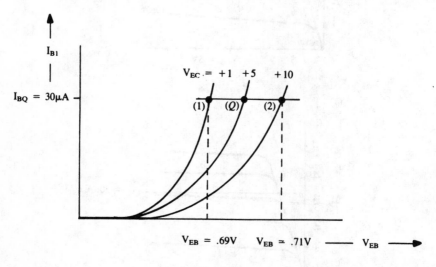

Fig. 9.16

Example (9.4). A horizontal line drawn through the curves at the Q point base current ($I_B = 30 \mu$A) would appear as shown in Fig. 9.15. Then recalling the definition of h_{12},

$$h_{12} = \frac{V_1}{V_2}\Bigg|_{I_1 = \text{const.}}$$

and for the common emitter configuration:

$$h_{re} = \frac{\Delta V_{EB}}{\Delta V_{EC}}\Bigg|_{I_B = \text{const.}} = \frac{V_{EB2} - V_{EB1}}{V_{EC2} - V_{EC1}}\Bigg|_{I_B = \text{const.}}$$

$$= \frac{0.71 - 0.69}{10 - 1}\Bigg|_{I_B = 30 \mu\text{A}}$$

$$= 0.0022 \text{ (dimensionless)}$$

Determination of h_{ie}
Finally, a tangent line drawn through the Q point yields the values given in Fig. 9.17.

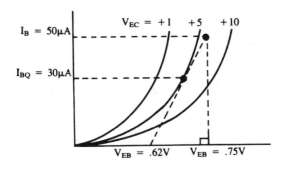

Fig. 9.17

Example (9.5). By drawing a right triangle using the tangent line as the hypotenuse of the triangle, we may determine h_{ie}:

$$h_{11} = \frac{V_1}{I_1}\bigg|_{V_2 = \text{const.}}$$

or for the common emitter configuration:

$$h_{ie} = \frac{\Delta V_{EB}}{\Delta I_B}\bigg|_{V_{EC} = \text{const.}} = \frac{V_{EB2} - V_{EB1}}{I_{B2} - I_{B1}}\bigg|_{V_{EC} = \text{const.}}$$
$$= \frac{0.75 - 0.62 \text{ V}}{0.05 - 0 \text{ mA}}\bigg|_{V_{EC} = +5 \text{ V}}$$

$$h_{ie} = 2.6 \text{ k}\Omega$$

9.1.3 Amplifier Analysis Using Approximation Formulas

Let us analyze the amplifier characteristics of the previous example, but noting that the numerical values of two of the h parameters (h_{re} and h_{oe}) were considerably less than the other parameters. Before we proceed with the analysis, it is desirable to acknowledge that numerous textbooks define the following:

$$\Delta^h = h_i\, h_o - h_f\, h_r$$

For a common emitter connection, an additional subscript e would be added. Thus,

$$\Delta^{he} = h_{ie}\, h_{oe} - h_{fe}\, h_{re}$$

For the previous example,

$$h_{fe} = 100,\ h_{re} = 0.0001,\ h_{oe} = 2(10^{-5})\ \Omega,\ h_{ie} = 3000\ \Omega$$

and, thus

$$\Delta^{he} = 0.05$$

The *voltage gain* was computed as follows:

$$A_V = \frac{-h_{fe}\, Z_L}{h_{ic} + \Delta^{he}\, Z_L} = \frac{-100\ (5000)}{3000 + 0.05\ (5)\ (10^3)} = -153.8$$

Example (9.6). If we had made the assumption that $\Delta^{he} \to 0$, then the approximation formula would be

$$A_V \doteq \frac{-h_{fe}\, Z_L}{h_{ie}} = \frac{-100\ (5000)}{3000} = -167$$

The percent error in calculating the voltage gain is

$$\%\ \text{error}\ A_V = \frac{-153.8 - 166.7}{-153.8} \times 100 = 8.38\%$$

Reviewing the *current gain*:

$$A_I = \frac{h_{fe}}{1 + h_{oe}\, Z_L} = \frac{100}{1 + 2(10^{-5})\ 5(10^3)} = 90.9$$

Example (9.7). By assuming that $h_{oe} \to 0$, we then have

$$A_I \doteq h_{fe} = 100$$

The percentage error in calculating the current gain is

$$\% \text{ error } A_I = \left| \frac{90.9 - 100}{90.9} \right| \times 100$$

$$\% \text{ error } A_I = 10\%$$

Reviewing the *input impedance* (transistor only, not including bias resistors):

$$Z_1 = h_{ie} - \frac{h_{re} h_{fe} Z_L}{1 + h_{oe} Z_L} = 3000 - \frac{10^{-4} (10^2) 5(10^3)}{1 + 2(10^{-5}) 5(10^3)} = 3000 - 45.5$$

$$Z_1 = 2954 \ \Omega$$

Example (9.8). By assuming that $h_{re} \to 0$, we then have

$$Z_1 \doteq 3000 \ \Omega$$

The percentage error in calculating the input impedance is

$$\% \text{ error } Z_1 = \left| \frac{2954 - 3000}{2954} \right| \times 100 = 1.54\%$$

Reviewing the *output impedance* (transistor only, not including collector resistor):

$$Z_2 = \frac{1}{h_{oe} - \dfrac{h_{re} h_{fe}}{h_{ie} + R_S}} = \frac{1}{2(10^{-5}) - \dfrac{(10^{-4}) (10^2)}{3000 + 900}} = 57.35 \text{ k}$$

Example (9.9). By assuming that $h_{re} \to 0$

$$Z_2 \doteq \frac{1}{h_{oe}} = \frac{1}{2(10^{-5})} = 50 \text{ k}\Omega$$

The percentage error in calculating the output impedance is

$$\% \text{ error } Z_2 = \left| \frac{57.35 - 50}{57.35} \right| \times 100 = 12.82 \ \%$$

We note that the greatest error in using the approximation formulas occurs in the computation of Z_2. However, we should also consider the fact that the output impedance of the amplifier stage Z_{out} is the transistor output impedance Z_2 in parallel with R_C:

$$Z_{out} = Z_2 \ /\!/ \ R_C$$

298

Because the collector resistor R_C value is usually considerably less than the resistance value of the transistor output impedance Z_2, a reasonable approximation of the amplifier output resistance is the value of the collector resistor, i.e.,

$$Z_{out} \doteq R_C \tag{9.23}$$

9.2 SMALL SIGNAL ANALYSIS OF COMMON BASE AMPLIFIER

Up to this point, we have developed the small-signal amplifier analysis formulas, applied them to the analysis of a common emitter stage, and defined various circuit parameters. To analyze a common base stage, we would need the h parameter values for a transistor connected in the common base connection. A manufacturer's specification sheet for a particular transistor may contain the h parameter values for a particular connection (e.g., common emitter, but not for common base). Table 9.4 enables us to make the conversion from one set of values to another.

Example (9.10). Let us use these formulas to determine the characteristics of a single stage, common base amplifier employing the same transistor as used in the previous common emitter example.

First, we must determine the common base hybird parameters. Both exact and approximate values are determined so that the reader can see the loss in accuracy in employing approximate formulas. From Table 9.4,

$$h_{ib} = \frac{h_{ie}}{(1 + h_{fe})(1 - h_{re}) + h_{ie} h_{oe}}$$

$$= \frac{h_{ie}}{1 - h_{re} + h_{fe} - h_{fe} h_{re} + h_{ie} h_{oe}}$$

$$= \frac{h_{ie}}{\Delta^{he} + h_{fe} + 1 - h_{re}}$$

$$h_{ib} = \frac{3000}{0.05 + 100 + 1 - 0.0001} = \frac{3000}{101.0499} = 29.688 \ \Omega$$

Noting that $\Delta^{he} \ll 1$ and $h_{re} \ll 1$,

$$h_{ib} \doteq \frac{h_{ie}}{1 + h_{fe}} = \frac{3000}{101} = 29.703 \ \Omega$$

$$h_{rb} = \frac{h_{ie} h_{oe} - h_{re}(1 + h_{fe})}{(1 + h_{fe})(1 - h_{re}) + h_{ie} h_{oe}}$$

$$= \frac{h_{ie}\, h_{oe} - h_{re} - h_{re}\, h_{fe}}{1 - h_{re} + h_{fe} - h_{fe}\, h_{re} + h_{ie}\, h_{oe}}$$

$$h_{rb} = \frac{\Delta^{he} - h_{re}}{\Delta^{he} - h_{re} + 1 + h_{fe}}$$

$$h_{rb} = \frac{0.05 - 0.0001}{0.05 - 0.0001 + 1 + 100} = 0.000494 \text{ (dimensionless)}$$

$$h_{rb} \doteq \frac{h_{ie}\, h_{oe}}{1 + h_{fe}} = -h_{re} = \frac{3000\,(2)\,(10^{-5})}{1 + 100} - 10^{-4}$$

$$h_{rb} = 0.000494 \text{ (dimensionless)}$$

$$h_{fb} = \frac{-h_{fe}\,(1 - h_{re}) - h_{ie}\, h_{oe}}{(1 + h_{fe})(1 - h_{re}) + h_{ie} h_{oe}} = \frac{-h_{fe} + h_{fe}\, h_{re} - h_{ie}\, h_{oe}}{1 - h_{re} + h_{fe} - h_{fe}\, h_{re} + h_{ie}\, h_{oe}}$$

$$= \frac{-h_{fe} - \Delta^{he}}{1 - h_{re} + h_{fe} + \Delta^{he}}$$

$$= \frac{-100 - 0.05}{1 - 10^{-4} + 100 + 0.05}$$

$$= -0.990 \text{ (dimensionless)}$$

$$h_{fb} \doteq \frac{-h_{fe}}{1 + h_{fe}}$$

$$h_{fb} \doteq -0.990 \text{ (dimensionless)}$$

$$h_{ob} = \frac{h_{oe}}{(1 + h_{fe})(1 - h_{re}) + h_{ie}\, h_{oe}} = \frac{h_{oe}}{1 + h_{fe} - h_{re} - h_{fe}\, h_{re} + h_{ie}\, h_{oe}}$$

$$= \frac{h_{oe}}{1 + h_{fe} - h_{re} + \Delta^{he}}$$

$$= \frac{2(10^{-5})}{101 - 0.0001 + 0.05} = 1.979\,(10^{-7})\ \Omega$$

$$h_{ob} \doteq \frac{h_{oe}}{h_{fe} + 1} = \frac{2(10^{-5})}{101} = 1.98\,(10^{-7})\ \Omega$$

Computing the amplifier characteristics for the common base configuration:

$$A_V = \frac{-h_{21}\, Z_L}{h_{11} + (h_{11}\, h_{22} - h_{12}\, h_{21})\, Z_L}$$

$$Z_L = 5(10^3)$$

$$h_{fb} = -0.99$$

$$= \frac{-h_{fb}\, Z_L}{h_{ib} + (h_{ib}\, h_{ob} - h_{rb}\, h_{fb})\, Z_L}$$

$$h_{rb} = 0.000494$$

$$h_{ib} = 29.7\ \Omega$$

$$= \frac{h_{fb}\, Z_L}{h_{ib} + \Delta^{hb}\, Z_L}$$

$$h_{ob} = 1.98\,(10^{-7})\ \Omega$$

Table 9.4

EXACT		APPROXIMATE	
Common-Emitter Configuration		**Common-Emitter Configuration**	

$$h_{ie} = \frac{h_{ib}}{(1 + h_{fb})(1 - h_{rb}) + h_{ob}\, h_{ib}} = h_{ic}$$

$$h_{re} = \frac{h_{ib}\, h_{ob} - h_{rb}\,(1 + h_{fb})}{(1 + h_{fb})(1 - h_{rb}) + h_{ob}\, h_{ib}} = 1 - h_{re}$$

$$h_{fe} = \frac{-h_{fb}\,(1 - h_{rb}) - h_{ob}\, h_{ib}}{(1 + h_{fb})(1 - h_{rb}) + h_{ob}\, h_{ib}} = -(1 + h_{fe})$$

$$h_{oe} = \frac{h_{ob}}{(1 + h_{fb})(1 - h_{rb}) + h_{ob}\, h_{ib}} = h_{oc}$$

$$h_{ie} \cong \frac{h_{ib}}{1 + h_{fb}} \cong \beta\, r_e$$

$$h_{re} \cong \frac{h_{ib}\, h_{ob}}{1 + h_{fb}} - h_{rb}$$

$$h_{fe} \cong \frac{-h_{fb}}{1 + h_{fb}} \cong \beta$$

$$h_{oe} \cong \frac{h_{ob}}{1 + h_{fb}}$$

Common-Base Configuration	Common-Base Configuration

$$h_{ib} = \frac{h_{ie}}{(1+h_{fe})(1-h_{re}) + h_{ie}h_{oe}} = \frac{h_{ie}}{h_{ie}h_{oe} - h_{fe}h_{re}}$$

$$h_{ib} \cong \frac{h_{ie}}{1+h_{fe}} \cong \frac{-h_{ie}}{h_{fe}} \cong r_e$$

$$h_{rb} = \frac{h_{ie}h_{oe} - h_{re}(1+h_{fe})}{(1+h_{fe})(1-h_{re}) + h_{ie}h_{oe}} = \frac{h_{fe}(1-h_{re}) + h_{ie}h_{oe}}{h_{ie}h_{oe} - h_{fe}h_{re}}$$

$$h_{rb} \cong \frac{h_{ie}h_{oe}}{1+h_{fe}} - h_{re} \cong h_{re} - 1 - \frac{h_{ie}h_{oe}}{h_{fe}}$$

$$h_{fb} = \frac{-h_{fe}(1-h_{re}) - h_{ie}h_{oe}}{(1+h_{fe})(1-h_{re}) + h_{ie}h_{oe}} = \frac{h_{re}(1+h_{fe}) - h_{ie}h_{oe}}{h_{ie}h_{oe} - h_{fe}h_{re}}$$

$$h_{fb} \cong \frac{-h_{fe}}{1+h_{fe}} \cong \frac{-(1+h_{fe})}{h_{fe}} \cong -\alpha$$

$$h_{ob} = \frac{h_{oe}}{(1+h_{fe})(1-h_{re}) + h_{ie}h_{oe}} = \frac{h_{oe}}{h_{ie}h_{oe} - h_{fe}h_{re}}$$

$$h_{ob} \cong \frac{h_{oe}}{1+h_{fe}} \cong \frac{-h_{oe}}{h_{fe}}$$

Common-Collector Configuration	Common-Collector Configuration

$$h_{ie} = \frac{h_{ib}}{(1+h_{fb})(1-h_{rb}) + h_{ob}h_{ib}} = h_{ie}$$

$$h_{ie} \cong \frac{h_{ib}}{1+h_{fb}} \cong \beta r_e$$

$$h_{re} = \frac{1+h_{fb}}{(1+h_{fb})(1-h_{rb}) + h_{ob}h_{ib}} = 1 - h_{re}$$

$$h_{re} \cong 1$$

$$h_{fe} = \frac{h_{rb} - 1}{(1+h_{fb})(1-h_{rb}) + h_{ob}h_{ib}} = -(1+h_{fe})$$

$$h_{fe} \cong \frac{-1}{1+h_{fb}} \cong -\beta$$

$$h_{oe} = \frac{h_{ob}}{(1+h_{fb})(1-h_{rb}) + h_{ob}h_{ib}} = h_{oe}$$

$$h_{oe} \cong \frac{h_{ob}}{1+h_{fb}}$$

but $\Delta^{hb} = h_{ib} h_{ob} - h_{rb} h_{fb}$
$$= 29.7 \, (1.98) \, (10^{-7}) - 0.000494 \, (-0.99) = 0.0004949$$

$$A_V = \frac{-(-0.99) \, 5(10^3)}{29.7 + 4.949 \, (5) \, 10^3 \, (10^{-4})} = \frac{4.95(10^3)}{29.7 + 2.4747}$$

$$= \frac{4.95 \, (10^3)}{32.17} = 153.8$$

Note: Common emitter gain was -153.8. Voltage gains are similar in value, but common base has no phase reversal.

$$A_I = \frac{h_{21}}{1 + h_{22} \, Z_L} = \frac{h_{fb}}{1 + h_{ob} \, Z_L} = \frac{-0.99}{1 + 1.98 \, (10^{-7}) \, 5(10^3)}$$

$$= -0.989 \text{ (dimensionless and a phase reversal)}$$

$$Z_1 = h_{11} - \frac{h_{12} \, h_{21} \, Z_L}{1 + h_{22} \, Z_L} = h_{ib} - \frac{h_{fb} \, h_{rb} \, Z_L}{1 + h_{ob} \, Z_L}$$

$$= 29.7 - \frac{-0.99 \, (4.94) \, (10^{-4})(5)(10^3)}{1 + 1.98 \, (10^{-7}) \, 5 \, (10^3)} = 29.7 + 2.443$$

$$Z_1 = 32.1 \, \Omega$$

$$Z_2 = \frac{1}{h_{22} - \dfrac{h_{12} \, h_{21}}{h_{11} + R_S}} = \frac{1}{h_{ob} - \dfrac{h_{fb} \, h_{rb}}{h_{ib} + R_S}}$$

$$= \frac{1}{1.98 \, (10^{-7}) - \dfrac{-0.99 \, (4.94) \, (10^{-4})}{29.7 + 900}} = \frac{1}{1.98 \, (10^{-7}) + 5.26(10^{-7})}$$

$$= 1.38 \, (10^6) \, \Omega$$

Because the source impedance appears to affect the output impedance significantly, let us see what happens if we reduce R_S to reflect the connection of a coaxial cable to the input ($Z_0 = 50 \, \Omega$). Thus, we have

$$R_S = R_1 \, /\!/ \, R_2 \, /\!/ \, R_{gen} = 100 \text{ k} \, /\!/ \, 10 \text{ k} \, /\!/ \, 50 \, \Omega = 49.73 \, \Omega$$

$$Z_2 = \frac{1}{1.98 \, (10^{-7}) - \dfrac{-0.99 \, (4.94) \, (10^{-4})}{29.7 + 49.7}} = \frac{1}{1.98 \, (10^{-7}) + 0.616 \, (10^{-7})}$$

$$= 3.85 \, (10^6) \, \Omega$$

Table 9.5 is provided for comparison purposes.

We can see from Table 9.5 that the voltage gains for both configurations (common emitter and common base) are similar, with the exception of the phase inversion that takes place within the common emitter (see Fig. 9.18 (a,b)).

Table 9.5

CONFIGURATION	A_V	A_I	Z_1	Z_2
Common Emitter	-153.8	$+90.9$	2954 Ω	57.4 K
Common Base	$+153.6$	-0.989	32.1 Ω	1.38 MΩ

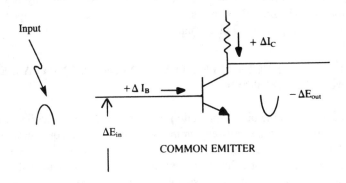

Fig. 9.18(a)

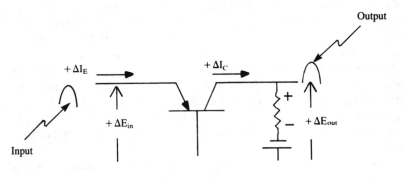

Fig. 9.18(b)

The current gain for the common emitter stage is considerable. Thus, a large power gain will be produced by the common emitter stage:

$$A_P = |A_V| \cdot |A_I| = 153.6(90.9) = 13,962$$

The common base stage, for all practical purposes, has unity current gain. Thus, the power gain for a common base stage is approximately equal to the voltage gain:

$$A_P = |A_V| \cdot |A_I| = 153.6(0.989) = 151.9$$

The input impedance is considerably greater for the common emitter stage. For capacitor coupling of cascaded stages, the common emitter connection more closely approximates the requirements of the maximum power transfer theorem, i.e., the output impedance of one stage is more nearly equal to the input impedance of a succeeding stage.

The low input impedance of the common base connection may have some desirability in RF amplifier circuitry, where the characteristic impedance of coaxial cable is typically 50 or 75 ohms.

9.3 THE COMMON EMITTER STAGE WITH AN UNBYPASSED EMITTER RESISTOR

In Chapter 4, on the bias stabilization, we discussed the unstabilized, emitter stabilized, and universal stabilized common emitter amplifier configurations. The effect of the presence of an emitter resistor was to provide circuit stability by the generation of an error voltage (or current). The error signal tended to oppose a spontaneously arising instability signal. If the instability signal was assumed to be ''positive going'' in nature, the correcting error signal that was ''fed back'' to cancel the error, was ''negative going.'' This stabilizing characteristic is given the special name *negative feedback*. Negative feedback, as incorporated into the dc bias circuitry of a transistor, can affect the ac signal characteristics of a transistor if we design our circuitry to incorporate this feature.

An amplifier stage with an emitter resistor that has a capacitor connected in parallel with it (see Fig. 9.19) may be analyzed from the standpoint of the transistor amplifier characteristics, as though the emitter were connected to ground. We state that the emitter is essentially at ground potential because we may choose a capacitor large enough that the reactance of the capacitor approaches zero ohms.

If the emitter resistor is left unbypassed, the negative feedback considerably modifies the amplifier characteristics. For example, the voltage gain may decrease appreciably. Our initial reaction to this may be that the loss in amplification is undesirable. However, with the loss in amplification, an increase in bandwidth is experienced by the amplifier stage. The increase in bandwidth can generally be interpreted to mean that one single-stage transistor can now amplify higher frequencies. A video amplifier in a TV receiver or TV camera must often be able to amplify, non-preferentially, signals from very low frequencies up to six megahertz. This frequency response would not be possible without negative feedback. An amplifier with negative feedback will also add less distortion to the amplified signal.

At this point, we may think that it might be desirable to sacrifice a small amount of amplification for a small increase in bandwidth. How can we accomplish this? To answer, let us examine the circuit in Fig. 9.19.

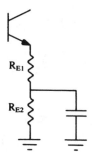

Fig. 9.19

We note that there are two resistors in series in the emitter circuit, R_{E1} and R_{E2}. The required dc bias emitter resistor value is simply the numerical value of R_{E1} added to R_{E2}. Unbypassed R_{E1}, on the other hand, provides the desired increase in bandwidth (and attendant loss in voltage gain).

The following are the analysis formulas for the common emitter amplifier with unbypassed emitter resistor. Because the derivations for A_V, A_I, Z_1, Z_2, are lengthy, only the beginning and end of each derivation are shown below. The complete derivations are provided in Appendix A.

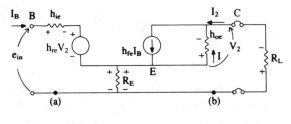

Fig. 9.20(a)   *Fig. 9.20(b)*

Common Emitter Analysis (with unbypassed emitter resistor)

Determine Z_1. Using Kirchoff's voltage law around the input loop:

at (a) c.c.w. $I_B R_E + I_2 R_E + h_{re} V_2 + I_B h_{ie} - e_{in} = 0$
at (b) c.w. $I_B R_E + I_2 R_E + V_2 + I_2 R_L = 0$

Using Kirchoff's current law at node E:

$$h_{fe} I_B = I_2 + I \quad \text{and} \quad V_2 = -\frac{I}{h_{oe}}$$

Solving, we have (see Appendix A for remainder of derivation):

$$Z_1 = h_{ie} + \left\{ \frac{-h_{fe} h_{re} Z_L + R_E [(1 + h_{fe})(1 - h_{re}) + h_{oe} Z_L]}{1 + h_{oe}(R_E + Z_L)} \right\}$$

(9.24)

If we let $R_E = 0$, then we have

$$Z_1 = h_{ie} - \frac{h_{fe} h_{re} Z_L}{1 + h_{oe} Z_L}$$

and we note that this is the equation derived for the common emmiter amplifier without an emitter resistor (i.e., $R_E = 0$) or a totally bypassed resistor via a large capacitor.

Example (9.11). Let us use the h parameter and resistor values used in the previous example, and assume $R_{E1} = 200 \; \Omega$:

$$Z_1 = h_{ie} + \left\{ \frac{-h_{fe} h_{re} Z_L + R_E [(1 + h_{fe})(1 - h_{re}) + h_{oe} Z_L]}{1 + h_{oe}(R_E + Z_L)} \right\}$$

$$= 3000 + \left\{ \frac{-100 \, (10^{-4}) \, 3(10^3) + 200 \, [(1 + 100)(1 - 10^{-4}) + 2(10^{-5}) \, 3(10^3)]}{1 + 2(10^{-5})(0.2 + 5) \, 10^3} \right\}$$

$$= 3000 + 18{,}270$$

$$Z_1 = 21{,}270 \; \Omega$$

We can see that the input impedance has increased approximately seven times by providing an unbypassed emitter resistor. If we had a high impedance signal source, which we did not wish to load, an unbypassed emitter resistor would accomplish this goal.

Because formula (9.24) is quite involved, let us make some assumptions in order to simplify its use.

Example (9.12). Let $h_{oe} = h_{re} = 0$. Thus,

$$Z_1 \doteq h_{ie} + R_E (1 + h_{fe})$$

Computing Z_1, we have

$Z_1 \doteq 3000 + 200 \,(101)$

$Z_1 \doteq 23{,}200 \; \Omega$

Let us investigate the amount of error in using the above approximation:

$$\% \text{ error} = \left| \frac{21{,}270 - 23{,}200}{21{,}270} \right| \times 100 = 9.1\%$$

We could further simplify the formula by assuming that $h_{fe} \gg 1$. Thus,

Therefore, $Z_1 \doteq h_{ie} + h_{fe}\,R_E = 3000 + 100\,(20)$

$$\doteq 23{,}000$$

$$\% \text{ error} = \left| \frac{21{,}270 - 23{,}000}{21{,}270} \right| \times 100 = 8.1\%$$

The formula for A_I is (see Appendix A for complete derivation):

$$A_I = \frac{h_{fe} - h_{oe}\,R_E}{1 + h_{oe}\,(Z_L + R_E)}$$

and, if $R_E = 0$, this reduces to

$$A_I = \frac{h_{fe}}{1 + h_{oe}\,Z_L}$$

which checks against the formula derived for the common emitter amplifier without an emitter resistor.

Example (9.13). Computing the value of A_I, we have

$$A_I = \frac{h_{fe} - h_{oe}\,R_E}{1 + h_{oe}\,(Z_L + R_E)} = \frac{100 - 2(10^{-5})\,200}{1 + 2(10^{-5})\,(5.2)\,(10^3)}$$
$$= 90.57$$

For the common emitter amplifier with no emitter resistor (or capacitor bypassed), we determined the current gain to be

$$A_I = 90.9$$

There is little difference between the two values.

The voltage gain formula is (see Appendix A for complete derivation):

$$A_V = \frac{-(h_{fe} - h_{oe} R_E) Z_L}{h_{ie} + \Delta^{he} (Z_L + R_E) + R_E (h_{fe} + 1 + h_{oe} Z_L - h_{re})}$$

Example (9.14). Computing the voltage gain for an unbypassed common emitter stage:

$$A_V = \frac{-[100 - 2(10^{-5}) 200] 5000}{3000 + 0.05 (5000 + 200) + 200 [101 + 2(10^{-5}) (5000) - 10^{-4}]}$$
$$= -21.29$$

If h_{oe} and h_{re} are assumed to be zero, then we have

$$A_V \doteq \frac{-h_{fe} Z_L}{h_{ie} + (h_{fe} + 1) R_E}$$

Thus,

$$A_V \doteq \frac{-100(5000)}{3000 + 101 (200)} = -21.55$$

The percentage error in using the approximation is

$$\% \text{ Error } A_V = \left| \frac{21.29 - 21.55}{21.29} \right| \times 100 = 1.22\%$$

Examining the approximation formula for A_V:

$$A_V \doteq \frac{-h_{fe} Z_L}{h_{ie} + (h_{fe} + 1) Z_E}$$

if we assume that

$$(h_{fe} + 1) R_E \gg h_{ie}$$

we have

$$A_V \doteq \frac{-h_{fe} Z_L}{(h_{fe} + 1) Z_L}$$

If we further assume that

$$h_{fe} \gg 1$$

then we have

$$A_V \doteq \frac{-h_{fe} Z_L}{h_{fe} Z_E} = \frac{Z_L}{Z_E} \tag{9.25}$$

The error in making the above assumptions can be as great as 30%. However, if a quick, "ball-park" value for the voltage gain is desired, the use of (9.25) may be appropriate.

Example (9.15). Determine the approximate voltage gain of the common emitter circuit shown in Fig. 9.21. Thus, we obtain

$$A_V \doteq \frac{R_C}{R_E} = \frac{1000}{200} = 5$$

$$A_{V\ exact} = 4.44$$

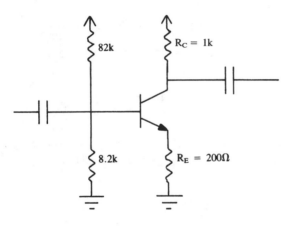

Fig. 9.21

9.4 AN EXPERIMENTAL METHOD FOR PREDICTING BIPOLAR TRANSISTOR AMPLIFIER PERFORMANCE

Recalling that the voltage gain formula for a common emitter amplifier with no emitter resistor or a bypassed emitter resistor was

$$A_V \doteq \frac{h_{fe} Z_L}{h_{ie}}$$

By examining a circuit we would know the value of Z_L, but we would have no idea of the values of h_{fe} and h_{ie}. Let us look at a typical common emitter amplifier (universally stabilized), which we examined in Chapter 4 on bias stabilization, as shown in Fig. 9.22.

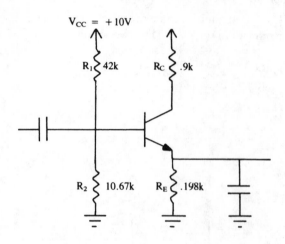

Fig. 9.22

Assuming that we could measure the voltage across the collector resistor and we determined that voltage to be $V_{RC} = 4.5$ V. Recall that

$$I_B = \frac{V_{CC} - V_{EB}\left(1 + \dfrac{R_1}{R_2}\right)}{R_1 + (\beta + 1) R_E \left(1 + \dfrac{R_1}{R_2}\right)}$$

and since, $I_B = \dfrac{I_C}{\beta}$

$$\frac{I_C}{\beta} = \frac{V_{CC} - V_{EB}\left(1 + \dfrac{R_1}{R_2}\right)}{R_1 + (\beta + 1) R_E \left(1 + \dfrac{R_1}{R_2}\right)}$$

and also, $V_{RC} = I_C R_C$

Therefore, $I_C = \dfrac{V_{RC}}{R_C}$

$$\frac{V_{RC}}{\beta R_C} = \frac{V_{CC} - V_{EB}\left(1 + \dfrac{R_1}{R_2}\right)}{R_1 + (\beta + 1) R_E \left(1 + \dfrac{R_1}{R_2}\right)}$$

Solving for β, we obtain

$$\beta = \frac{V_{RC}\left[R_1 + R_E\left(1 + \dfrac{R_1}{R_2}\right)\right]}{R_C V_{CC} - (R_C V_{EB} + V_{RC} R_E)\left(1 + \dfrac{R_1}{R_2}\right)}$$

Example (9.16). For the schematic shown in Fig. 9.22, we have

$$\beta_{dc} = \frac{4.5\,[42\,k + 0.2\,k\left(1 + \dfrac{42\,k}{10.67\,k}\right)}{0.9\,k(10) - [0.9\,k(0.6) + 4.5(0.2\,k)]\left(1 + \dfrac{42\,k}{10.67\,k}\right)} = 100$$

If we assume that the ac β is approximately equal to the dc β, i.e.,

$h_{fe} \doteq h_{fe}\,(= \beta_{dc})$ then $h_{fe} \doteq 100$

Now the only term in our voltage gain formula that remains to be determined is h_{ie}. It can be demonstrated by solid state physics that

$$h_{ie} \doteq h_{fe}\frac{26\,mV}{I_E}$$

for our particular example:

$$I_C = \frac{V_{RC}}{R_C} = \frac{4.5}{0.9\,k} = 5\,mA$$

If we assume that $I_E \doteq I_C$, then

$$h_{ie} = 100 \frac{26\,\text{mV}}{5\,\text{mA}} = 520\,\Omega$$

Therefore,

$$A_V \doteq \frac{h_{fe} Z_L}{h_{ie}} = \frac{100\,(0.9\,\text{k})}{0.52\,\text{k}} = 173$$

In addition, we can also predict that

$$Z_1 \doteq h_{ie} \doteq 520\,\Omega$$

and

$$Z_{in} = R_1 \mathbin{/\!/} R_2 \mathbin{/\!/} Z_1 = 42\,\text{k} \mathbin{/\!/} 10.68\,\text{k} \mathbin{/\!/} 0.52\,\text{k}$$
$$Z_{in} = .49\,\text{k}$$

Also,

$$Z_{out} \doteq R_C = 0.9\,\text{k}$$

and

$$A_I \doteq h_{fe} = 100$$

Thus, with a single dc voltage measurement, we have been able to predict most of the ac amplifier characteristics!

The voltage gain computed for the common emitter amplifier without an emitter resistor (or R_E bypassed) was

$$A_V = -153.8$$

We can see that there is a considerable loss in voltage gain with the insertion of a 200 Ω emitter resistor. (See Appendix A for complete derivation).

Computing Z_2:

$$Z_2 = \frac{h_{ie} + R_S + R_E(1 + h_{fe} + h_{oe} R_S + \Delta^{he} - h_{re})}{\Delta^{he} + h_{oe}(R_S + R_E)}$$
$$Z_2 = \frac{3000 + 900 + 200\,(1 + 100 + 2(10^{-5})\,900 + 0.05 - 10^{-4})}{0.05 + 2(10^{-5})\,(900 + 200)}$$
$$= 335{,}000\,\Omega$$

For approximation purposes, let us see what happens when we assume that

$h_{re}, h_{oe}, \Delta^{he} \to 0.$

Thus,

$$Z_2 \doteq \frac{h_{ie} + R_S + R_E(1 + h_{ie})}{0} \to \infty$$

Recalling that our previous computation for the output impedance (without an unbypassed emitter resistor) was 57,353.56, we can conclude that *an unbypassed emitter resistor increases the output impedance.*

Let us examine the output impedance formula when $R_E = 0$:

$$Z_2 = \frac{h_{ie} + R_S + R_E(1 + h_{fe} + h_{oe}R_S + \Delta^{he} - h_{re})}{\Delta^{he} + h_{oe}(R_S + R_E)}$$

at $R_E = 0$

$$Z_2 = \frac{h_{ie} + R_S}{\Delta^{he} + h_{oe}R_S} = \frac{1}{\dfrac{h_{ie}h_{oe} - h_{fe}h_{re} + h_{oe}R_S}{h_{ie} + R_S}}$$

$$= \frac{1}{\dfrac{h_{oe}(h_{ie} + R_S)}{h_{ie} + R_S} - \dfrac{h_{fe}h_{re}}{h_{ie} + R_S}} = \frac{1}{h_{oe} - \dfrac{h_{fe}h_{re}}{h_{ie} + R_S}}$$

which we note is the formula derived for the common emitter circuit without an emitter resistor (or an emitter resistor that has a parallel connected bypass capacitor).

Table 9.6 enables the reader to compare the electrical characteristics of the common emitter amplifier circuit with and without an emitter resistor.

where

$h_{oe} = 2(10^{-5})$
$h_{fe} = 100$
$h_{re} = 10^{-4}$
$h_{ie} = 3000$
$R_S = 900$
$Z_L = 5000$

Table 9.6

R_E	A_V	A_I	Z_1	Z_2
0 Ω	−153.8	90.9	2,954	57,350 Ω
200 Ω	−21.9	90.57	21,270	334,900 Ω

9.4.1 The Emitter Follower (Common Collector Circuit)

As we learned during our study of dc electricity, if maximum power is to be transferred from a source to a load, the resistance value of the load impedance must be equal to the resistance value of the source impedance (maximum power transfer theorem).

One method of accomplishing impedance matching is by employing transformers. A difficulty that arises in the use of transformers is that economically priced units have limited frequency response.

Also, for optimum operation, coaxial cables that interconnect a signal source to a load and carry very high frequency signals should have signal source and load impedances that are equal to the characteristic impedance of the cable. If all impedances are not equal, high frequency signals will be reflected in the coaxial line. As an example, a TV antenna connected to a TV receiver that is mismatched to the 300 ohms down-lead, could cause a "ghost" in the picture.

A common transistor circuit, which is employed where a high impedance source is converted to a low impedance source, is the emitter follower (common collector circuit). Instead of taking the output signal from the collector as is done in the common emitter and common base circuits, the output signal is taken from the emitter. This circuit has the properties of supplying a considerable current gain, approximately unity voltage gain, a high input impedance, and low output impedance.

With a low output impedance, the emitter follower is a convenient circuit for feeding a signal to a coaxial cable.

At this point, it is appropriate to define the electrical parameters, which were not defined previously for this particular circuit, Z_3, Z_{out}.

To determine Z_3, "stand" at the emitter, "look" toward the transistor, and determine the impedance measured to ground. Z_3 will be comprised of h parameters and R_S (previously defined). Ignore all resistances "behind you." The impedance "behind you" would consist of R_E in parallel with the impedance "seen" when "looking into" the coaxial cable (usually 50 or 75 ohms). To determine Z_E, "stand" at the emitter, "look away" from the transistor, and measure all impedances to ground (R_E in parallel with the coaxial cable input impedance). Ignore all impedances "behind you," i.e., Z_3.

Biasing of the emitter follower circuit was covered in a previous chapter. Therefore, only the ac parameters are of interest to us at this point. We will want to determine the input impedance of the transistor, Z_1, excluding the effect of the bias resistors. After determining this value, we may add in parallel either R_B or R_1 and R_2 to determine the total input impedance, Z_{in}. See Fig. 9.24 (a,b).

We should also want to determine the formulas for the voltage gain A_V, current gain A_I, and output impedance Z_3.

We note that the input impedance formula for the emitter follower, should be the same as the formula for the common emitter circuit (with emitter resistor)

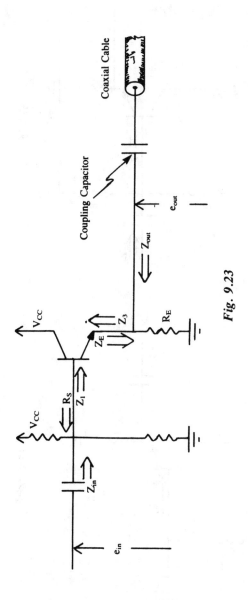

Fig. 9.23

316

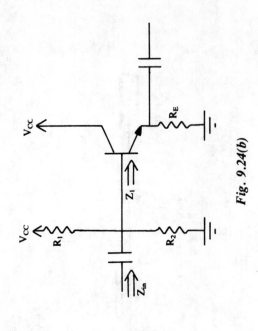

Fig. 9.24(b)

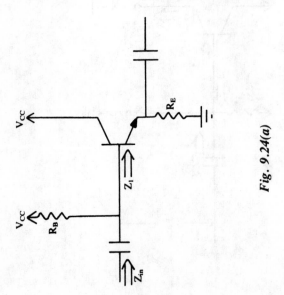

Fig. 9.24(a)

with the exception that $Z_L = 0$. Z_L is equal to zero ohms because the collector is connected directly to V_{CC} and we know that V_{CC} is at ac ground potential.

Thus, for the common emitter circuit (with emitter resistor):

$$Z_1 = h_{ie} + \left\{ \frac{-h_{fe} h_{re} Z_L + R_E \left[(1 + h_{fe}) (1 - h_{re}) + h_{oe} Z_L \right]}{1 + h_{oe} [R_E + Z_L]} \right\}$$

For the emitter follower, $Z_L = 0$ and R_E must be replaced with Z_E. Therefore, we have

$$Z_1 = h_{ie} + \left\{ \frac{Z_E \left[(1 + h_{fe}) (1 - h_{re}) \right]}{1 + h_{oe} Z_L} \right\} \tag{9.26}$$

Using Figs. 9.25 and 9.26, let us develop the derivation of the input impedance and see if it agrees with the above formula (9.26).

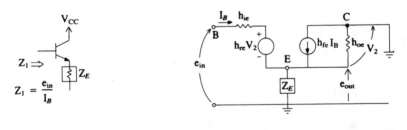

Fig. 9.25   **Fig. 9.26**

Redrawing the circuit, we have the configuration of Fig. 9.27.

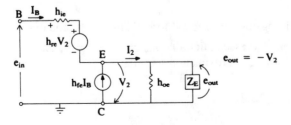

Fig. 9.27

Employing Kirchoff's voltage law around the input loop: at C c.c.w.

$$e_{out} + h_{re} V_2 + I_B h_{ie} - e_{in} = 0$$
$$e_{out} - h_{re} e_{out} + I_B h_{ie} = e_{in}$$
$$e_{out} (1 - h_{re}) + I_B h_{ie} = e_{in} \qquad (9.27)$$

Employing Kirchoff's current law at the emitter node, E:

$$I_2 = h_{fe} I_B + I_B = I_B (h_{fe} + 1)$$

Also by employing Ohm's law:

$$e_{out} = \left(\frac{\dfrac{Z_E}{h_{oe}}}{\dfrac{1}{h_{oe}} + Z_E} \right) I_2$$

$$e_{out} = \frac{\dfrac{Z_E}{h_{oe}}}{\dfrac{1}{h_{oe}} + Z_E} (I_B (h_{fe} + 1)) \qquad (9.28)$$

Using (9.27) and (9.28):

$$\frac{Z_E I_B (h_{fe} + 1) (1 - h_{re})}{1 + Z_E h_{oe}} + I_B h_{ie} = e_{in}$$

$$e_{in} = I_B h_{ie} + I_B \frac{(h_{fe} + 1) (1 - h_{re})}{1 + h_{oe} Z_E}$$

Therefore, $\dfrac{e_{in}}{I_B} = Z_1 = h_{ie} + \dfrac{(h_{fe} + 1) (1 - h_{re})}{1 + h_{oe} Z_E}$ $\qquad (9.29)$

which checks against our original formula (using the common emitter formula).

If we make the assumption that $h_{oe} \doteq 0$, $h_{re} \doteq 0$, and $h_{fe} >> 1$, then we have the following approximation formula:

$$Z_1 \doteq h_{ie} + h_{fe} Z_E \qquad (9.30)$$

In some instances, if the transistor ac β is a large value ($h_{fe} > 100$), Z_E is appreciable, and h_{ie} is a small value, then a further approximation of the emitter follower input impedance is given by neglecting h_{ie}. Thus,

$$Z_1 \doteq h_{fe} Z_E \tag{9.31}$$

Example (9.18). From formula (9.31), we see that the emitter follower acts like a transformer, "stepping-up" the load impedance Z_E (which includes the emitter bias resistor) by the β (ac) of the transistor. As an example, suppose we had an emitter follower with an emitter resistor of 1,000 Ω and a 50 Ω coaxial cable connected to the emitter, as shown in Fig. 9.28.

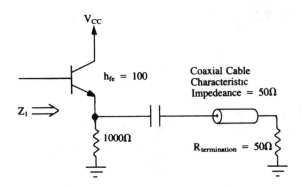

Fig. 9.28

Thus,

$$Z_1 \doteq h_{fe} Z_E = 100(1000 /\!/ 50) \doteq 100(50) = 5000 \ \Omega$$

The Emitter Follower Current Gain

Redrawing Fig. 9.29(a,b), we obtain Fig. 9.30, where $e_0 = -V_2$.
Employing Kirchoff's voltage law around the input loop at node C (c.c.w.):

$$I_o Z_E + h_{re} V_2 + I_B h_{ie} - e_{in} = 0$$
$$I_o Z_E - h_{re} e_{out} + I_B h_{ie} - e_{in} = 0$$
$$I_o Z_E - h_{re} (I_o Z_E) + I_B h_{ie} - I_B Z_1 = 0$$
$$I_o Z_E - I_o h_{re} Z_E + I_B h_{ie} - I_B \left[h_{ie} + \frac{(h_{fe} + 1)(1 - h_{re}) Z_E}{1 + h_{oe} Z_E} \right] = 0$$
$$I_o Z_E (1 - h_{re}) + I_B \left[h_{ie} - h_{ie} + \frac{(h_{fe} + 1)(1 - h_{re}) Z_E}{1 + h_{oe} Z_E} \right] = 0$$
$$I_o Z_E (1 - h_{re}) = I_B \left[- \frac{(h_{fe} + 1)(1 - h_{re}) Z_E}{1 + h_{oe} Z_E} \right]$$
$$\frac{I_o}{I_B} = A_I = \frac{h_{fe} + 1}{1 + h_{oe} Z_E} \tag{9.32}$$

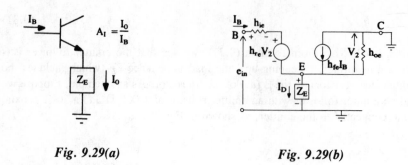

Fig. 9.29(a)   Fig. 9.29(b)

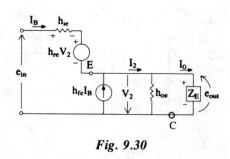

Fig. 9.30

If we make the assumption $h_{oe} \doteq 0$, then we have

$$A_I \doteq h_{fe} + 1 \tag{9.33}$$

If we further assume that $h_{fe} \gg 1$, then we have

$$A_I \doteq h_{fe} \tag{9.34}$$

Emitter Follower Voltage Gain

From Fig. 9.31, we would expect that the output voltage (ac) would be very similar to the input voltage because the base-to-emitter junction is a forward-biased diode, i.e., if the voltage at the base should increase by a small amount Δe_{in}, then the emitter voltage should "track" the base and increase by almost (but not quite) the same amount. Therefore, the numerical value for the voltage gain should be almost unity (see Fig. 9.31(a,b)).

Employing Kirchoff's voltage law around the input loop at C c.c.w.

$$e_{out} + h_{re} V_2 + I_B h_{ie} - e_{in} = 0$$
$$e_{out} - h_{re} e_{out} + I_B h_{ie} = e_{in}$$

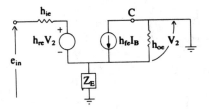

Fig. 9.31(a)

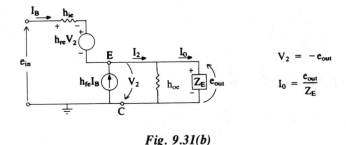

Fig. 9.31(b)

$$V_2 = -e_{out}$$

$$I_0 = \frac{e_{out}}{Z_E}$$

Using the current divider rule:

$$I_o = \frac{\dfrac{1}{h_{oe}}}{\dfrac{1}{h_{oe}} + Z_E} I_2$$

Employing Kirchoff's current law at node E

$$I_2 = h_{fe} I_B + I_B = I_B (h_{fe} + 1)$$

Combining equations:

$$I_o = \frac{\dfrac{1}{h_{oe}}}{\dfrac{1}{h_{oe}} + Z_E} I_B (h_{fe} + 1)$$

Therefore, $I_B = \dfrac{I_o\left(\dfrac{1}{h_{oe}} + Z_E\right)}{\dfrac{1}{h_{oe}}(h_{fe} + 1)}$

or $I_B = \dfrac{e_{out}\left(\dfrac{1}{h_{oe}} + Z_E\right)}{Z_E\,\dfrac{1}{h_{oe}}(h_{fe} + 1)}$

$ = \dfrac{e_{out}(1 + h_{oe}\,Z_E)}{(h_{fe} + 1)\,Z_E}$

Substituting the value of I_B determined on the right:

$$e_{out}(1 - h_{re}) + \frac{e_{out}(1 + h_{oe}\,Z_E)\,h_{ie}}{(h_{fe} + 1)\,Z_E} = e_{in}$$

$$e_{out}\left[1 - h_{re} + \frac{1 + h_{oe}\,Z_E)\,h_{ie}}{(h_{fe} + 1)\,Z_E}\right] = e_{in}$$

$$\frac{e_{out}}{e_{in}} = \frac{1}{1 - h_{re} + \dfrac{(1 + h_{oe}\,Z_E)\,h_{ie}}{(h_{fe} + 1)\,Z_E}}$$

$$\frac{e_{out}}{e_{in}} = \frac{1}{\dfrac{(h_{fe} + 1)(1 - h_{re})\,Z_E + (1 + h_{oe}\,Z_E)\,h_{ie}}{(h_{fe} + 1)\,Z_E}}$$

$$= \frac{1}{\dfrac{(h_{fe} - h_{fe}\,h_{re} + 1 - h_{re})\,Z_E + h_{ie} + h_{oe}\,h_{ie}\,Z_E}{(h_{fe} + 1)\,Z_E}}$$

$$= \frac{1}{\dfrac{((h_{fe} + 1) + h_{ie}\,h_{oe} - h_{fe}\,h_{re} - h_{re})\,Z_E + h_{ie}}{(h_{fe} + 1)\,Z_E}}$$

$$= \frac{1}{\dfrac{(h_{fe} + 1)\,Z_E}{(h_{fe} + 1)\,Z_E} + \dfrac{\Delta^{he}\,Z_E - h_{re}\,Z_E}{(h_{fe} + 1)\,Z_E} + \dfrac{h_{ie}}{(h_{fe} + 1)\,Z_E}}$$

$$A_V = \frac{1}{1 + \dfrac{\Delta^{he} - h_{re}}{h_{fe} + 1} + \dfrac{h_{ie}}{(h_{fe} + 1)\,Z_E}} \tag{9.35}$$

The numerical value of the terms,

$$\frac{\Delta^{he} - h_{re}}{h_{fe} + 1} \quad \text{and} \quad \frac{h_{ie}}{(h_{fe} + 1) Z_E}$$

are both small compared to unity, so we can see that the voltage gain approaches unity.

If we assume that $h_{oe} \doteq 0$ and $h_{re} \doteq 0$, then we have

$$A_V \doteq \frac{1}{1 + \dfrac{h_{ie}}{(h_{fe} + 1) Z_E}} \tag{9.36}$$

Emitter Follower Output Impedance

We must recall that to find the Thevenin equivalent circuit for a complex circuit with voltage and current sources in addition to series and parallel connected resistors, we shorted all voltage sources and opened all current sources.

Thus, to determine the output impedance of an emitter follower, we must short out the signal source connected to the transistor base. We *must*, however, keep the generator impedance connected to ground (see Fig. 9.32(a,b,c)).

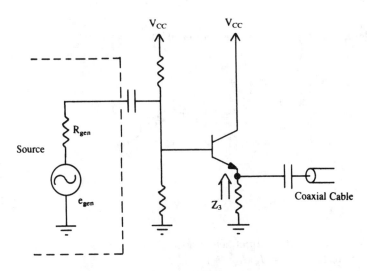

Fig. 9.32(a)

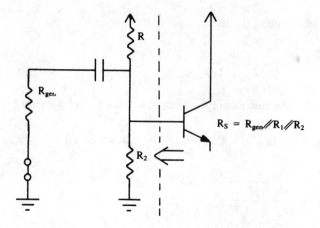

Fig. 9.32(b)

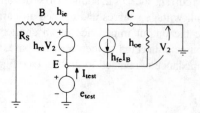

Fig. 9.32(c)

Redrawing Fig. 9.32, we obtain Fig. 9.33.

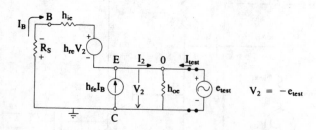

Fig. 9.33

Employing Kirchoff's current law at node E:

$$I_2 = h_{fe} I_B + I_B = I_B (h_{fe} + 1)$$

also at node O:

$$I_2 + I_{test} = e_{test} \, h_{oe}$$

Therefore, $I_B \, (h_{fe} + 1) + I_{test} = e_{test} \, h_{oe}$

$$I_B \, (h_{fe} + 1) = e_{test} \, h_{oe} - I_{test}$$

$$I_B = \frac{e_{test} \, h_{oe} - I_{test}}{h_{fe} + 1}$$

Employing Kirchoff's voltage law around the input loop: at C c.c.w.

$$e_{test} + h_{re} \, V_2 + I_B \, h_{ie} + I_B \, R_S = 0$$

$$e_{test} - h_{re} \, e_{test} + I_B \, (h_{ie} + R_S) = 0$$

$$e_{test} \, (1 - h_{re}) + \left[\frac{e_{test} \, h_{oe} - I_{test}}{h_{fe} + 1} \right] (h_{ie} + R_S) = 0$$

$$e_{test} \, (h_{fe} + 1) \, (1 - h_{re}) + (e_{test} \, h_{oe} - I_{test}) \, (h_{ie} + R_S) = 0$$

$$e_{test} \, (h_{fe} + 1) \, (1 - h_{re}) + e_{test} \, h_{oe} \, (h_{ie} + R_S) = I_{test} \, (h_{ie} + R_S)$$

$$e_{test} \, [(h_{fe} + 1) \, (1 - h_{re}) + h_{oe} \, (h_{ie} + R_S)] = I_{test} \, (h_{ie} + R_S)$$

Thus,

$$\frac{e_{test}}{I_{test}} = Z_3 = \frac{h_{ie} + R_S}{(h_{fe} + 1) \, (1 - h_{re}) + h_{oe} \, (h_{ie} + R_S)}$$

$$= \frac{h_{ie} + R_S}{h_{fe} - h_{fe} \, h_{re} + 1 - h_{re} + h_{oe} \, h_{ie} + h_{oe} \, R_S}$$

Therefore,

$$Z_3 = \frac{h_{ie} + R_S}{h_{fe} + 1 + \Delta^{he} - h_{re} + h_{oe} \, R_S} \tag{9.37}$$

If we assume that Δ^{he}, h_{re} and $h_{oe} \to 0$, then

$$Z_3 = \frac{h_{ie} + R_S}{h_{fe} + 1 + \Delta^{he} - h_{re} + h_{oe} \, R_S}$$

reduces to

$$Z_3 \doteq \frac{h_{ie} + R_S}{h_{fe} + 1} \tag{9.38}$$

and if the transistor β is much greater than unity:

$$Z_3 \doteq \frac{h_{ie} + R_S}{h_{fe}} \tag{9.39}$$

Thus, we can see that for approximation purposes we can determine the transistor output impedance by dividing the source resistance and h_{ie} by the transistor β.

Let us determine some typical values for the electrical characteristics of an emitter follower stage.

Example (9.19). From the previous examples,

$$\Delta^{he} = 0.05, R_S = 900, R_E = 200, h_{oe} = 2(10^{-5}), h_{ie} = 3000, h_{re} = 10^{-4},$$
$$h_{fe} = 100$$

Let us also assume that there is a 50 Ω coaxial cable connected to the emitter. For this circumstance,

$$Z_E = R_E \,/\!/\, 50 \ \Omega \ (\text{coax})$$
$$= 200 \,/\!/\, 50 = 40 \ \Omega$$

$$A_V = \cfrac{1}{1 + \cfrac{\Delta^{he} - h_{re}}{h_{fe} + 1} + \cfrac{h_{ie}}{(h_{fe} + 1)\, Z_E}}$$

$$= \cfrac{1}{1 + \cfrac{0.05 - 0.0001}{101} + \cfrac{3000}{101\,(40)}} = \cfrac{1}{1 + \cfrac{0.0499}{101} + \cfrac{3000}{4040}}$$

$$A_V = 0.574$$

or by using the approximation formula:

$$A_V \doteq \cfrac{1}{1 + \cfrac{h_{ie}}{h_{fe}\, Z_E}} = \cfrac{1}{1 + \cfrac{3000}{100(40)}} = 0.571$$

Computing the current gain A_I, we have

$$A_I = \frac{h_{fe} + 1}{1 + h_{oe}\, Z_E} = \frac{101}{1 + 2(10^{-5})40} = \frac{101}{1 + 8(10^{-5})} = 100.92$$

or by use of the approximation formula:

$$A_I \doteq h_{fe} + 1 = 101$$

Computing the input impedance Z_1, we have

$$Z_1 = h_{ie} = \frac{(h_{fe} + 1)\,(1 - h_{re})\, Z_E}{1 + h_{oe}\, Z_E} = 3000 + \frac{101\,(1 - 10^{-4})\,40}{1 + 40(2)\,(10^{-5})}$$
$$Z_1 = 7036 \ \Omega$$

or by using the approximation formula:

$$Z_1 \doteq h_{ie} + h_{fe} Z_E = 3000 + 100(40) = 7000$$

Computing the output impedance Z_3, we have

$$Z_3 = \frac{h_{ie} + R_S}{h_{fe} + 1 + \Delta^{he} - h_{re} + h_{oe} R_S}$$
$$= \frac{3000 + 900}{101 + 0.05 - 0.0001 + 2(10^{-5}) \, 900}$$
$$Z_3 = 38.6 \, \Omega$$

or by using the approximation formula:

$$Z_3 \doteq \frac{h_{ie} + R_S}{h_{fe}} = \frac{3000 + 900}{100} = 39 \, \Omega$$

Computed emitter follower characteristics are given in Table 9.7.

Table 9.7

A_V	A_I	Z_1	Z_3
0.574	100.9	7036 Ω	38.6 Ω

Summarizing, we can see that the emitter follower has a fairly high input impedance, even with a low impedance load (50 Ω) connected to its output. It has a low impedance output (38.6 Ω), which nearly matches the characteristic impedance of the load (50 Ω coaxial cable). There is a slight loss of the signal voltage from passing through the stage (57% of input is present at output terminals), and the signal current is amplified by an amount approximately equal to the transistor β ($\beta = 100$).

9.4.2 Common Emitter with Collector Feedback

In the chapters dealing with dc bias (Ch. 3–4), we noted the effect on stability of connecting a resistor from collector to base. Let us see what this connection does to the ac characteristics of the transistor.

Due to the length of the derivations for A_V, A_I, Z_1 and Z_2, only the initial steps or the final result for each amplifier characteristic will be presented in the main body of the text. For a complete derivation, the reader is referred to Appendix A.

328

First, let us determine the input impedance, Z_1 (see Fig. 9.34):

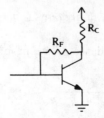

Fig. 9.34(a)

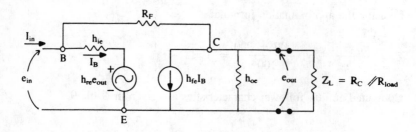

Fig. 9.34(b)

Employing Kirchoff's current law at node B:

$$-I_{in} + \frac{e_{in} - e_{out}}{R_F} + I_B = 0$$

and at node C:

$$h_{fe} I_B + \frac{e_{out} - e_{in}}{R_F} + e_{out} h_{oe} + \frac{e_{out}}{Z_L}$$

Employing Kirchoff's voltage law around the input loop at E c.c.w.

$$h_{re} e_{out} + I_B h_{ie} - e_{in} = 0$$

Solving these equations simultaneously for e_{in}/I_{in} we obtain

$$Z_1 = \frac{h_{ie}(R_F + Z_L) + \Delta^{he} R_F Z_L}{h_{ie} + \Delta^{he} Z_L + Z_L(h_{fe} + 1 - h_{re}) + R_F(1 + h_{oe} Z_L)} \qquad (9.40)$$

If we wish to employ a less involved version of the formula (9.40), we can assume that $h_{oe} \rightarrow 0$ and $h_{re} \rightarrow 0$,

$$Z_1 \doteq \frac{h_{ie}(Z_L + R_F)}{h_{ie} + Z_L(h_{fe} + 1) + R_F} \tag{9.41}$$

The equation for the voltage gain is

$$A_V = \frac{-\left(h_{fe} - \dfrac{h_{ie}}{R_F}\right)}{h_{ie}\left(\dfrac{1}{R_F} + \dfrac{1}{Z_L}\right) + \Delta^{he}} = \frac{-\left(h_{fe} - \dfrac{h_{ie}}{R_F}\right)Z_L}{h_{ie}\left(1 + \dfrac{Z_L}{R_F}\right) + \Delta^{he} Z_L} \tag{9.42}$$

As a check on the correctness of formula (9.42), we can see that as R_F tends toward infinity, the formula should approach that for the simple common emitter, or

$$A_V = \frac{-(h_{fe} - 0)Z_L}{h_{ie} + \Delta^{he} Z_L}$$

which checks!

If $h_{re} \rightarrow 0$ and $h_{oe} \rightarrow 0$, then, we have

$$A_V \doteq \frac{-\left(h_{fe} - \dfrac{h_{ie}}{R_F}\right)Z_L}{h_{ie}\left(1 + \dfrac{Z_L}{R_F}\right)} \tag{9.43}$$

The equation for the current gain A_I is

$$A_I = \frac{h_{fe} - \dfrac{h_{ie}}{R_F}}{\dfrac{h_{ie}}{Z_L} + \dfrac{Z_L}{R_F}(h_{fe} + 1 + \Delta^{he} - h_{re}) + 1 + h_{oe} Z_L} \tag{9.44}$$

As a check on the correctness of this equation, we may let $R_F \rightarrow \infty$, which then reduces the circuit to a simple common emitter circuit, or

$$A_I = \frac{h_{fe}}{1 + h_{oe} Z_L}$$

which checks!

A less cumbersome version of Equation (9.44) will result if we allow $h_{re} \rightarrow 0$ and $h_{oe} \rightarrow 0$, then, we have

$$A_I \doteq \frac{h_{fe} - \dfrac{h_{ie}}{R_F}}{\dfrac{h_{ie}}{R_F} + \dfrac{Z_L}{R_F}(h_{fe} + 1) + 1}$$

$$A_I \doteq \frac{h_{fe} R_F - h_{ie}}{h_{ie} + (h_{fe} + 1) Z_L + R_F} \tag{9.45}$$

Let us see what happens to the output impedance by using collector feedback. Again, we must remember that we are essentially determining the Thevenin equivalent impedance of the transistor; therefore, we must short out all voltage sources ($e_{gen} = 0$) and we cannot disregard the generator internal impedance, R_{gen}. For our analysis, let us use R_S instead of R_{gen}, since there may be other resistors to consider (in addition to R_{gen}):

$$Z_2 = \frac{h_{ie} R_F + R_S R_F + R_S h_{ie}}{h_{ie} + R_F \Delta^{he} + R_S[(1 + h_{fe})(1 - h_{re}) + h_{oe} R_F]} \tag{9.46}$$

If we let $h_{re} \rightarrow 0$ and $h_{oe} \rightarrow 0$, then, we have

$$Z_2 \doteq \frac{h_{ie} R_F + R_S R_F + R_S h_{ie}}{h_{ie} + R_S(1 + h_{fe})} \tag{9.47}$$

If we let $R_F \rightarrow \infty$, then Equation (9.47) should be the same as was derived for the output impedance of a simple common emitter circuit.

Dividing (9.46) by R_F, we obtain

$$Z_2 = \frac{h_{ie} + R_S + \dfrac{R_S h_{ie}}{R_F}}{\dfrac{h_{ie}}{R_F} + \Delta^{he} + R_S\left[\dfrac{(1 + h_{fe})(1 - h_{re})}{R_F} + h_{oe}\right]}$$

If $R_F \rightarrow \infty$, then

$$Z_2 = \frac{h_{ie} + R_S}{\Delta^{he} + h_{oe} R_S} = \frac{1}{\dfrac{h_{ie} h_{oe} - h_{fe} h_{re} + h_{oe} R_S}{h_{ie} + R_S}}$$

$$Z_2 = \cfrac{1}{\cfrac{h_{oe}(h_{ie} + R_S) - h_{fe}h_{re}}{h_{ie} + R_S}} = \cfrac{1}{h_{oe} - \cfrac{h_{fe}h_{re}}{h_{ie} + R_S}}$$

which checks!

Example (9.20). Let us evaluate the electrical characteristics of a common emitter circuit with collector feedback. We will use the h parameter and resistor values previously used, and the value for R_F will be comparable to that used in Chapters 3 and 4 on dc bias design and analysis ($R_F = 100$ k):

$$Z_1 = \frac{h_{ie}(R_F + Z_L) + \Delta^{he}R_F Z_L}{h_{ie} + \Delta^{he}Z_L + Z_L(h_{fe} + 1 - h_{re}) + R_F(1 + h_{oe}Z_L)}$$

$$= \frac{3(10^3)\,100(10^3) + 5(10^3) + 0.05(10^5)\,5(10^3)}{3(10^3) + 0.05(5)(10^3) + 5(10^3)}$$

$$\times \frac{1}{(101 - 0.0001) + 100(10^3)[1 + 2(10^{-5})\,5(10^3)]}$$

$$Z_1 = 550\ \Omega$$

$$Z_1 \doteq \frac{h_{ie}(Z_L + R_F)}{h_{ie} + Z(h_{fe} + 1) + R_F} = \frac{3(10^3)\,[5(10^3) + 100(10^3)]}{3(10^3) + 5(10^3)(101) + 100(10^3)}$$

$$\doteq 518\ \Omega$$

Computing the voltage gain, we have

$$A_V = \cfrac{-\left(h_{fe} - \cfrac{h_{ie}}{R_F}\right)}{h_{ie}\left(\cfrac{1}{R_F} + \cfrac{1}{Z_L}\right) + \Delta^{he}} = \cfrac{-\left[100 - \cfrac{3(10^3)}{100(10^3)}\right]}{3(10^3)\left[\cfrac{1}{100(10^3)} + \cfrac{1}{5(10^3)}\right] + 0.05}$$

$$A_V = -147$$

Computing the current gain, we have

$$A_I = \cfrac{h_{fe} - \cfrac{h_{ie}}{R_F}}{\cfrac{h_{ie}}{R_F} + \cfrac{Z_L}{R_F}(h_{fe} + 1 - h_{re} + \Delta^{he}) + 1 + h_{oe}Z_L}$$

$$= \cfrac{100 - \cfrac{3(10^3)}{100(10^3)}}{\cfrac{3(10^3)}{100(10^3)} + \cfrac{5(10^3)}{100(10^3)}[100 + 1 + 0.05 - 0.0001]}$$

$$\times \frac{1}{+ 1 + 2(10^{-5})\,5(10^3)}$$

$$A_I = 16.17$$

Computing the output impedance, we have

$$Z_2 = \frac{h_{ie} R_F + R_S R_F + R_S h_{ie}}{h_{ie} + R_F \Delta^{he} + R_S [(1 + h_{re}) + h_{oe} R_F]}$$

$$= \frac{3(10^3) 100(10^3) + 0.9(10^3) 100(10^3) + 0.9(10^3) 3(10^3)}{3(10^3) + 100(10^3) 0.05 + 0.9(10^3)}$$

$$\times \frac{1}{[(1 + 100)(1 - 0.0001) + 2(10^{-5}) 100(10^3)]}$$

$$Z_2 \doteq 3.9 \text{ k}\Omega$$

Comparing these computed values as listed in Table 9.8 for the common emitter amplifier with collector feedback with the values for the simple common emitter amplifier, Table 9.6, we see that the input impedance is approximately one-sixth of the common emitter value, the voltage gain is comparable, the current gain is approximately one-sixth, and the output impedance is approximately one-fifteenth of the common emitter value.

Table 9.8

A_V	A_I	Z_1	Z_2
-147	16.17	550 Ω	3.9 KΩ

If we wished to retain the dc stability, which is inherent in the common emitter amplifier with collector feedback, but without the changes in ac characteristics, the circuit shown below in Fig. 9.35 would operate in this manner.

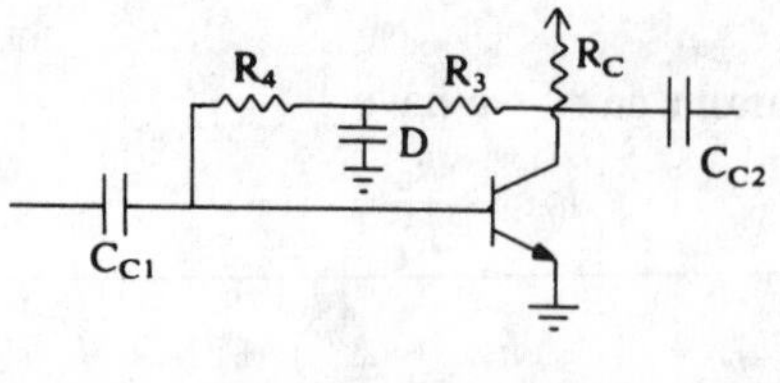

Fig. 9.35

The decoupling capacitor D would allow dc feedback, but prevent ac feedback by passing any signal from the collector to ground, thereby preventing the negative feedback signal from entering the base lead.

PROBLEMS

1. For the transistor amplifier shown below, determine the following: Z_1, Z_2, A_V, A_I, Z_{in}, Z_{out}. Given $h_{fe} = 49$, $h_{re} = .001$, $h_{ie} = 1$ k Ω, $h_{oe} = 2(10^{-5})\ \Omega$.

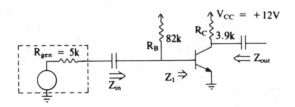

Fig. 9.36

Use the exact formulas first and then employ the approximate formulas. What is the percent error between the value determined for Z_1, *et cetera* using the exact and approximate formulas? Which parameter exhibits the greatest error between exact and approximate formulas?

2. For the transistor amplifier shown below, determine the following: Z_1, Z_2, A_V, A_I, Z_{in}, Z_{out}. Given $h_{fe} = 79$, $h_{re} = .001$, $h_{ie} = 1.5$ kΩ, $h_{oe} = 4(10^{-5})\ \Omega$.

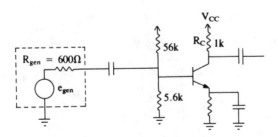

Fig. 9.37

Use the approximate formulas.

3. For the common emitter input and output curves shown below, compute the four h parameters at the operating point.

4. Referring to Problem 3 in the chapter on graphical analysis (Ch. 7), use the given curves to determine h_{ie}, h_{fe}, h_{oe}. Use the h parameter values to determine A_V, A_I, Z_1 and Z_{in}. What is the percent difference between

the values determined graphically and by use of the h parameters? *Compare Z_1 with $r_{in\ ac}$, A_V, A_I.*

5. For the common base amplifier shown below, determine the following: A_V, A_I, Z_1, Z_2, Z_{in}, Z_{out}. Given $h_{ib} = 30\ \Omega$, $h_{fb} = -0.98$, $h_{rb} = 5(10^{-4})$, $h_{ob} = 10^{-7}\ \Omega$, $V_{BE} = 0.3V$.

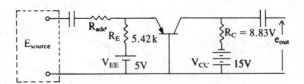

Fig. 9.38

6. The circuit shown below can be analyzed as a common emitter amplifier from a dc bias standpoint. However, from an ac amplifier standpoint, it is a common base circuit. We note that the signal is "fed into" the emitter and "detected" at the collector. We also note that the base is at ac ground potential due to the by-pass capacitor. This circuit requires only a single dc power source, V_{CC}, thus reducing circuit complexity and cost. *Compute A_V, A_I, Z_1, Z_{in}, Z_2, Z_{out}.* Given $h_{ib} = 50\ \Omega$, $h_{ob} = 2(10^{-7})$ Ω, $h_{fb} = -0.98$, $h_{rb} = .0003$.

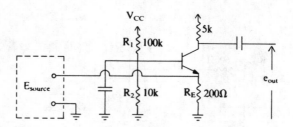

Fig. 9.39

7. Referring to Problem 1 in the chapter on graphical analysis (Ch. 7), use the given curves to determine h_{ib}, h_{rb}, h_{fb}, h_{ob}. Use these h parameter values to determine A_V, A_I, Z_1 and Z_{in}. *Compare* these results with those determined graphically, however, *compare $r_{in\ ac}$ with Z_1.*

8. Reanalyze Problem 1 with a 200 Ω resistor (unbypassed) in the emitter to ground circuit. *Compare* the amplifier characteristics with those obtained in Problem 1. Determine the percent difference (use Problem 1 results as the reference data). Use the approximate formulas.

9. For the circuit shown below, determine A_V, A_I, Z_1, Z_{in}, Z_3, Z_{out}. Given $h_{FE} = 75$, $V_{EB} = 0.7$ V, $h_{fe} = 50$, $h_{ie} = 2$ kΩ, $h_{re} = .001$, $h_{oe} = 0.2(10^{-4})$ Ω.

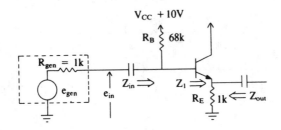

Fig. 9.40

10. For the circuit shown below, determine A_V, A_I, Z_1, Z_{in}, Z_3, Z_{out}.

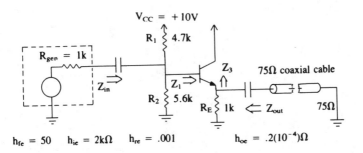

Fig. 9.41

11. For the circuit shown below, determine A_v, A_I, Z_1, Z_{in}, Z_2, Z_{out}. Given $h_{FE} = 49$, $h_{fe} = 49$, $h_{re} = .001$, $h_{ie} = 1$ k, $h_{oe} = 2(10^{-5})$ Ω.

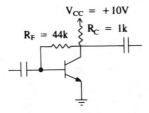

Fig. 9.42

Compare these results with the results of Problem 1 by computing the percent difference. Use the results of Problem 1 as the reference values.

12. Referring to Problem 11, what would happen to A_V, A_I, Z_1, Z_{in}, Z_2, Z_{out} if we altered the circuit as shown below:

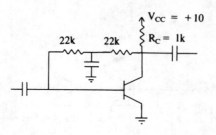

Fig. 9.43

What is the difference in the dc bias conditions between Problems 11 and 12?

Chapter 10

Single-Stage Small-Signal Amplifier Frequency Considerations

10.1 AMPLIFIER FREQUENCY CONSIDERATIONS

Up to this point, we have dealt with voltage amplification of an ac signal without considering the effect that the input signal frequency might have upon the amplifier performance.

Both inductances and capacitances are frequency dependent devices. All circuits have inductances and capacitances, although the inductances and capacitances might not be discrete components that were designed into the circuit. For example, a wire that lies close to a metal chassis forms a capacitance. At very high frequencies, a wire connecting one component to another "acts" like an inductance. Thus, we can see that amplifier performance is frequency dependent.

We can lump amplifiers into two general categories: *untuned* and *tuned*. Most student experiments associated with a course on devices are oriented toward untuned amplifiers. The reason for introducing the student to untuned rather than tuned amplifiers is that amplifier principles can be readily understood by studying untuned amplifiers. These concepts can then be applied to the more complex concepts involved in tuned amplifier design and analysis.

A *bandpass curve* is a pictoral representation of the amplification of a particular amplifier at various operating frequencies. Because we are interested in the amount of amplification that takes place at various frequencies, the frequency is usually plotted as the independent variable on the horizontal axis and amplification is plotted on the vertical axis. Because the test frequencies applied to an amplifier may vary over many decades (multiples of 10), it is quite common to use a logarithmic scale along the horizontal axis. If the amplification varies greatly (over many decades), it is also possible to use a logarithmic scale for the vertical axis. However, a more common approach for expressing amplification is to define a special unit of measurement, the *decibel* (dB). The amplification in decibels is given by the following equation:

$$A_{dB} = 20 \log_{10} \frac{V_{out}}{V_{in}} (= 20 \log_{10} A_V) \tag{10.1}$$

Example (10.1). A common emitter amplifier stage with a two-millivolt input signal applied to its base has a 70-millivolt signal at its collector. What is its absolute voltage amplification and its amplification expressed in decibels? Hence, we derive

$$A = \frac{V_{out}}{V_{in}} = \frac{70}{2} = 35$$
$$A_{dB} = 20 \log_{10} A_V = 20 \log 35 = 30.9 \text{ dB}$$

Example (10.2). Two common emitter amplifier stages are connected in tandem, i.e., one "feeds" the other. A one-millivolt input signal into the first amplifier results in a one-volt signal at the output of the second stage. What is the magnitude of the overall voltage gain? Express this value in decibels, thereby obtaining:

$$A_V = \frac{V_{out}}{V_{in}} = \frac{1 \text{ V}}{1 \text{ mV}} = \frac{1000}{1} = 1000$$
$$A_{VdB} = 20 \log_{10} A_V = 20 \log_{10} 1000 = 60 \text{ dB}$$

Although the absolute voltage amplification for each of the examples varied greatly, we can see that the difference in decibels varies only by approximately two to one. Thus, we can use a linear scale for the vertical (amplification) axis. Special graph paper is available that has a logarithmic scale on the horizontal axis and a linear scale on the vertical axis. It is referred to as "semi-log graph paper," or simply "semi-log paper."

If we were to plot voltage gain for a typical R-C coupled audio amplifier, a gain *versus* frequency curve might appear as shown in Fig. 10.1.

The name *R-C coupled* is derived from the fact that only discrete *resistors* and *capacitors* (no inductors), in conjunction with an active device, are used to construct the amplifier, whereby the capacitor's function is to couple the signal from one stage to the next or from an amplifier stage to a load.

We note that the amplifier stage amplifies all frequencies by the same amount for each input frequency from approximately 20 Hz to 30 kHz. The curve is essentially a straight line from 20 Hz to 30 kHz and is said to have a "flat" frequency response over the range from 20 Hz to 30 kHz. The frequency where the amplification begins to decrease is referred to as the "break frequency" or "cut-off frequency." The decline in the voltage gain occurs at the rate of six decibels per octave (for a single-stage amplifier). An octave is a doubling (or halving) in frequency.

By definition, at the cut-off frequency, the voltage amplification is 70.7% of the mid-band amplification.

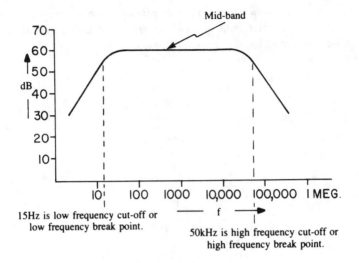

Fig. 10.1

10.2 AMPLIFIER FREQUENCY CONSIDERATIONS

In performing amplifier small-signal analysis, we mentioned that capacitors are used in two general circumstances. They are:

1. *To couple a signal from one stage to another or from an amplifier to a load* (e.g., a loudspeaker, a picture tube, *et cetera*). In performing this task the capacitor serves to "pass" the ac signal from one stage to another; it also "blocks" the dc level from the collector of one stage from affecting the dc level at the transistor base of a succeeding stage. This blocking action is also employed in isolating the drain potential of a FET stage from affecting the dc level at the gate of a succeeding stage (see Fig. 10.2).

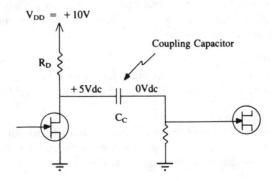

Fig. 10.2

340

2. *To bypass a signal to ground.* For example, when using self-bias, the emitter of a bipolar transistor and the source of a FET, should be at ac ground potential for many applications.

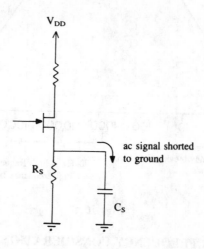

Fig. 10.3

What value capacitor should be selected for a particular application? To answer this question, it is appropriate to review capacitor ratings. They are:

1. The working voltage;
2. The peak voltage;
3. The capacitance value.

1.) The Working Voltage

If the collector of one stage is operating at 8 V above ground and the base of a succeeding stage is at 2 V above ground, the minimum working voltage rating of the capacitor must be 6 V. Naturally, a voltage of slightly higher rating would be chosen because changes in the operating points of either amplifier stage are possible.

2.) The Peak Voltage

If, in addition to the six volts of dc potential across a capacitor, the collector swung an additional 4 V positive, a peak rating of 10 V would be required.

3.) The Capacitance Value

The capacitance value is not as simply computed as the working voltage and the peak voltage rating. Let us discuss the bypass capacitor value first, since this value is more readily computed. Refer to Fig. 10.4.

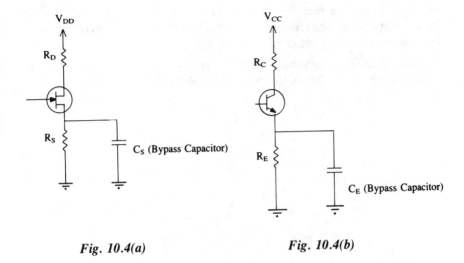

Fig. 10.4(a)   Fig. 10.4(b)

We know that the capacitive reactance is determined by the formula:

$$X_C = \frac{1}{2 \pi f C_E} \qquad \left(\text{or } X_C = \frac{1}{2 \pi f C_S} \right)$$

and that the reactance increases with decreasing frequency. The purpose of the bypass capacitor is to short an ac signal to ground, or *bypass* the ac signal around an emitter or source resistor to ground. It is common design procedure to choose a value of the capacitive reactance equal to one-tenth of the emitter or source resistor value. Because the reactance of a capacitor is greatest at the lowest frequency, the capacitor value is computed at the lowest operating frequency. At higher operating frequencies, it becomes an "even better short." Thus,

$$X_C = .1 \, R_E \quad \text{or} \quad X_C = .1 \, R_S$$

$$\frac{1}{2 \pi f C_E} = .1 \, R_E$$

Therefore,

$$C_E = \frac{1}{2 \pi f (.1) \, R_E} = \frac{10}{2 \pi f R_E} = \frac{5}{\pi f_{low} R_E} \qquad (10.2)$$

$$\text{or } C_S = \frac{5}{\pi f_{low} R_S} \qquad (10.3)$$

where f_{low} is the low frequency break-point. Quite often, f_{low} is selected as 20 Hz for an R-C coupled audio amplifier, since 20 Hz is approximately the lowest frequency that humans are capable of hearing.

Example (10.3). A universal stabilized common-emitter amplifier appears as shown in Fig. 10.5. Design the bypass capacitor for operation over the audio range.

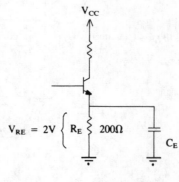

Fig. 10.5

$$C_E = \frac{5}{\pi f_{low} R_E} = \frac{5}{\pi (20)(200)} = 398 \ \mu F$$

Choose a capacitor of 400 μF at 5 W Vdc.

Selection of the coupling capacitor between two stages involves choosing the correct working voltage and the correct capacitance value.

To determine the correct capacitance value, let us first make use of the voltage divider rule to determine A_V mid-band for the circuit shown in Fig. 10.6.

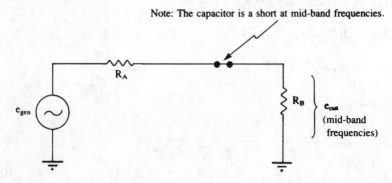

Fig. 10.6

Therefore, we obtain

$$e_{out\ MIDBAND} = e_{gen} \frac{R_B}{R_A + R_B}$$

Thus,

$$\frac{e_{out\ MIDBAND}}{e_{gen}} = A_{V\ MIDBAND} = \frac{R_B}{R_A + R_B} \qquad (10.4)$$

If we have the circuit of Fig. 10.7 at low frequencies, we derive

$$e_{out\ low} = e_{gen} \frac{R_B}{R_A + R_B - j X_C}$$

$$\frac{e_{out\ low}}{e_{gen}} = \frac{R_B}{R_A + R_B - j X_C} \qquad (10.5)$$

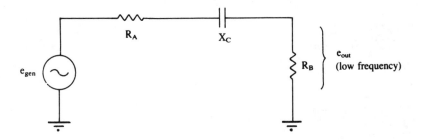

Fig. 10.7

Thus, R_A may be considered to be all of the resistance to the left of the capacitor, or the Thevenin source resistance. R_B consists of all the resistances connected to ground (in parallel) to the right of the coupling capacitor. Hence, dividing (10.5) by (10.4), we have,

$$\frac{A_{V\ low}}{A_{V\ MID}} = \frac{\dfrac{R_B}{R_A + R_B - j X_C}}{\dfrac{R_b}{R_A + R_B}} = \frac{R_A + R_B}{R_A + R_B - j X_C}$$

Dividing the numerator and denominator by $R_A + R_B$, we obtain

$$\frac{A_{V\ low}}{A_{V\ MID}} = \frac{1}{1 - \dfrac{j\,X_C}{R_A + R_B}}$$

(10.6)

Suppose:

$$\frac{A_{V\ low}}{A_{V\ MID}} = \frac{1}{1 - j\,1}$$

$$= \frac{1}{\sqrt{2}\ \underline{/-45^\circ}}$$

(10.7)

Then,

$$\frac{A_{V\ low}}{A_{V\ MID}} = \frac{1}{\sqrt{2}} = .707$$

and we know that *by definition* the magnitude of the amplifier voltage gain at the low frequency "break point" is 70.7% of the mid-band frequency amplification. Therefore, from (10.6), at f_{LCO}, we have

$$f_{LCO}, \frac{X_C}{R_A + R_B} = 1$$

or $X_C = R_A + R_B$

but $X_C = \dfrac{1}{2\,\pi\,f_{LBP}\,C_C} = R_A + R_B$

Thus,

$$C_C = \frac{1}{2\,\pi\,f_{LBP}\,(R_A + R_B)}$$

(10.8)

where

f_{LCO} = low frequency cut-off,
f_{LBP} = low frequency break point.

Example (10.4). Determine the coupling capacitor value for the circuit shown in Fig. 10.8, where $h_{ie2} = 2$ k. Thus,

$$R_A = Z_{O1} = R_{C1} \, /\!/ \, Z_{21} \doteq R_{C1} \, /\!/ \, h_{oe1} \doteq R_{C1} = 4 \text{ k}$$
$$R_B = Z_{in2} = R_1 \, /\!/ \, R_2 \, /\!/ \, Z_{12} \doteq R_1 \, /\!/ \, R_2 \, /\!/ \, h_{ie2}$$
$$\quad = 100 \text{ k} \, /\!/ \, 10 \text{ k} \, /\!/ \, 4 \text{ k}$$
$$R_B = 1.64 \text{ k}$$
$$C_C = \frac{1}{2 \, \pi \, f_{LBP} \, (R_A + R_B)} = \frac{1}{2 \, \pi \, (20)(4 \text{ k} + 1.64 \text{ k})} = 1.41 \; \mu F$$

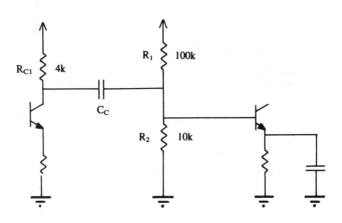

Fig. 10.8

Example (10.5). Determine the coupling capacitor value for the circuit shown in Fig. 10.9. Thus,

$$R_A = Z_{O1} = R_{D1} \, /\!/ \, V_{D1} \doteq R_{D1} = 4 \text{ k}$$
$$R_B = Z_{in2} = R_G = 100 \text{ k}$$
$$C_C = \frac{1}{2 \, \pi \, f_{LBP} \, (R_A + R_B)} = \frac{1}{2 \, \pi \, (20) \, (4 \text{ k} + 100 \text{ k})} = .0765 \; \mu F$$

Choose .1 μF at 20 W Vdc.

We see that with the high input impedance of the FET, a much lower value of C_C is required. Thus, C_C is also physically smaller.

10.2.1 High Frequency Considerations

Let us examine the simple circuit shown in Fig. 10.10 and discuss the effect upon the output voltage if we maintain a constant input signal voltage amplitude but vary the input signal frequency.

For low and mid-band frequencies, the reactance of C_{Hi} is negligible compared with the circuit resistances, and the output voltage remains constant with increasing frequency. At a given frequency, i.e., the high frequency cut-off, the reactance of C_{Hi} begins to decrease and, therefore, the output signal decreases as a result of voltage divider action. At a sufficiently high signal frequency, the reactance approaches zero, and no output voltage is present.

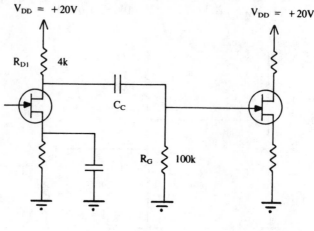

Fig. 10.9

According to the voltage divider rule, we can write the following:

At mid-band frequencies:

$$e_{out\ MID} = \frac{R_{Load}}{R_{gen} + R_{Load}} \cdot e_{gen}$$

$$\frac{e_{out\ MID}}{e_{gen}} = A_{V\ MID} = \frac{R_{Load}}{R_{gen} + R_{Load}}$$

(10.9)

where $X_{CHi} \to \infty$.

At high frequencies:

$$e_{out\ Hi} = e_{gen} \frac{R_{Load} \mathbin{/\mkern-5mu/} (-j\,X_{C\ Hi})}{R_{gen} + R_{Load} \mathbin{/\mkern-5mu/} (-jX_{C\ Hi})}$$

$$\frac{e_{out\ Hi}}{e_{gen}} = A_{V\ Hi} = \frac{\dfrac{R_L\,(-j\,X_C)}{R_L - j\,X_C}}{R_g + \dfrac{R_L\,(-j\,X_C)}{R_L - j\,X_C}}$$

$$= \frac{R_L \, (-j \, X_C)}{R_g \, (R_L \, - \, j \, X_C) \, + \, R_L \, (-j \, X_C)}$$

$$= \frac{R_L \, (-j \, X_C)}{R_g \, R_L \, + \, R_g \, (-j \, X_C) \, + \, R_L \, (-j \, X_C)}$$

$$= \frac{R_L \, (-j \, X_C)}{R_g \, R_L \, + \, (-j \, X_C)(R_g \, + \, R_L)}$$

$$= \frac{R_L}{\dfrac{R_g \, R_L}{-j \, X_C} \, + \, (R_g \, + \, R_L)}$$

$$= \frac{R_L}{(R_g \, + \, R_L) \, + \, \dfrac{R_g \, R_L}{-j \, X_C} \left[\dfrac{R_g \, + \, R_L}{R_g \, + \, R_L} \right]}$$

$$A_{V \, Hi} = \frac{R_L}{(R_g \, + \, R_L) \left[1 \, + \, j \, \dfrac{R_g \, R_L}{X_C \, (R_g \, + \, R_L)} \right]}$$

$$(10.10)$$

Dividing (10.10) by (10.9), we have

$$\frac{A_{V \, Hi}}{A_{V \, MID}} = \frac{\dfrac{R_L}{R_g \, + \, R_L \left(1 \, + \, j \, \dfrac{R_g \, R_L}{X_C \, (R_g \, + \, R_L)} \right)}}{\dfrac{R_L}{R_g \, + \, R_L}}$$

$$\frac{A_{V \, Hi}}{A_{V \, MID}} = \frac{1}{1 \, + \, j \, \dfrac{R_g \, R_L}{X_C \, (R_g \, + \, R_L)}}$$

$$(10.11)$$

Suppose:

$$\frac{A_{V \, Hi}}{A_{V \, MID}} = \frac{1}{1 \, + \, j \, 1}$$

$$(10.12)$$

Thus,

$$\frac{A_{V \, Hi}}{A_{V \, MID}} = \frac{1}{\sqrt{2} \, \underline{/+ \, 45°}}$$

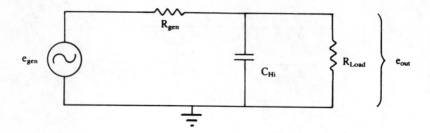

Fig. 10.10

and

$$\frac{A_{V\,Hi}}{A_{V\,MID}} = \frac{1}{\sqrt{2}} = .707$$

We know *by definition*, the magnitude of the amplifier voltage gain at the high frequency break point is 70.7% of the mid-band frequency amplification. Thus, from (10.11) and (10.12), we obtain at the high frequency cut-off:

$$\frac{R_G\,R_L}{X_C\,(R_G + R_L)} = 1$$

or at break point:

$$\frac{R_G\,R_L}{R_G + R_L} = X_C = \frac{1}{2\,\pi\,f_{Hi\,BP}\,C_{Hi}}$$

Therefore,

$$f_{Hi\,BP} = \frac{1}{2\,\pi\,C_{Hi}\left(\dfrac{R_G\,R_L}{R_G + R_L}\right)} \qquad (10.13)$$

Example (10.6). Compute the high frequency cut-off for a circuit as shown in Fig. 10.11(a). Thus,

$$f_{Hi} = \frac{1}{2\,\pi\,C_{Hi}\,\dfrac{R_G\,R_L}{R_G + R_L}}$$

$$= \frac{1}{2 \pi (20)(10^{-12}) \left[\dfrac{3 \text{ k } (100 \text{ k})}{3 \text{ k } + 100 \text{ k}}\right]} = 2.73 \text{ MHz}$$

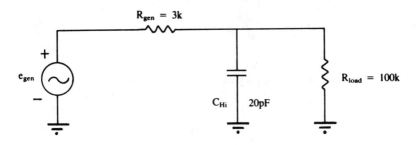

Fig. 10.11(a)

The circuit of Fig. 10.11(a) could be the equivalent circuit for one FET feeding another, as shown in Fig. 10.11(b), where C_{Hi} consists of various capacitances to ground, such as $C_{wiring} + C_{out1} + C_{in2}$.

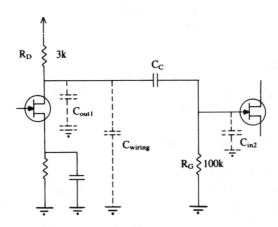

Fig. 10.11(b)

Example (10.7). Compute the high frequency cut-off for the circuit of Fig. 10.12:

$$f_{Hi\ BP} = \cfrac{1}{2\ \pi\ C_{Hi}\left(\cfrac{R_G\ R_L}{R_G\ +\ R_L}\right)}$$

$$= \cfrac{1}{2\ \pi\ (20)(10^{-12})\left[\cfrac{3\ k\ (1.64\ k)}{3\ k\ +\ 1.64\ k}\right]} = 4.85\ \text{MHz}$$

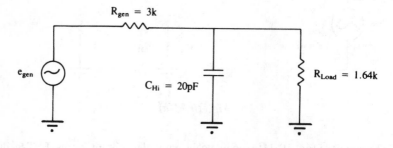

Fig. 10.12

The circuit shown in Fig. 10.12 could be the equivalent circuit for one bipolar transistor "feeding" another, as shown in Fig. 10.13, where C_{Hi} consists of various capacitances to ground, such as $C_{wiring}\ +\ C_{out1}\ +\ C_{in2}$, and

$$R_{gen} = Z_{out1} = R_{C1}\ /\!/\ Z_{21} \doteq R_{C1}\ /\!/\ \frac{1}{h_{oe2}} \doteq R_{C1} = 3\ k$$

$$R_{Load} = Z_{in2} = R_1\ /\!/\ R_2\ /\!/\ Z_{12} \doteq R_1\ /\!/\ R_2\ /\!/\ h_{ie2}$$
$$= 100\ k\ /\!/\ 10\ k\ /\!/\ 2\ k = 1.64\ k$$

10.2.2 The Evaluation of Device Capacitance

In the previous section we referred to the output and input capacitances of either a bipolar or field-effect transistor stage. The inter-electrode capacitances are specified by the manufacturer on the appropriate data sheets. "Inter-electrode capacitance" is a fancy term for the capacitance as measured between two device leads, such as the base and emitter or gate and source. The inter-electrode capacitance is to some extent a function of the dc voltage between the two electrodes and, therefore, a manufacturer will quite often specify the conditions that existed when the capacitance measurements were recorded.

For a bipolar transistor, the values of the following three capacitances would be desirable for predicting the high frequency performance (high frequency cut-off) of an amplifier stage: C_{BC}, C_{CE} and C_{EB} (see Fig. 10.14).

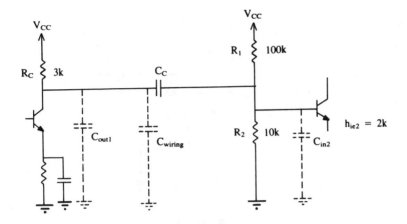

Fig. 10.13

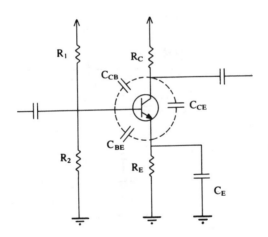

Fig. 10.14

It is customary to show each of the inter-electrode capacitances with dotted lines. This notation signifies that discrete capacitors do not exist between these points, but rather the capacitances arise from the proximity of two conducting surfaces or media.

For FETs, the inter-electrode capacitances would be as shown in Fig. 10.15 below.

The output capacitance of an amplifier stage is simply the capacitance "seen" when looking backwards into the output of an amplifier stage. Referring to the schematic of Fig. 10.16, we note that the only capacitance to ground when we "look back" into the output of the FET amplifier stage is the capacitance from source to drain.

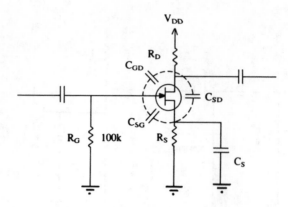

Fig. 10.15

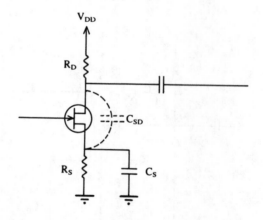

Fig. 10.16

We must remember that the source is at ac ground as a result of the "shorting effect" of the source resistor bypass capacitor, C_S.

The wiring capacitances, as it is called, consist of all the capacitances caused by wires, printed circuit runs, or components that interconnect two stages and lie close to a "ground plane." A "ground plane" is a fancy term for a chassis, or other conductor, which is at ground potential.

Depending upon the length of the inter-stage connections—be they wires, printed circuit runs, or components—and their proximity to a ground plane, the wiring capacitance will vary greatly. See Fig. 10.17.

C_{wiring} is quite often estimated in order to predict an amplifier's high frequency cut-off. Values of 5 to 20 pF are not uncommon.

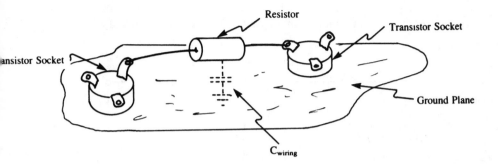

Fig. 10.17

If we examine the schematic for a FET amplifier stage shown in Fig. 10.18 and ask what capacitance we "see" when looking into the input terminals of the stage, the answer might simply be the capacitance from gate to source. We may be inclined to assume that the gate-to-drain capacitance is not a capacitance to ground because the drain resistor isolates the drain from ground.

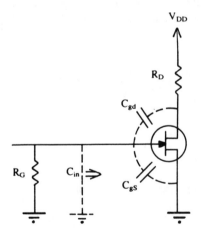

Fig. 10.18

We know from our previous discussion of FETs that the gate-to-source junction is essentially a back-biased junction for a JFET, and the gate to substrate contains an insulating layer for a MOSFET. Thus, the primary component of the input impedance, particularly at high frequencies is the capacitance reactance due to C_{gd}

An equivalent circuit for a FET can thus be shown schematically (Fig. 10.19) as an open circuit with a shunt capacitance from gate to ground.

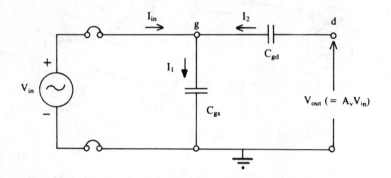

Fig. 10.19

We cannot ignore the gate-to-drain capacitance because this reactance provides a feedback path for some of the output signal fed from the output terminals (drain-to-ground) to the input terminals (ground-to-gate).

We may analyze the circuit shown in Fig. 10.19 as follows:

1.) By Ohm's law:

$$I_2 = (V_{out} - V_{in}) Y_{gd} = (A_V V_{in} - V_{in}) Y_{gd}$$
$$= V_{in} (A_V - 1) Y_{gd}$$

2.) By Ohm's law:

$$I_1 = Y_{in} Y_{gs}$$

3.) Employing Kirchoff's current law at node g:

$$\Sigma I_{out} = \Sigma I_{in}$$
$$I_1 = I_{in} + I_2$$

Thus,

$$I_{in} = I_1 - I_2$$
$$= V_{in} V_{gs} - V_{in} (A_V - 1) Y_{gd}$$
$$= V_{in} [Y_{gs} - (A_V - 1) Y_{gd}]$$

$$I_{in} = V_{in} [Y_{gs} + (1 - A_V) Y_{gd}]$$

Thus,

$$\frac{I_{in}}{V_{in}} = Y_{gs} + (1 - A_V) Y_{gd}$$
$$Y_{in} = Y_{gs} + (1 - A_V) Y_{gd}$$
$$j\omega C_{in} = j\omega (G_s + (1 - A_V) C_{gd})$$

Therefore, for a common source amplifier, we have

$$C_{in} = C_{gs} + (1 - A_V) C_{gd} \qquad (10.14)$$

or

$$C_{in} = C_{be} + (1 - A_V) C_{bc} \qquad (10.15)$$

for a common emitter amplifier.

We note that the input capacitance is a function of the stage gain. This is referred to as the "Miller effect."

Example (10.8). For the FET amplifier stage shown in Fig. 10.20, compute the input capacitance, given

$$g_m = 2.2 \text{ mmho}$$
$$C_{gs} = 4 \text{ pF}$$
$$C_{gd} = 2.5 \text{ pF}$$
$$C_{ds} = 1.2 \text{ pF}$$

Also,

$$A_V \doteq G_m R_D$$
$$= -2.2 \, (3) = -6.6$$

where

$$C_{in} = C_{gs} + (1 - A_V) C_{gd}$$
$$C_{in} = 4 \text{ pF} + [1 - (-6.6)] \, 2.5 \text{ pF}$$
$$C_{in} = 23 \text{ pF}$$

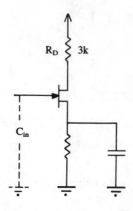

R_D 3k

Fig. 10.20

Let us see what the input capacitance is for a source-follower amplifier stage. Refer to the schematic and equivalent circuits shown in Fig. 10.21(a,b,c).

Writing Kirchoff's current law at node g, we have

$$\Sigma I_{in} = \Sigma I_{out}$$
$$I_{in} = I_1 + I_2$$
$$\text{and } I_2 = (V_{in} - V_{out}) Y_{sg}$$
$$= (V_{in} - A_V V_{in}) T_{sg}$$
$$= V_{in} (1 - A_V) Y_{sd}$$
$$I_{in} = V_{in} Y_{gd} + V_{in} (1 - A_V) Y_{sg}$$
$$= V_{in} [Y_{gd} + (1 - A_V) Y_{sg}]$$
$$\text{Also } I_1 = V_{in} Y_{sd}$$
$$\frac{I_{in}}{V_{in}} = Y_{in} = Y_{gd} + (1 - A_V) Y_{sg}$$
$$j \omega C_{in} = j \omega C_{gd} + (1 - A_V) \omega C_{sg}$$

Therefore, for a source follower, we obtain

$$C_{in} = C_{gd} + (1 - A_V) C_{sg} \tag{10.16}$$

or

$$C_{in} = C_{BC} + (1 - A_V) C_{CB} \tag{10.17}$$

for an emitter follower.

357

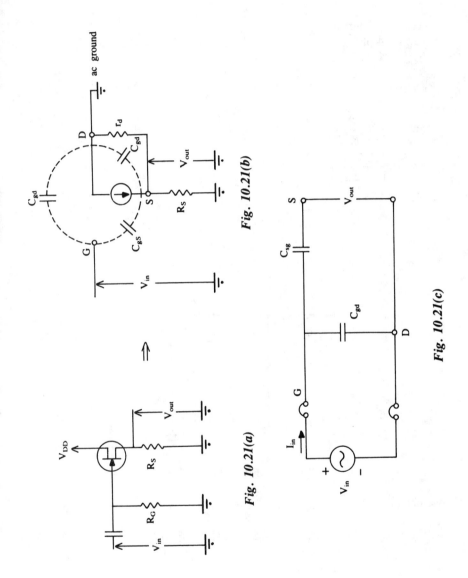

Fig. 10.21(b)

Fig. 10.21(a)

Fig. 10.21(c)

Example (10.9).   Determine the input capacitance for the source-follower stage shown in Fig. 10.22. Given

$$G_m = 2.2 \text{ mmho}$$
$$C_{sg} = 4 \text{ pF}$$
$$C_{sd} = 1.2 \text{ pF}$$
$$C_{gd} = 2.5 \text{ pF}$$

where

$$A_V \doteq \frac{g_m R_S}{1 + g_m R_S}$$
$$= \frac{2.2 (1)}{1 + 2.2 (1)}$$
$$A_V \doteq .688$$

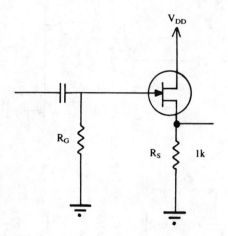

Fig. 10.22

we obtain

$$C_{in} = C_{gd} + (1 - A_V) C_{sg} = 2.5 + (1 - .688) 4$$

Thus,

$$C_{in} = 3.75 \text{ pF}$$

The input capacitance for a common-source connected FET is considerably greater than for a source follower. For a source follower, the Miller effect is minimized as a result of the term $1 - A_V$ tending toward zero. A_V is a positive value slightly less than unity, which is, in turn, subtracted from unity.

PROBLEMS

1. A common emitter amplifier has 3 mV applied to its base. An output voltage of 0.6 V is measured at its collector. What is its absolute voltage gain and its voltage gain measured in decibels?

2. An emitter follower amplifier has a one-volt signal applied to its base and one-half volt signal is measured at its emitter. What is its absolute voltage gain and its voltage gain expressed in decibels?

3. A source follower amplifier has a two volt signal applied to its gate. A 0.4 V signal is measured at its source. What is the amplifier's absolute voltage gain and its voltage gain expressed in decibels?

4. For a two stage amplifier, the input to the first stage is one-half millivolt and the output of the first stage is 40 mV. The output of the second stage is measured at 2.5 V. What is the absolute voltage gain of each stage and the voltage gain of each stage expressed in decibels?

5. At 25 Hz the voltage output of an amplifier has decreased to 70.7% of the mid-band output voltage. Express this ratio in decibels.

$$\frac{E_{out} \ (25 \ \text{Hz})}{E_{out} \ (\text{Mid})} = ?$$

6. For the emitter network shown below (Fig. 10.23), select the correct value for both the capacitance and the operating voltage of the bypass capacitor if the low frequency cut-off is 25 Hz, given $I_{CQ} = 5$ mA.

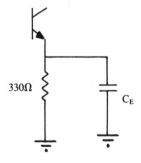

330Ω C_E

Fig. 10.23

360

7. For the source network shown below (Fig. 10.24), select the correct value for both the capacitance and the operating voltage of the bypass capacitor if the low frequency cut-off is 30 Hz, given I_{DQ}.

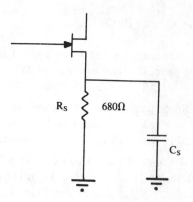

Fig. 10.24

8. Determine the coupling capacitor value for the circuit shown below (Fig. 10.25). Specify the minimum working voltage for the capacitor, where f_{LBP} = 25 Hz.

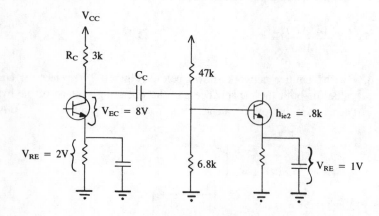

Fig. 10.25

9. Determine the coupling capacitor value for the circuit shown below (Fig. 10.26). Specify the minimum working voltage for the capacitor, where f_{LBP} = 30 Hz.

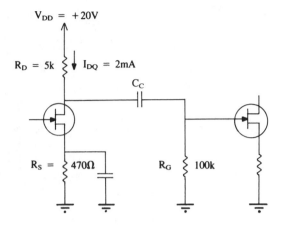

Fig. 10.26

10. Compute the high frequency cut-off, for the input, interstage and output networks of the two stage FET amplifier shown below (Fig. 10.27). Given

$$C_{sg1} = C_{sg2} = 4.2 \text{ pF}, \ C_{sd1} = C_{sd2} = 1.2 \text{ pF}, \ C_{gd1} = C_{gd2} = 2.5 \text{ pF}$$

where $\quad g_{m1} = g_{m2} = 4$ mmho

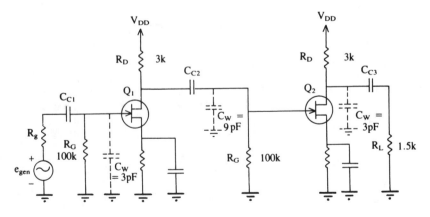

Fig. 10.27

11. Compute the high frequency cutoff, for the input, interstage and output network of the two stage bipolar transistor amplifier shown below (Fig. 10.28). Given

$C_{be1} = C_{be2} = 30 \text{ pF}, C_{ec1} = C_{ec2} = 5 \text{ pF}, C_{bc1} = C_{bc2} = 1.2 \text{ pF}$
where
$h_{ie1} = 2 \text{ k}, h_{ie2} = 1 \text{ k}, h_{fe1} = 200, h_{fe2} = 125$

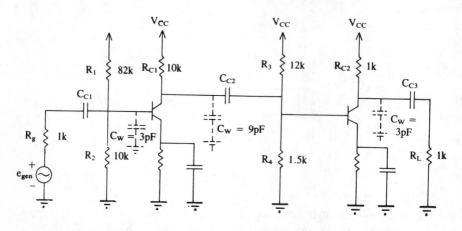

Fig. 10.28

12. Compute the high frequency cut-off for the input, interstage, and output networks of the two-stage FET amplifier shown below (Fig. 10.29).

Given
$C_{sg1} = C_{sg2} = 6 \text{ pF}, C_{sd1} = C_{sd2} = .8 \text{ pF}, e_{gd1} = C_{gd2} = 2.4 \text{ pF}$
where
$g_{m1} = g_{m2} = 10 \text{ mmho}$

13. Compute the high frequency cut-off for the input, interstage, and output networks of the two-stage bipolar transistor amplifier shown below (Fig. 10.30).

Given
$C_{be1} = C_{be2} = 30 \text{ pF}, C_{ec1} = C_{ec2} = 5 \text{ pF}, C_{bc1} = C_{bc2} = 1.2 \text{ pF}$
where
$h_{ie1} = 2 \text{ k}, h_{ie2} = 1 \text{ k}, h_{fe1} = 200, h_{fe2} = 125$

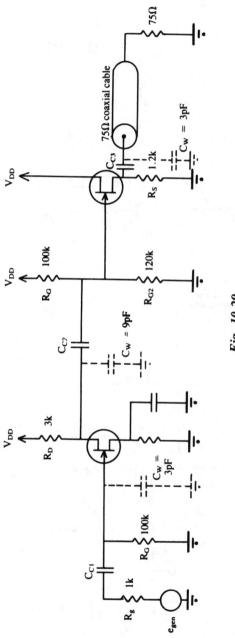

Fig. 10.29

364

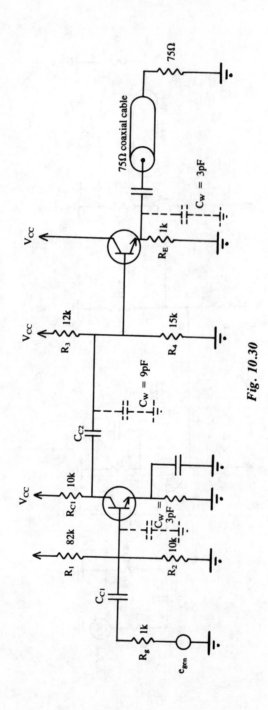

Fig. 10.30

Chapter 11

Multistage Amplifier Small-Signal Considerations

11.1 MULTISTAGE AMPLIFIER VOLTAGE GAIN

The electromagnetic wave intercepted by a radio antenna may generate a signal of only microvolts (millionths of a volt) in the initial amplifier input circuit of a radio. Eventually, this signal must be amplified sufficiently to drive a loudspeaker. If the loudspeakers are part of a high quality stereo system, they may require an input of 100 W per channel. The method by which we develop 100 W from a signal of only microvolts is to connect numerous amplifier stages in tandem, as shown in Fig. 11.1.

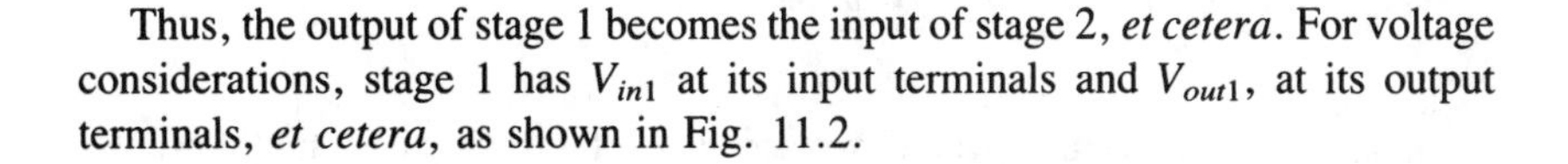

Signal In $\longrightarrow$ A_1 $\longrightarrow$ A_2 $\longrightarrow$ A_3 $\longrightarrow$ Signal Out

Fig. 11.1

Thus, the output of stage 1 becomes the input of stage 2, *et cetera*. For voltage considerations, stage 1 has V_{in1} at its input terminals and V_{out1}, at its output terminals, *et cetera*, as shown in Fig. 11.2.

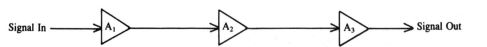

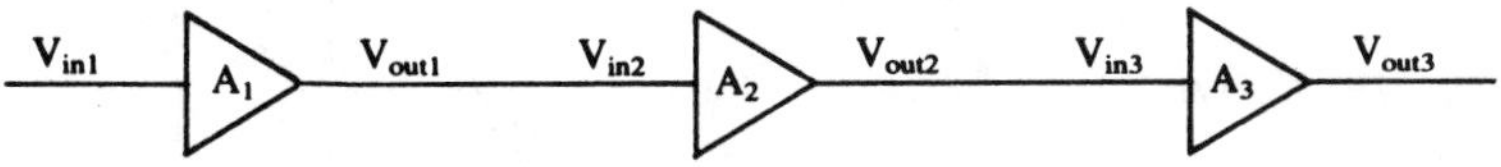

Fig. 11.2

We also observe that the individual stage voltage gain, as previously defined, is

$$A_{V1} = \frac{V_{out1}}{V_{in1}}, \quad A_{V2} = \frac{V_{out2}}{V_{in2}}, \quad A_{V3} = \frac{V_{out3}}{V_{in3}}$$

365

and the overall voltage gain is

$$A_{V\ (overall)} = \frac{V_{out\ (overall)}}{V_{in\ (overall)}} \tag{11.1}$$

and we note that $V_{out\ (overall)}$ is equal to V_{out3} and $V_{in\ (overall)}$ is equal to V_{in1}
Thus

$$A_{V\ (overall)} = \frac{V_{out3}}{V_{in1}} \tag{11.2}$$

If we were to multiply the individual stage voltage gains together, we would
have

$$A_{V1} \cdot A_{V2} \cdot A_{V3} = \frac{V_{out1}}{V_{in1}} \cdot \frac{V_{out2}}{V_{in2}} \cdot \frac{V_{out3}}{V_{in3}}$$

However,

$$V_{in2} = V_{out1},\ V_{in3} = V_{out2}$$

Substituting:

$$A_{V1} \cdot A_{V2} \cdot A_{V3} = \frac{V_{in2}}{V_{in1}} \cdot \frac{V_{in3}}{V_{in2}} \cdot \frac{V_{out3}}{V_{in3}} = \frac{V_{out3}}{V_{in1}}$$

and we see that

$$A_{V1} \cdot A_{V2} \cdot A_{V3} = \frac{V_{out3}}{V_{in1}} = A_{V\ (overall)}$$

Thus, to determine the overall voltage gain of a multistage amplifier, we
simply multiply each stage gain together:

$$A_{V\ (overall)} = A_{V1} \cdot A_{V2} \cdot A_{V3} \ldots \tag{11.3}$$

By the same process, we have

$$A_{I\ (overall)} = A_{I1} \cdot A_{I2} \cdot A_{I3} \ldots \tag{11.4}$$

The power gain is defined as the ratio of the output power divided by the
input power:

$$A_P = \frac{P_{out}}{P_{in}} = \frac{V_{out} \, I_{out}}{V_{in} \, I_{in}} = \left(\frac{V_{out}}{V_{in}}\right) \cdot \left(\frac{I_{out}}{I_{in}}\right) = A_V \cdot A_I \qquad (11.5)$$

If we wish to express the overall voltage gain in terms of decibels, we would compute the gain as follows:

$$
\begin{aligned}
A_{V \, (overall)} \, (dB) &= 20 \log_{10} A_{V \, (overall)} \\
&= 20 \log_{10} (A_{V1} \cdot A_{V2} \cdot A_{V3} \cdot \ldots) \\
&= 20 \, (\log_{10} A_{V1} + \log_{10} A_{V2} + \log A_{V3} + \ldots) \\
&= 20 \log_{10} A_{V1} + 20 \log A_{V2} + \ldots \\
A_{V \, (overall)} \, (dB) &= A_{V1} \, (dB) + A_{V2} \, (dB) + \ldots \qquad (11.6)
\end{aligned}
$$

Thus, the overall voltage gain expressed in decibels is simply the sum of each of the stage gains expressed in decibels.

Similarly, for the overall current gain:

$$A_{I \, (overall)} \, (dB) = A_{I1} \, (dB) + A_{I2} \, (dB) + \ldots \qquad (11.7)$$

and

$$
\begin{aligned}
A_{P \, (overall)}(dB) &= 10 \, [\log_{10} (A_{P1} \cdot A_{P2} \cdot A_{P3} \ldots)] \\
&= 10 \, [\log_{10} A_{P1} + \log A_{P2} + \log A_{P3} + \ldots] \\
&= 10 \log A_{P1} + 10 \log A_{P2} + 10 \log A_{P3} + \ldots \\
A_{P \, (overall)}(dB) &= A_{P1dB} + A_{P2dB} + A_{P3dB} + \ldots \qquad (11.8)
\end{aligned}
$$

Let us use formula (11.3) and the voltage gain formulas developed in a previous chapter to determine the overall voltage gain of a two-stage bipolar transistor amplifier.

Example (11.1). Determine the overall voltage gain of the two stage amplifier shown in Fig. 11.3.

Starting with the output stage, we compute the stage voltage gain:

$$A_{V2} \doteq \frac{-h_{fe2} \, Z_{L2}}{h_{ie2} + (h_{fe2} + 1) \, R_{E2}} = \frac{-50 \, (.5)}{.5 + 51 \, (.18)} = 2.58$$

$$A_{V1} \doteq \frac{-h_{fe1} \, Z_{L1}}{h_{ie1}} = \frac{-200 \, (10 \text{ k} \, /\!/ \, 56 \text{ k} \, /\!/ \, 5.6 \text{ k} \, /\!/ \, 9.68 \text{ k})}{5 \text{ k}}$$

$$\doteq -100$$

$$A_{V \, (overall)} \doteq A_{V1} \cdot A_{V2} = 258.2$$

368

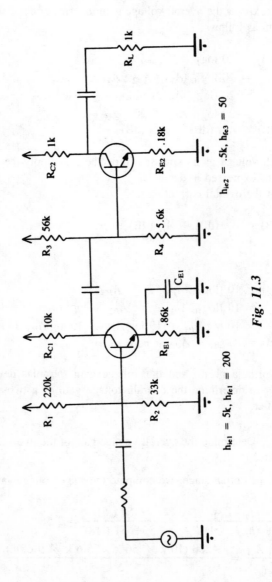

Fig. 11.3

$h_{ie2} = .5k, h_{fe3} = 50$

$h_{ie1} = 5k, h_{fe1} = 200$

Example (11.2).   Determine the overall voltage gain of the two-stage amplifier shown in Fig. 11.4.

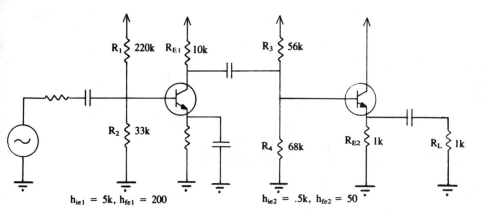

Fig. 11.4

Once again, starting at the output stage, we compute:

$$A_{V2} \doteq \frac{1}{1 + \dfrac{h_{ie1}}{(h_{fe1} + 1) Z_{E2}}} = \frac{1}{1 + \dfrac{.5}{51\,(.5)}} = .98$$

$$Z_{12} \doteq h_{ie2} + (h_{fe2} + 1) Z_{E2} = .5 + 51\,(.5) = 26 \text{ k}$$

$$Z_{L1} = R_{C1} \,/\!/\, R_3 \,/\!/\, R_4 \,/\!/\, Z_{12} \doteq 10\text{ k} \,/\!/\, 56\text{ k} \,/\!/\, 68\text{ k} \,/\!/\, 26\text{ k}$$
$$\doteq 5.85 \text{ k}$$

$$A_{V1} \doteq \frac{h_{fe1}\, Z_{L1}}{h_{ie1}} = \frac{200\,(5.85)}{5} = -233.9$$

$$A_{V\ (overall)} = A_{V1} \cdot A_{V2} \doteq -233.9\,(.98) = -229$$

Example (11.3).   Determine the voltage gain of the two-stage amplifier shown in Fig. 11.5.

We calculate:

$$A_{V2} = \frac{-g_{m2}\, Z_{L2}}{1 + \dfrac{R_{S2}}{r_{d2}}(1 + g_{m2}\, r_{d2}) + \dfrac{Z_{L2}}{r_{d2}}} = \frac{-2.5\,(2.5)}{1 + \dfrac{.2}{100}\,[1 + 2.5(100)] + \dfrac{2.5}{100}}$$
$$= -4.09$$

370

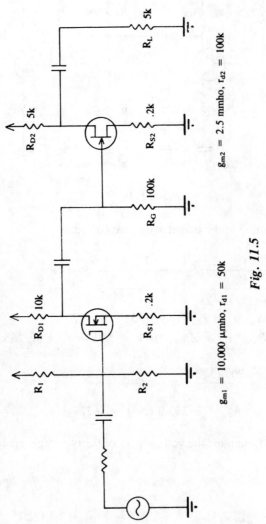

$g_{m1} = 10,000 \ \mu mho, \ r_{d1} = 50k$

$g_{m2} = 2.5 \ mmho, \ r_{d2} = 100k$

Fig. 11.5

$$A_{V1} = \frac{g_{m1} Z_{L1}}{1 + \dfrac{R_{S1}}{r_{d1}} (1 + g_{m1} r_{d1}) + \dfrac{Z_{L1}}{r_{d1}}}$$

$$= \frac{-10 \, (10 \,/\!/\, 100)}{1 + \dfrac{.2}{50} [1 + 10(50)] + \dfrac{10 \,/\!/\, 100}{50}}$$

$$= -28.5$$

$$A_{V \, (overall)} = A_{V1} \cdot A_{V2} = -28.5 \, (-4.09) = 117$$

Example (11.4). Determine the overall voltage gain of the two-stage amplifier shown in Fig. 11.6.

We calculate:

$$A_{V2} = \frac{1}{1 + \dfrac{1 + Z_{S2}/r_{d2}}{g_{m2} Z_{S2}}} = \frac{1}{1 + \dfrac{1 + .5/100}{2.5 \, (.5)}} = .554$$

$$A_{V1} = \frac{-g_{m1} Z_{L1}}{1 + \dfrac{R_{S1}}{r_{d1}} (1 + g_{m1} r_{d1}) + \dfrac{Z_{L1}}{r_{d1}}} = \frac{-10 \, (10 \,/\!/\, 120 \,/\!/\, 100)}{1 + 0 + \dfrac{10 \,/\!/\, 120 \,/\!/\, 100}{50}}$$

$$= -84.5$$

$$A_{V \, (overall)} = -84.5 \, (.554) = -46.8$$

11.2 CURRENT GAIN AND POWER GAIN

To this point, we have considered voltage gain to the exclusion of current and power gain. For bipolar transistors, the signal current is increased from one stage to the succeeding stage. For FETs, however, because, in theory, no current flows into the gate lead, it is somewhat meaningless to discuss current gain per stage. Overall current gain for a multistage FET amplifier, on the other hand, does have some meaning because a weak signal source will deliver some signal current into the input transistor gate resistor, and signal current certainly will flow in the FET-output stage drain and load resistors. Thus, a ratio of output signal current to input signal current can be computed.

A more meaningful expression for current gain is possible for multistage bipolar transistor amplifiers. It is important to emphasize that dc bias currents are not considered in the computation of signal current gain. The only possible interest that a designer may have in dc bias currents for signal current analysis is that the peak signal current cannot exceed the dc bias current because the transistor will be driven into cut-off.

372

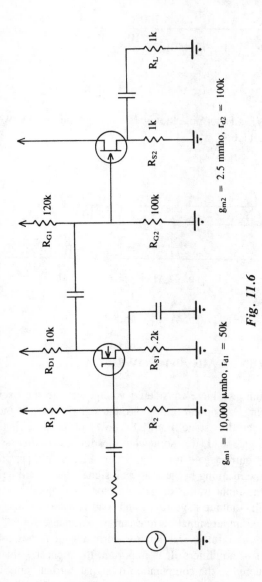

$g_{m1} = 10{,}000\ \mu\text{mho},\ r_{d1} = 50\text{k}$

$g_{m2} = 2.5\ \text{mmho},\ r_{d2} = 100\text{k}$

Fig. 11.6

It is important to emphasize that current gain is the ratio of *output signal current to input signal current*. Wherever resistor networks form current dividers, signal current will be lost (see Fig. 11.7).

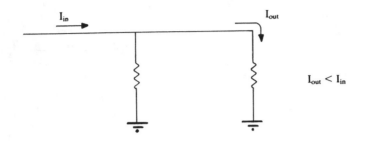

Fig. 11.7

Let us examine a single-stage bipolar transistor amplifier and determine its current gain.

Example (11.5). Determine the overall current gain for the amplifier shown in Fig. 11.8.

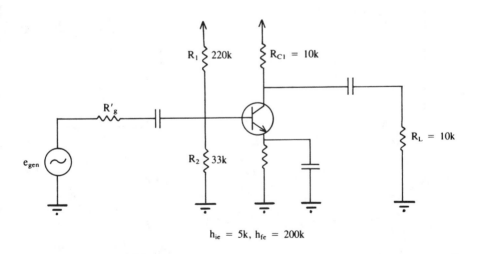

$h_{ie} = 5k$, $h_{fe} = 200k$

Fig. 11.8

Redrawing the ac equivalent circuit, we have the schematic of Fig. 11.9. Redrawing again, we obtain the schematic of Fig. 11.10.

374

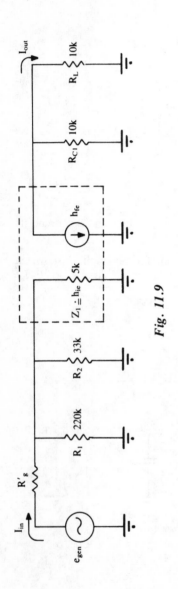

Fig. 11.9

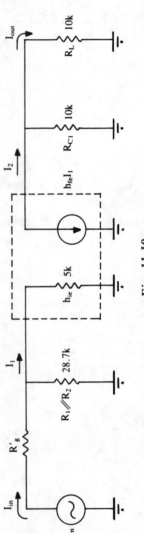

Fig. 11.10

We note in the first equivalent circuit (Fig. 11.9) that the top ends of R_1 and R_{C1} are shown at ac ground potential. (The positive terminal of the power supply is at ac ground potential.)

By use of the current divider rule, let us determine an expression for I_1. We derive

$$I_1 = I_{in} \frac{R_1 \mathbin{/\mkern-5mu/} R_2}{R_1 \mathbin{/\mkern-5mu/} R_2 + h_{ie}} =$$

or $\dfrac{I_1}{I_{in}} = \dfrac{R_1 \mathbin{/\mkern-5mu/} R_2}{R_1 \mathbin{/\mkern-5mu/} R_2 + h_{ie}} = \dfrac{28.7}{28.7 + 5} = 0.85$

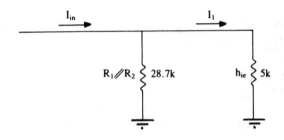

I_{in} I_1

$R_1/\!/R_2$ ⌇ 28.7k h_{ie} ⌇ 5k

Fig. 11.11

We know that the ratio of I_2 to I_1 is simply the current gain of the transistor by itself, or

$$A_I = \frac{I_2}{I_1} \doteq h_{fe} = 200$$

By use of the current divider rule, let us compute the value of I_{out}. We derive

$$I_{out} = I_2 \frac{R_{C1}}{R_{C1} + R_2}$$

Rewriting:

$$\frac{I_{out}}{I_2} = \frac{R_{C1}}{R_{C1} + R_2} = \frac{10}{10 + 10} = 0.5$$

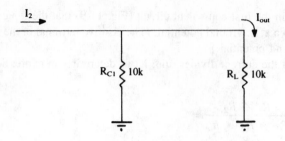

Fig. 11.12

We also know that

$$A_{I \text{ (overall)}} = A_{I1} \cdot A_{I2} \cdot A_{I3}$$
$$= \frac{I_1}{I_{in}} \cdot \frac{I_2}{I_1} \cdot \frac{I_{out}}{I_2} = \frac{I_{out}}{I_{in}}$$
$$= 0.85 \ (200) \ (0.5)$$
$$A_{I \text{ (overall)}} = 85$$

While we are involved in the analysis of this stage, let us compute the voltage gain:

$$A_V \doteq \frac{-h_{fe} \, Z_L}{h_{ie}} = \frac{-200 \ (5)}{5} = -200$$

Restating the voltage gain:

$$A_V = \frac{V_{out}}{V_{in}}$$

and for our example,

$$V_{out} = I_{out} \, R_L \text{ and } V_{in} = I_{in} \, Z_{in}$$

and $Z_{in} = R_1 \mathbin{/\mkern-5mu/} R_2 \mathbin{/\mkern-5mu/} h_{ie} = 220 \text{ k} \mathbin{/\mkern-5mu/} 33 \text{ k} \mathbin{/\mkern-5mu/} 5 \text{ k}$
$$= 4.26 \text{ k}$$

If

$$A_V = \frac{V_{out}}{V_{in}} = \frac{-I_{out} \, R_L}{I_{in} \, Z_{in}}$$

then we can rewrite:

$$A_V = -\left(\frac{I_{out}}{I_{in}}\right)\frac{R_L}{Z_{in}} = -A_I\frac{R_L}{Z_{in}} \tag{11.9}$$

and for our example,

$$A_V = -85\frac{10\text{ k}}{4.26\text{ k}} = -200$$

This checks against our computation for A_V using the formula:

$$A_V = \frac{-h_{fe}\,Z_L}{h_{ie}}$$

11.2.1 Calculation of the Overall Current Gain

Let us compute the overall current gain of a two-stage amplifier. We will use the circuit from a previous example, Example (11.2), to compute its current gain. With the computed current gain value, we will then compute the voltage gain and compare the result against the previously determined value of A_V.

Example (11.6). Compute the current gain and use this value to determine the voltage gain for the two-stage amplifier shown in Fig. 11.13.

Redrawing Fig. 11.13, we obtain the schematic of Fig. 11.14.

According to the current divider rule:

$$I_1 = I_{in}\frac{R_1 /\!/ R_2}{R_1 /\!/ R_2 + h_{ie1}}$$

or

$$\frac{I_1}{I_{in}} = \frac{R_1 /\!/ R_2}{R_1 /\!/ R_2 + h_{ie1}} = \frac{28.7}{28.7 + 5} = .852$$

$$\frac{I_2}{I_1} \doteq h_{fe1} = 200$$

By the current divider rule:

$$I_3 = I_2\frac{R_{C1} /\!/ R_3 /\!/ R_4}{R_{C1} /\!/ R_3 /\!/ R_4 + Z_{12}}$$

or

$$\frac{I_3}{I_2} = \frac{R_{C1} \parallel R_3 \parallel R_4}{R_{C1} \parallel R_3 \parallel R_4 + Z_{12}} = \frac{3.37}{3.37 + 9.68} = .258$$

where

$$Z_{12} = h_{ie2} + (h_{fe2} + 1) R_{E2} = .5 + 51 (.18) = 9.68 \text{ k}$$
$$\frac{I_4}{I_3} \doteq h_{fe2} = 50$$

Also,

$$I_{out} = I_4 \frac{R_{C2}}{R_{C2} + R_L}$$

or

$$\frac{I_{out}}{I_4} = \frac{R_{C2}}{R_{C2} + R_L} = \frac{1}{1 + 1} = .5$$

$$A_{I \text{ (overall)}} = A_{I1} \cdot A_{I2} \cdot A_{I3} \cdot A_{I4} \cdot A_{I5}$$
$$= \frac{I_1}{I_{in}} \cdot \frac{I_2}{I_1} \cdot \frac{I_3}{I_2} \cdot \frac{I_4}{I_3} \cdot \frac{I_{out}}{I_4}$$
$$A_{I \text{ (overall)}} = .852 \ (200) \ (.258) \ (50) \ (.5) = 1100$$
$$A_{V \text{ (overall)}} = \frac{V_{out}}{V_{in}} = \frac{I_{out} R_L}{I_{in} Z_{in}} = \left(\frac{I_{out}}{I_{in}}\right) \frac{R_L}{Z_{in}} = A_I \frac{R_L}{Z_{in}}$$

$$A_{V \text{ (overall)}} = 1100 \cdot \frac{1 \text{ k}}{4.26 \text{ k}} = 258.3$$

which checks against our previous results.

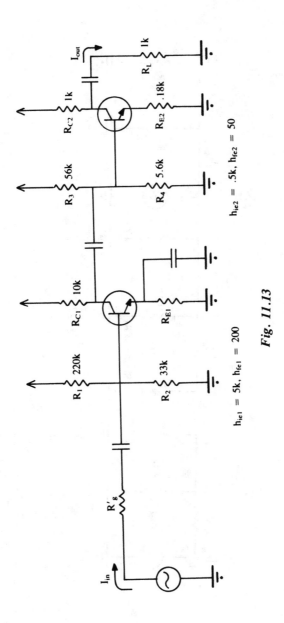

Fig. 11.13

380

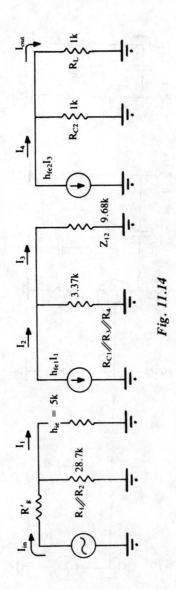

Fig. 11.14

PROBLEMS

1. The voltage gain of each of three FET amplifier stages connected in tandem is given by the diagram of Fig. 11.15.

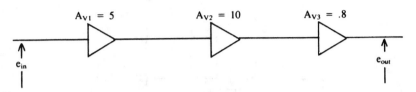

Fig. 11.15

If an input signal of three millivolts (3 mV) is applied to the network, what is the value of the output voltage?

2. What is the overall voltage gain of the amplifier of Problem 1? Express this value in decibels (dB).

3. If the overall current gain is equal to the voltage gain, what is the overall power gain expressed as an absolute value and in terms of decibels?

4. For the circuit shown in Fig. 11.16:
   (a) compute the overall voltage gain;
   (b) compute the overall current gain;
   (c) use the results of (b) to compute the voltage gain. How does this compare with (a)?

5. Repeat the steps of Problem 4, for the circuit shown in Fig. 11.17.

6. Determine the overall voltage gain for the circuit shown in Fig. 11.18. Express the result in decibels.

7. Determine the overall voltage gain for the circuit shown in Fig. 11.19. Express the result in decibels.

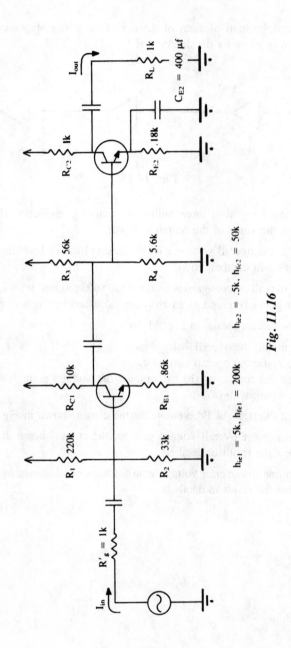

Fig. 11.16

383

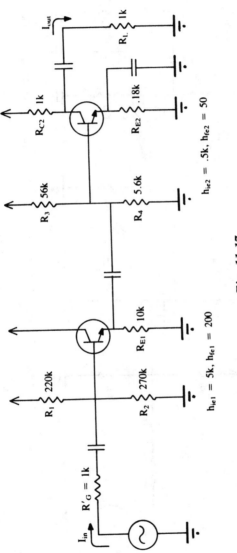

Fig. 11.17

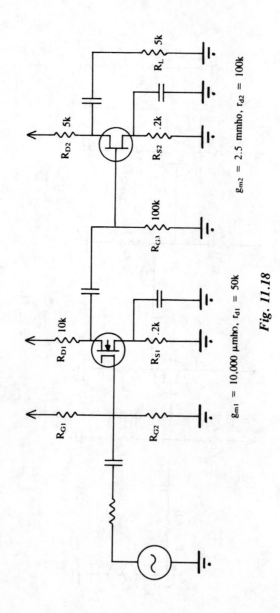

Fig. 11.18

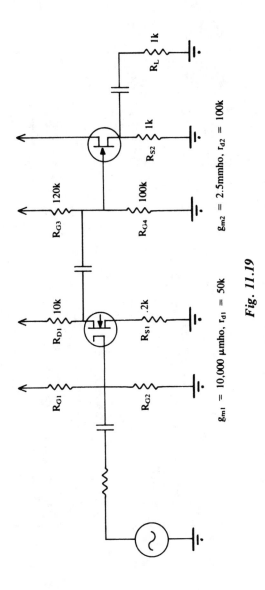

$g_{m1} = 10,000 \ \mu mho, \ r_{d1} = 50k$ $g_{m2} = 2.5mmho, \ r_{d2} = 100k$

Fig. 11.19

Chapter 12

Multistage Amplifier Frequency Considerations

12.1 FREQUENCY RESPONSE

12.1.1 Two-Stage Bipolar Transistor (Common-Emitter Feeding Common-Emitter)

In this chapter we will utilize the formulas developed in previous chapters to both analyze and design the high- and low-frequency characteristics of multistage, small-signal amplifiers.

Example (12.1). Let us compute the voltage gain of a two-stage bipolar transistor amplifier, and a few unspecified component values, and the high frequency cutoff of the input, interstage, and output networks. See Fig. 12.1.

Given:

$$h_{fe1} = 100, \; h_{ie1} = 1000 \; \Omega, \; h_{fe2} = 50, \; h_{ie2} = 500 \; \Omega$$

where

$$C_{be1} = C_{be2} = 8 \; \text{pF}, \; C_{bc1} = C_{bc2} = 1.2 \; \text{pF}, \; C_{Ce1} = C_{Ce2} = 3 \; \text{pF}$$

Determine:

$$C_{C1}, \; C_{E1}, \; C_{C2}, \; C_{E2}, \; C_{C3}$$

Design each capacitor for a low frequency cut-off of 20 Hz. Also determine $A_{V(overall)}$ and the high frequency cut-off of the input, interstage, and output networks. Thus,

$$A_{V1} \doteq \frac{h_{fe1} Z_{L1}}{h_{ie1}} \qquad Z_{L1} = R_{C1} \; /\!/ \; R_3 \; /\!/ \; R_4 \; /\!/ \; Z_{12}$$

$$\doteq R_{C1} \; /\!/ \; R_e \; /\!/ \; R_4 \; /\!/ \; h_{ie2}$$

387

$$A_{V1} \doteq \frac{100\,(.382)}{1} \qquad Z_{L1} \doteq 5k \,/\!/\, 22k \,/\!/\, 2.7k \,/\!/\, 0.5k$$

$$\doteq -38.2 \qquad\qquad\quad \doteq 0.382k$$

$$A_{V2} \doteq \frac{h_{fe2}\, Z_{L2}}{h_{ie2}} \qquad Z_{L2} = R_{C2} \,/\!/\, R_L = 1.2k \,/\!/\, 2.4k$$

$$\doteq \frac{50(0.8)}{0.5} \qquad\quad Z_{L2} = 0.8k$$

$$\doteq -80$$

$$A_{V\,(overall)} = A_{V1} \cdot A_{V2} = -38.2\,(-80) = 3058$$

For a coupling capacitor, we have

$$C_{C1} = \frac{1}{2\,\pi\, f_{LOW}\,(R_A + R_B)} = \frac{1}{2\,\pi\, f_{LOW}\,(R_g + Z_{in1})}$$

$$= \frac{1}{2\,\pi\,(20)\,(1\,k + 0.9\,k)}$$

where

$$Z_{in1} = R_1 \,/\!/\, R_2 \,/\!/\, Z_{11}$$
$$\doteq R_1 \,/\!/\, R_2 \,/\!/\, h_{ie1}$$
$$\doteq 100\,k \,/\!/\, 10\,k \,/\!/\, 1\,k$$
$$\doteq 0.9\,k$$
$$C_{C1} = 4.18\ \mu F$$

and for the first bypass capacitor:

$$C_{E1} = \frac{5}{\pi\, f_{LOW}\, R_E} = \frac{5}{\pi\,(20)\,400} = 199\ \mu F$$

for the second coupling capacitor:

$$C_{C2} = \frac{1}{2\,\pi\, f_{LOW}\,(R_A + R_B)}$$

$$= \frac{1}{2\,\pi\,(20)\,(5 + 0.413)k}$$

$$C_{C2} = 1.47 \ \mu\text{F}$$

where

$$
\begin{aligned}
R_A &= Z_{21} \doteq R_{C1} = 5k \\
R_B &\doteq R_3 \ /\!/ \ R_4 \ /\!/ \ Z_{12} \\
&\doteq R_3 \ /\!/ \ R_4 \ /\!/ \ h_{ie2} \\
&\doteq 22k \ /\!/ \ 2.7k \ /\!/ \ 0.5k \\
&\doteq 0.413k
\end{aligned}
$$

and for the second bypass capacitor:

$$C_{E2} = \frac{5}{\pi \ f_{LOW} \ R_{E2}} = \frac{5}{\pi \ (20) \ 86} = 925 \ \mu\text{F}$$

and for the third coupling capacitor:

$$
\begin{aligned}
C_{C3} &= \frac{1}{2 \ \pi \ f_{LOW} \ (R_A + R_B)} \\
&= \frac{1}{2 \ \pi \ (20) \ (1.2k + 2.4k)} \\
C_{C3} &= 2.2 \ \mu\text{F}
\end{aligned}
$$

where
$$
\begin{aligned}
R_A &= R_{C2} \ /\!/ \ Z_{22} \\
R_B &= R_L = 2.4k
\end{aligned}
$$

The high frequency cut-off of the input network can be computed as follows, but first let us determine C_{in1}:

$$
\begin{aligned}
C_{in1} &= C_{be1} + (1 - A_{V1}) \ C_{cb1} = 8 \ \text{pF} + [1 - (-36.4)] \ 1.2 \ \text{pF} \\
C_{in1} &= 52.88 \ \text{pF}
\end{aligned}
$$

$$C_{Hi \ input} = \frac{1}{2 \ \pi \ C_{Hi} \dfrac{(R_g \ R)}{R_g + R}} = \frac{1}{2 \ \pi \ C_{Hi} \dfrac{(R_g \ Z_{in1})}{R_g + Z_{in1}}}$$

But $Z_{in1} = R_1 \ /\!/ \ R_2 \ /\!/ \ Z_{11} \doteq R_1 \ /\!/ \ R_2 \ /\!/ \ h_{ie1}$
$$\doteq 100k \ /\!/ \ 10k \ /\!/ \ 1k = 0.9k$$

$$f_{Hi \ input} = \frac{1}{2 \ \pi \ (52.88) \ (10^{-12}) \left[\dfrac{1k \ (0.9k)}{1k + 0.9k} \right]} = 5.8 \ \text{MHz}$$

$$C_{in2} = C_{be2} + (1 - A_{V2}) C_{cb2} = 8 + [1 - (-80)] \qquad (1.2)$$
$$= 105.2 \text{ pF}$$
$$C_{out1} = C_{bc1} = 1.2 \text{ pF}$$
$$C_w \text{ (interstage)} = 10 \text{ pF}$$
$$C_{Hi} = C_{out1} + C_w \text{ (interstage)} + C_{in2} = 1.2 + 10 + 105.2 + 116.4 \text{ pF}$$
$$f_{Hi} \text{ (interstage)} = \cfrac{1}{2\pi \, C_{Hi} \cfrac{(R_g \, R)}{R_g + R}}$$

where $R_g = Z_{o1} = R_{c1} /\!/ Z_{21} \doteq R_{C1} /\!/ h_{oe1} \doteq R_{C1} = 5\text{k}$

$$R = Z_{in2} = R_3 /\!/ R_4 /\!/ Z_{in2} \doteq R_3 /\!/ R_4 /\!/ h_{ie2}$$
$$\doteq 22\text{k} /\!/ 2.7\text{k} /\!/ 0.5\text{k} = 0.414\text{k}$$

$$f_{Hi} \text{ (interstage)} = \cfrac{1}{2\pi \, (116.4) \, (10^{-12}) \left[\cfrac{5\text{k} \, (0.414\text{k})}{5\text{k} + 0.414\text{k}} \right]}$$

$$f_{Hi} \text{ (interstage)} = 3.58 \text{ MHz}$$

For the output network:

$$C_{Hi} = C_{out2} + C_w = C_{ce2} + C_w = 3\text{pF} + 4\text{pF} = 7\text{pF}$$
$$f_{Hi} \text{ (output)} = \cfrac{1}{2\pi \, C_{Hi} \cfrac{(R_g \, R)}{R_g + R}}$$
$$= \cfrac{1}{2\pi \, (7) \, (10^{-12}) \left[\cfrac{1.2\text{k} \, (2.4\text{k})}{1.2\text{k} + 2.4\text{k}} \right]}$$

$$f_{Hi} \text{ (output)} = 89.3 \text{ MHz}$$

where $R_g = R_{C2} /\!/ Z_{22}$
$$\doteq R_{C2} = 1.2\text{k}$$
$$R = R_L = 2.4\text{k}$$

391

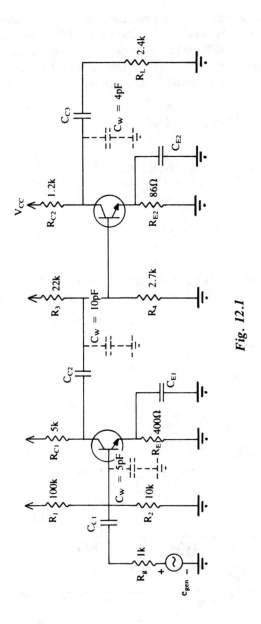

Fig. 12.1

12.1.2 Common-Emitter Feeding Emitter-Follower

Let us analyze another circuit, a common-emitter stage feeding an emitter-follower stage. From the standpoint of performing a laboratory experiment, this circuit may be preferable because the previous two-stage common emitter circuit with its high voltage gain may have a tendency to be unstable and, thus, oscillate, particularly if great care is not taken in the circuit layout and component placement.

Example (12.2). Refer to Fig. 12.2.
Given:

$$h_{fe1} = 100, \ h_{ie1} = 1k, \ h_{fe2} = 50, \ h_{ie2} = 0.5k$$
$$C_{be1} = C_{be2} = 8pF, \ C_{bc1} = C_{bc2} = 1.2pF, \ C_{ce1} = C_{ce2} = 3pF$$

Determine:

$$C_{C1}, \ C_{E1}, \ C_{C2}, \ C_{C3}$$

Design each capacitor for a low frequency cut-off of 20 Hz. Also determine $A_{V \ (overall)}$ and the high frequency cut-off of the input, interstage, and output networks. Thus,

$$A_{V1} \doteq \frac{-h_{fe1} \ Z_{L1}}{h_{ie1}}$$

$$\doteq \frac{-100 \ (3.57)}{1}$$

$$= -357$$

where $Z_{L1} = R_{C1} \ /\!/ \ R_3 \ /\!/ \ R_4 \ /\!/ \ Z_{12}$

$$\doteq R_{C1} \ /\!/ \ R_3 \ /\!/ \ R_4 \ /\!/ \ [h_{ie2} + (h_{fe2} + 1) \ Z_{E2}]$$

$$\doteq 5k \ /\!/ \ 33k \ /\!/ \ 39k \ /\!/ \ [0.5k + 51 \ (0.8k)]$$

$$\doteq 3.57k$$

$$A_{V2} \doteq \frac{1}{1 + \dfrac{h_{ie2}}{(h_{fe2} + 1)Z_{E2}}} = \frac{1}{1 + \dfrac{0.5}{51 \ (0.8)}} = 0.988$$

$$A_{V \ (overall)} = A_{V1} \cdot A_{V2} = -357 \ (0.988) = -353$$

$$C_{C1} = \frac{1}{2\pi f_{LOW} \ (R_A + R_B)} = \frac{1}{2\pi f_{LOW} \ (R_{gen} + Z_{in1})} = 4.18 \ \mu F$$

from previous example:

$$C_{C2} = \frac{1}{2\pi f_{LOW} (R_A + R_B)} = \frac{1}{2\pi f_{LOW} (R_{C1} \,/\!/\, Z_{21} + R_3 \,/\!/\, R_4 \,/\!/\, Z_{12})}$$

$$C_{C2} = \frac{1}{2\pi (20) [5k \,/\!/\, \infty + 33k \,/\!/\, 39k \,/\!/\, (0.5k + 51(0.8))]} = 0.455 \ \mu F$$

$$C_{C3} = \frac{1}{2\pi f_{LOW} (R_A + R_B)} = \frac{1}{2\pi f_{LOW} (Z_3 \,/\!/\, R_E + R_L)}$$

$$= \frac{1}{2\pi f_{LOW} \left[\dfrac{R_{C1} \,/\!/\, R_3 \,/\!/\, R_4 + h_{ie2}}{h_{fe2} + 1} \,/\!/\, R_E + R_L \right]}$$

$$C_{C3} = \frac{1}{2\pi (20) \left[\dfrac{5k \,/\!/\, 33k \,/\!/\, 39k + 0.5k}{51} \,/\!/\, 1.2k + 2.4k \right]} = 3.2 \ \mu F$$

The high frequency cut-off of the input network can be computed as follows, but first let us determine C_{in1}:

$$C_{in1} = C_{be1} + (1 - A_{V1}) C_{cb1} = 8 + [1 - (-357)] 1.2 = 437.6 pF$$
$$C_{Hi} = C_{in1} + C_w = 437.6 + 5 + 442.6 pF$$
$$f_{Hi} = \frac{1}{2\pi C_{Hi} \left(\dfrac{R_g R}{R_g + R} \right)} = \frac{1}{2\pi C_{Hi} \left(\dfrac{R_g Z_{in1}}{R_g + Z_{in1}} \right)}$$
$$Z_{in1} = R_1 \,/\!/\, R_2 \,/\!/\, Z_{11} \doteq R_1 \,/\!/\, R_2 \,/\!/\, h_{ie1} = 100k \,/\!/\, 10k \,/\!/\, 1k = 0.9k$$

$$f_{Hi \ (input)} = \frac{1}{2\pi (442.6) (10^{-12}) \left[\dfrac{1k (0.9k)}{1k + 0.9k} \right]} = 759 \ kHz$$

The high frequency cut-off of the interstage network can be computed as follows, but first let us compute C_{in2}:

$$C_{in2} = C_{bc2} + (1 - A_{V2}) C_{be2} = 1.2 + (1 - 0.988) 8 = 1.3 pF$$
$$C_{Hi \ (interstage)} = C_{out1} + C_{wiring} + C_{in2} = C_{ce1} + C_w + C_{in2}$$
$$= 3pF + 10pF + 1.3pF = 14.3 pF$$
$$f_{Hi \ (interstage)} = \frac{1}{2\pi C_{Hi} \left[\dfrac{R_g R}{R_g + R} \right]}$$

394

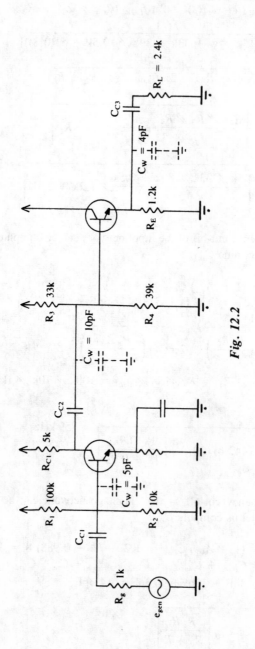

Fig. 12.2

where $R_g = Z_{o1} = R_{C1} /\!/ Z_{21} \doteq R_{C1} = 5\text{k}$

$R = Z_{in2} = R_3 /\!/ R_4 /\!/ Z_{12} \doteq R_3 /\!/ R_4 /\!/ [h_{ie2} + (h_{fe2} + 1)Z_2]$

$\doteq 33\text{k} /\!/ 39\text{k} /\!/ [0.5\text{k} + 51\,(0.8\text{k})] = 12.5\text{k}$

therefore

$$f_{Hi} = \cfrac{1}{2\pi\,(14.3)\,(10^{-12})\left[\cfrac{5\text{k}\,(12.5\text{k})}{5\text{k} + 12.5\text{k}}\right]} = 3.12\,\text{MHz}$$

Computing the high frequency cut-off of the output network:

$$C_{o2} = C_{ce2} = 3\text{pF}$$

and

$$C_{Hi} = C_{o2} + C_{wiring} = 3\text{pF} + 4\text{pF} = 7\text{pF}$$

$$f_{Hi} = \cfrac{1}{2\pi\,C_{Hi}\left(\cfrac{R_g R}{R_g + R}\right)} = \cfrac{1}{2\pi\,C_{Hi}\left(\cfrac{R_g R_L}{R_g + R_L}\right)}$$

where

$$R_g = Z_3 /\!/ R_E$$

and

$$Z_3 \doteq \frac{h_{ie2} + R_{32}}{h_{fe2} + 1} = \frac{h_{ie2} + R_{C1} /\!/ R_3 /\!/ R_4}{h_{fe2} + 1}$$

$$\doteq \frac{0.5\text{k} + 5\text{k} /\!/ 33\text{k} /\!/ 39\text{k}}{51} = 0.0864\text{k}$$

$$R_g = Z_3 /\!/ R_E = 0.0864\text{k} /\!/ 1.2\text{k} = 0.806\text{k}$$

$$f_{Hi} = \cfrac{1}{2\pi\,(7)\,(10^{-12})\left[\cfrac{0.0806\text{k}\,(2.4\text{k})}{0.0806\text{k} + 2.4\text{k}}\right]} = 291\,\text{MHz}$$

To complete our analysis, we must determine the overall current gain. Drawing an equivalent circuit, we have the schematic of Fig. 12.3.

Redrawing Fig. 12.3 and combining resistors to facilitate use of the current divider rule, we derive Fig. 12.4,

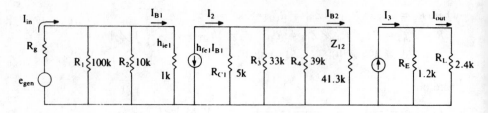

Fig. 12.3

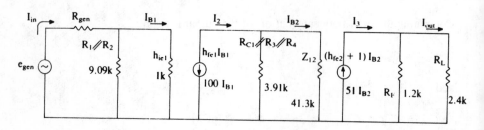

Fig. 12.4

where

$$Z_{12} = h_{ie2} + (h_{fe2} + 1) Z_E = 0.5 + 51(0.8) = 41.3\text{k}$$

We note that the arrows representing the direction of current flow in the second current generator (Fig. 12.4) are pointing upward. The reason for this direction is that an increase in base current causes an increase in the emitter current flowing *into* the emitter and load resistors.

According to the current divider rule:

$$I_{B1} = I_{in} \frac{R_1 /\!/ R_2}{R_1 /\!/ R_2 + h_{ie1}}$$

therefore

$$\frac{I_{B1}}{I_{in}} = \frac{R_1 /\!/ R_2}{R_1 /\!/ R_2 + h_{ie1}} = \frac{9.09\text{k}}{9.09\text{k} + 1\text{k}}$$

$$\frac{I_{B1}}{I_{in}} = 0.9$$

Also,

$$I_2 = -h_{fe} I_{B1}$$

and the ratio of I_2 to I_{B1} is

$$\frac{I_2}{I_{B1}} = \frac{-h_{fe1} I_{B1}}{I_{B1}} = -h_{fe1} = -100$$

By use of the current divider rule:

$$I_{B2} = I_2 \frac{R_{C1} /\!/ R_3 /\!/ R_4}{R_{C1} /\!/ R_3 /\!/ R_4 + Z_{12}}$$

Thus,

$$\frac{I_{B2}}{I_2} = \frac{R_{C1} /\!/ R_3 /\!/ R_4}{R_{C1} /\!/ R_3 /\!/ R_4 + Z_{12}} = \frac{3.91k}{3.91k + 41.3k} = 0.0864$$

The ratio of I_3 to I_{B2} is

$$\frac{I_3}{I_{B2}} = \frac{(h_{fe2} + 1) I_{B2}}{I_{B2}} = 51$$

By the current divider rule:

$$I_{out} = I_3 \frac{R_E}{R_E + R_L}$$

or

$$\frac{I_{out}}{I_3} = \frac{R_E}{R_E + R_L} = \frac{1.2k}{1.2k + 2.4k} = 0.333$$

Therefore,

$$A_{I\,(overall)} = \frac{I_{B1}}{I_{in}} \times \frac{I_2}{I_{B1}} \times \frac{I_{B2}}{I_2} \times \frac{I_3}{I_{B2}} \times \frac{I_{out}}{I_3}$$
$$= 0.9 \times (-100) \times (0.0864) \times (51) \times (0.333)$$
$$A_{I\,(overall)} = -132.4$$

Let us use the overall current gain value to compute the overall voltage gain (as a check against our earlier work):

$$A_{V\,(overall)} = \frac{V_{out}}{V_{in}} = \frac{I_{out} Z_L}{I_{in} Z_{in}} = A_I \frac{Z_L}{Z_{in}}$$

First, let us compute Z_{in}:

$$Z_{in} = R_1 /\!/ R_2 /\!/ Z_{11} \doteq R_1 /\!/ R_2 /\!/ h_{ie1} = 100k /\!/ 10k /\!/ 1k = 0.9k$$

$$A_{V \, (overall)} = -132.4 \left(\frac{2.4k}{0.9k}\right) = -353$$

which checks!

12.2 OVERALL FREQUENCY RESPONSE

In previous chapters, we discussed the effects of coupling and bypass capacitors on low frequency response and the effects of inter-electrode and wiring capacitance on the high frequency response. We also noted that the low frequency response was a parameter which we could control. If we wanted superior low frequency response, we simply increased the selected bypass and coupling capacitor values. We also noted that our control over high frequency response was more restricted because the wiring capacitance could be minimized, but not eliminated. The inter-electrode capacitance could also be minimized by selecting an appropriate transistor. A high β or transconductance could also be utilized to improve the high frequency performance.

What effect do these various capacitances have upon the overall frequency response curve? Perhaps an example would be appropriate to answer this question.

Example (12.3). For the two-stage amplifier shown in Fig. 12.5, design for a low frequency break point of 20 Hz for the coupling capacitors, and sketch the overall frequency response curve.

The break-point frequency of the bypass capacitors may be designed considerably below 20 Hz, and then the coupling capacitors' break-point frequency may be designed for 20 Hz. The effect of each coupling capacitor is to make the voltage gain decrease by 6 dB per octave below the cut-off frequency (20 Hz in this example), and because we have three coupling capacitors, each selected for a 20 Hz low frequency break point, the total loss in amplification below 20 Hz would be 18 dB per octave. In reality, the cut-off frequency will no longer be 20 Hz for the overall two-stage amplifier, consisting of three coupling capacitors.

The effect of each R-C coupling network must be considered as follows. We know that at the cut-off frequency the ratio of the output voltage to the input voltage is 70.7%. If we consider only two coupling capacitor networks connected in series, the output of the first is 70.7% of its input (at the break-point frequency, f_{low}). This voltage becomes the input to the second coupling capacitor network and its output may be determined by following the diagram of Fig. 12.6.

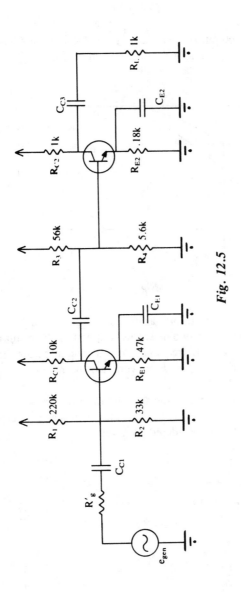

Fig. 12.5

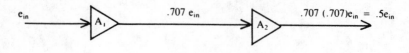

Fig. 12.6

If we sketch the output (normalized) *versus* the input for both one and two coupling networks, we would have the diagram of Fig. 12.7.

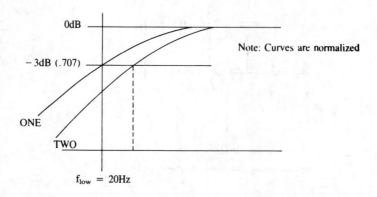

Fig. 12.7

Thus, the low frequency cut-off has increased from 20 Hz for a single network to a slightly higher value for two networks. Implicit in this result is the idea that if we desire a 20 Hz cut-off frequency for our two-stage amplifier with three coupling capacitor networks, *each* network will be required to have a cut-off frequency *below* 20 Hz.

From Equation (10.6), we know that the ratio of the output voltage to the mid-band voltage at any frequency, f_n, where the capacitive reactance in a coupling network is to be considered, is given by

$$\frac{e_{out}(f_n)}{e_{out\,(mid)}} = \frac{1}{1 - j\dfrac{X_C}{R_A + R_B}}$$

However,

$$X_C = \frac{1}{2\pi f_n C_C}$$

Thus,

$$\frac{e_{out}(f_n)}{e_{out\,(mid)}} = \frac{1}{1 - j\dfrac{1}{2\pi f_n C_C (R_A + R_B)}}$$

From Equation (10.8), we know that

$$f_{LOW} = \frac{1}{2\pi C_C (R_A + R_B)}$$

Substituting, we have

$$\frac{e_{out}(f_n)}{e_{out\,(mid)}} = \frac{1}{1 - j\dfrac{f_{LOW}}{f_n}}$$

and the *magnitude* of this ratio is

$$\left|\frac{e_{out}(f_n)}{e_{out\,(mid)}}\right| = \frac{1}{\sqrt{1 + \left(\dfrac{f_{LOW}}{f_n}\right)^2}}$$

If we had three coupling networks, each with an identical cut-off frequency of

$$\left.\frac{e_{out}(f_n)}{e_{out\,(mid)}}\right|_{overall} = \left[\frac{e_{out}(f_n)}{e_{out\,(mid)}}\right]^3 = \left[\frac{1}{\sqrt{1 + \left(\dfrac{f_{LOW}}{f_n}\right)^2}}\right]^3$$

and if we wanted this value to be equal to a -3 dB decrease, then

$$\left.\frac{e_{out}(f_n)}{e_{out\,(mid)}}\right|_{overall} = \frac{1}{\sqrt{2}}\left[\frac{1}{\sqrt{1 + \left(\dfrac{f_{LOW}}{f_n}\right)^2}}\right]^3$$

[*Note*: Decrease in magnitude for cut-off (or break-point) frequency $= \dfrac{1}{\sqrt{2}}$ (or .707) of midband frequency.]

or $2^{1/2} = \left[1 + \left(\dfrac{f_{LOW}}{f_n} \right)^2 \right]^{-3/2}$

or $2 = \left[1 + \left(\dfrac{f_{LOW}}{f_n} \right)^2 \right]^3$

$1 + \left(\dfrac{f_{LOW}}{f_n} \right)^2 = 2^{1/3}$

$\left(\dfrac{f_{LOW}}{f_n} \right)^2 = 2^{1/3} - 1$

$\dfrac{f_{LOW}}{f_n} = \sqrt{2^{1/3} - 1}$

therefore $f_n = \dfrac{f_{LOW}}{\sqrt{2^{1/3} - 1}}$

and for our example,

$$f_n = \frac{20}{\sqrt{2^{1/3} - 1}} = 39.2 \, \text{Hz}$$

Thus, if we used a low frequency cut-off of 20 Hz for each of three coupling networks, the overall cut-off frequency would be 39 Hz. If we rewrite the above equation:

$$f_{LOW} = f_n \sqrt{2^{1/3} - 1}$$

and substitute 20 Hz for the overall cut-off frequency, we derive

$$f_{LOW} = 20 \sqrt{2^{1/3} - 1} = 20(0.51) = 10.2 \, \text{Hz}$$

and, thus, *each* coupling network must have a cut-off frequency of 10.2 Hz if we are to have an overall cut-off frequency of 20 Hz. A more general expression for the relationship is as follows:

$$f_{LOW} = f_n \sqrt{2^{1/n} - 1} \tag{12.1}$$

where n is the number of coupling networks with identical low frequency cut-offs.

For a two-transistor amplifier with three coupling networks, each with a 20 Hz cut-off, as shown in Example (12.3), the bandpass curve might appear as given in Fig. 12.8.

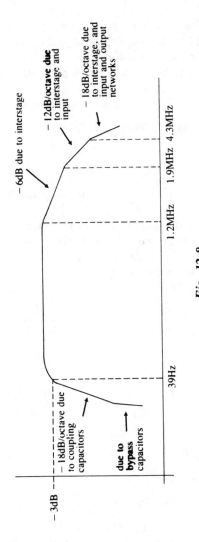

Fig. 12.8

If, by chance, the upper cut-off frequency of each stage were identical, a similar decrease in the overall high frequency cut-off would occur.

PROBLEMS

1. Compute the mid-band voltage gain of the two-stage bipolar transistor amplifier of Fig. 12.9, the coupling and bypass capacitor values, and the input, interstage, and output network high frequency break points. Assume that the low frequency break point occurs at 25 Hz.

   Given:

   $$h_{fe1} = 120 \ h_{ie1} = 2000 \ \Omega \qquad h_{fe2} = 50 \qquad h_{ie1} = 500 \ \Omega$$

   where

   $$C_{be1} = C_{be2} = 15pF, \ C_{bc1} = C_{bc2} = 0.8pF, \ C_{ce1} = c_{ce2} = 3pF$$

2. Compute the mid-band voltage gain of the two-stage FET amplifier of Fig. 12.10, the coupling and bypass capacitor values, and the input, interstage, and output network high frequency break points. Assume that the low frequency break point occurs at 30 Hz.

   Given

   $$g_{m1} = 2.5 \ \text{mmho}$$
   $$r_{d1} = 100k$$
   $$g_{m2} = 4 \ \text{mmho}$$
   $$r_{d2} = 80k$$

   where

   $$C_{gs1} = C_{gs2} = 8pF$$
   $$C_{sd1} = C_{sd2} = 0.8pF$$
   $$C_{gd1} = C_{gd2} = 2.8pF$$

3. Repeat Problem 1 for the circuit shown in Fig. 12.11.

   Given $h_{fe1} = 120$, $h_{ie1} = 2000 \ \Omega$, $h_{fe2} = 50$, $h_{ie2} = 0.5k$, where $C_{be1} = C_{be2} = 15pF$, $C_{bc1} = C_{bc2} = 0.8pF$, $C_{ce1} = C_{ce2} = 3pF$.

4. Repeat Problem 2 for the circuit shown in Fig. 12.12.

   Given $g_{m1} = 2.5 \ \text{mmho}$, $r_{d1} = 100k$, $g_{m2} = 4 \ \text{mmho}$, $r_{d2} = 80k$, where $C_{gs1} = C_{gs2} = 8pF$, $C_{sd1} = C_{sd1} = 0.8pF$, $C_{gd1} = C_{gd2} = 2.8pF$.

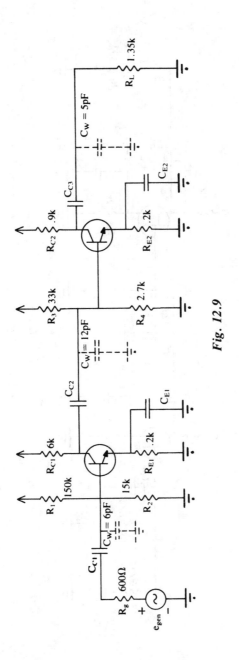

Fig. 12.9

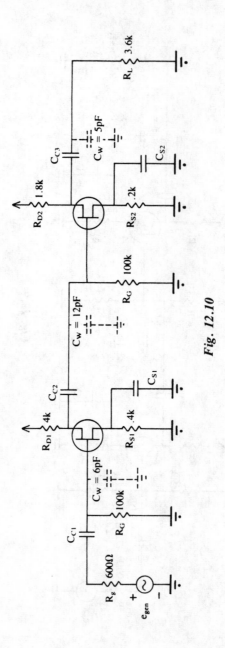

Fig. 12.10

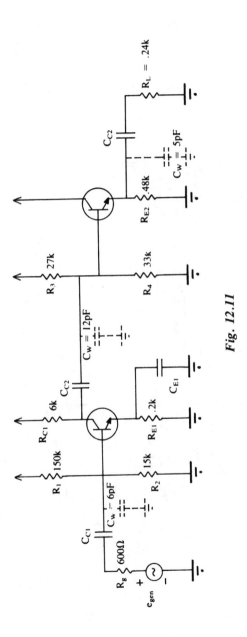

Fig. 12.11

408

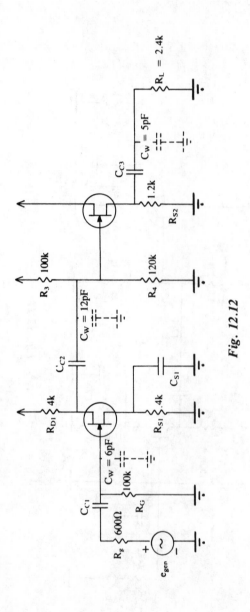

Fig. 12.12

5. For the circuit shown in Problem 1, compute the mid-band overall current gain. Use this value to check the mid-band voltage gain computed in Problem 1.

6. For the circuit shown in Problem 3, compute the mid-band overall current gain. Use this value to check the mid-band voltage gain computed in Problem 3.

Chapter 13

Operational Amplifiers

With the growth of semiconductor technology came the ability to package an ever-increasing number of transistors and support components in a single encapsulated unit.

A large number of transistors and support components grown on a single monolithic silicon wafer is referred to as an *integrated circuit*. A special purpose integrated circuit (IC) is the *operational amplifier* (OP AMP). An OP AMP is a versatile IC that may be utilized to perform many mathematical and control operations—hence, the name operational amplifier.

An operational amplifier exhibits very large open-loop voltage gain, low output impedance, and high input impedance values. An OP AMP consists of a series of cascaded differential amplifier stages, which may be thought of as an extremely high voltage gain, single-stage differential amplifier.

A differential amplifier is a unit that typically has two input terminals and one or two output terminals. The differential amplifier derives its name from the fact that the two inputs enable the amplifier to perform an algebraic summing function.

The non-inverting input terminal (see Fig. 13.1) (+) of a differential amplifier will accept an input signal, in which, after passing through the amplifier, the output signal will appear as an amplified replica of the input. See Figure 13.2 (a–d).

DIFFERENTIAL AMPLIFIER SCHEMATIC SYMBOL

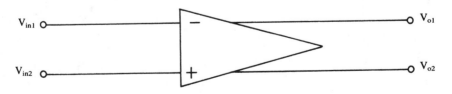

Fig. 13.1 Differential Amplifier Stage

411

412

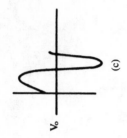

(a)

(b)

(c)

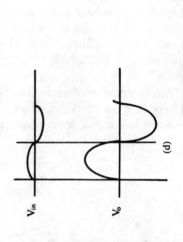

(d)

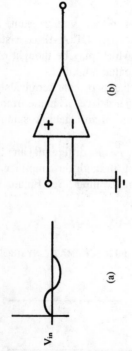

Fig. 13.2 Single-Output Differential Amplifier

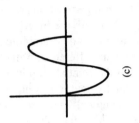

(a)

(b)

(c)

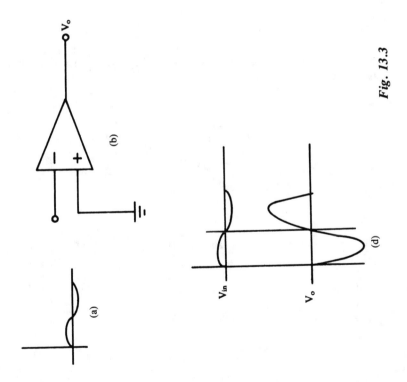

(d)

Fig. 13.3

If the inverting terminal $(-)$ is utilized and the non-inverting terminal $(+)$ is grounded, we have the results shown in Fig. 13.3 (a–d).

We note that the output signal is something of a replica of the input signal, but differs in that it is both inverted and amplified. By the superposition principal, we can see that if the same input signal were applied to both the inverting and non-inverting input terminals, the output signal (assuming equal amplification for both channels) would be the algebraic sum (arithmetic difference) of the two input signals, or (in theory) zero volts. In a practical differential amplifier, the amplification that can be anticipated for each channel (input 1 to output and input 2 to output) will not be exactly the same magnitude. Thus, the output signal will be almost (but usually not) equal to zero volts. The voltage gain of a differential amplifier is defined by two formulas: the differential gain and common-mode gain. The differential gain is simply the voltage gain that would be computed by dividing the measured value of the output voltage by the difference between the measured values of the two input voltages, or

$$A_d = \frac{V_{out\ d}}{V_{in\ d}} \qquad\qquad (13.1)$$

Example (13.1). Determine the differential voltage gain for the differential amplifier shown in Fig. 13.4.

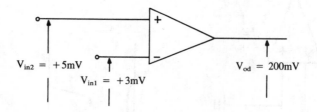

Fig. 13.4

We calculate

$$V_{in\ d} = +5\ mV - (+3\ mV) = +2\ mV$$

Thus,

$$A_d = \frac{V_{out\ d}}{V_{in\ d}} = \frac{200\ mV}{2\ mV} = 100$$

If we rewrite Equation (13.1), we have

$$A_d = \frac{V_{out\ d}}{V_{in\ d}}$$

hence,

$$V_{out\ d} = A_d\ V_{in\ d}$$

and if

$$V_{in\ d} = 0$$

then

$$V_{out\ d} = 0$$

Example (13.2) Redraw the differential amplifier of Example (13.1) and compute the output voltage if

$$V_{in\ 1} = +3\ mV \text{ and } V_{in\ 2} = +3\ mV$$

See Fig. 13.5.

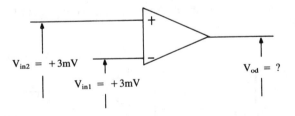

Fig. 13.5

Given from Example (13.1):

$$A_d = 100$$

therefore

$$V_{out\ d} = A_d\ V_{in\ d}$$

and

$$V_{in\ 2} = +3\ \text{mV} - (+3\ \text{mV}) = 0 \quad \text{or} \quad V_{out\ d} = 100(0) = 0\ \text{V}$$

Earlier it was stated that, although the input signals may be identically equal, a small signal would be measured at the output terminal. This output signal voltage, when divided by the input signal voltage, is given a special designation—the *common-mode voltage gain*, A_C:

$$A_C = \frac{V_{C\ out}}{V_{C\ in}} \tag{13.2}$$

In most instances, the common-mode voltage gain is a measured value. To predict this value from theoretical considerations would be a highly involved undertaking. Desirably, the common-mode voltage gain A_C should be a very small value (ideally, zero) and the difference voltage gain should be a large value. A small value for the common-mode voltage gain may be indicative of a "good match" of the transistors within the differential amplifier. A commonly accepted "figure of merit" for a differential amplifier is the ratio of the difference voltage gain to the common mode voltage gain. This ratio is referred to as the *common-mode rejection ratio*:

$$CMRR = \frac{A_d}{A_C} \tag{13.3}$$

Example (13.3). Compute the differential mode voltage gain for the amplifier shown in Fig. 13.6.

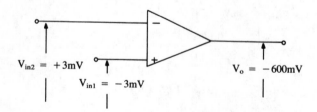

$$V_{in2} = +3\text{mV}$$
$$V_{in1} = -3\text{mV}$$
$$V_o = -600\text{mV}$$

Fig. 13.6

Thus,

$$A_d = \frac{V_{out\ d}}{V_{in\ d}} = \frac{-600\ \text{mV}}{-3\ \text{mV} - (+3\ \text{mV})} = 100$$

Example (13.4). Compute the common-mode voltage gain for the amplifier shown in Fig. 13.7.

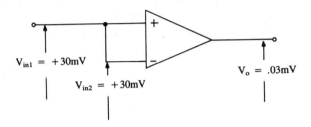

$V_{in1} = +30mV$

$V_{in2} = +30mV$

$V_o = .03mV$

Fig. 13.7

Thus,

$$A_C = \frac{V_{out\ C}}{V_{in\ C}} = \frac{0.03\ mV}{30\ mV} = 0.001$$

Example (13.5). Compute the common-mode rejection ratio for the amplifier of Examples (13.3) and (13.4). Thus,

$$CMRR = \frac{A_d}{A_C} = \frac{100}{0.001} = 100,000$$

The common mode rejection ratio is usually a large value, therefore, it is not uncommon to express the CMRR in decibels:

$$CMRR_{dB} = 20\ \log_{10} CMRR \qquad (13.4)$$

Example (13.6). Express the value of the CMRR determined in Example (13.5) in decibels. Thus,

$$\begin{aligned} CMRR_{dB} &= 20\ \log_{10} CMRR \\ &= 20\ \log_{10} 100,000 \\ &= 100\ dB \end{aligned}$$

13.1 DESIGN AND ANALYSIS CONSIDERATIONS

To this point, the differential amplifier has been designated schematically as a triangular symbol with only input and output terminals shown.

418

Although differential amplifiers are readily available in integrated circuit chip form, a brief discussion of the circuit design and analysis considerations for a differential amplifier constructed of discrete transistors and resistors will enable us to gain an understanding of the basic operations of most differential amplifiers, manufactured by discrete or integrated circuit techniques.

A single-stage differential amplifier could be constructed with two bipolar transistors as shown in Fig. 13.8.

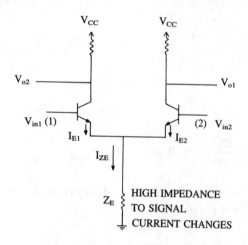

Fig. 13.8

If a positive gaining signal $+\Delta E$ is applied to terminal 1, the base current I_{B1} would increase and, thus, the collector current I_{C1} would increase. A corresponding decrease in the ground-to-collector voltage would be noted. We further note that $I_{ZE} = I_{E1} + I_{E2}$. If both transistors were identical, for quiescent conditions: $I_{E1} = I_{E2}$. If the impedance in the emitter circuit Z_E was chosen as an extremely large value, then I_{ZE} would remain essentially constant, regardless of changes in the individual emitter currents. Therefore, if the emitter current of Q_1 increased by $+\Delta I_{E1}$, to compensate for this increase, the emitter current of Q_2 would of necessity decrease by a corresponding amount $-\Delta I_{E2}$. Thus,

$$| \Delta I_{E1} | = | \Delta I_{E2} |$$

or

$$\Delta I_{E1} = -\Delta I_{E2}$$

If the emitter current of Q_2 decreased, then we would anticipate a corresponding decrease in the collector current of Q_2, I_{C2}. Therefore, the change in the collector current of Q_1 would equal the change in collector current in Q_2. Therefore, we have

$$| \Delta I_{C1} | = | \Delta I_{C2} |$$

therefore

$$+\Delta I_{C1} = -\Delta I_{C2}$$

If the collector resistors were equal in value, then the output voltages measured from ground to each collector would be equal:

$$| \Delta V_{C1} | = | \Delta V_{C2} |$$
$$+ \Delta V_{C1} = -\Delta V_{C2}$$

or

$$V_{out\ 1} = -V_{out\ 2}$$

This result would establish proper differential amplifier operation.

We may have a nagging concern about the feasability of using a very high impedance in the emitter of the differential transistor pair. We may reason that to have a very large resistor in the emitter pair, a very large dc bias voltage would be required in order to have a sufficient quiescent current for proper bias of the transistors. This is correct reasoning. What is needed in the emitter circuit is an impedance that is low in value at low frequencies, i.e., at direct current (zero frequency), and high in value with respect to what the emitters of Q_1 and Q_2 "see."

An inductor meets these requirements to some extent, but has the disadvantage that it is frequency dependent and will resonate with stray capacitances at specific frequencies. Therefore, we must select a different device for an *effective* large value emitter impedance. Let us examine a bipolar transistor common-emitter characteristic curves, as shown in Fig. 13.9.

We note that at low base-current values the output curves are essentially flat, i.e., changes in emitter-to-collector voltage V_{EC} result in little or no change in the collector current I_C, if the base current I_B is held constant. Therefore, once we have selected a desired emitter current for Q_1 and Q_2, the collector current for Q_3, I_{C3} is set at $2 I_{E1}$ ($= 2 I_{E2}$). See Fig. 13.10.

420

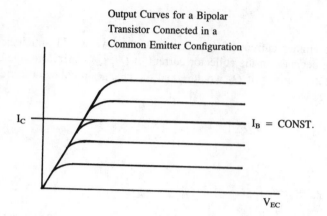

Output Curves for a Bipolar
Transistor Connected in a
Common Emitter Configuration

I_C

I_B = CONST.

V_{EC}

Fig. 13.9 Output Curves for a Bipolar Transistor Connected in a Common Emitter Configuration

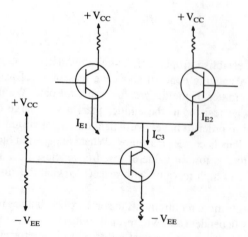

+V_{CC}

+V_{CC}

+V_{CC}

I_{E1}

I_{E2}

I_{C3}

−V_{EE}

−V_{EE}

Fig. 13.10

With this arrangement, the ac impedance "looking into" the collector terminal of Q_3 is a very high value. If the base bias arrangement of Fig. 13.10 is changed to that shown below (Fig. 13.11), the resistor/Zener diode pair will provide a superior constant source of base current for Q_3.

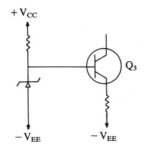

Fig. 13.11

With I_{B3} "more constant," then I_{C3} will be "more constant." This circuit is essentially a common base configuration. We know that the output curves for a common base connection are very flat with consequent very high ac output impedance. Recall that output impedance is determined by determining the slope of the tangent line to a curve as given by Fig. 13.12.

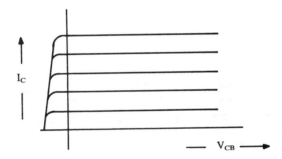

Fig. 13.12 Common Base Output Curves

A differential amplifier could also be constructed by employing field-effect transistors. Certain advantages and disadvantages are attendant with their use. For example, the voltage gain will be considerably less than for a bipolar stage. However, the range of input voltage values that may be tolerated before serious distortion occurs is considerably greater for a FET stage. If a FET is used as Q_3, it exhibits very flat common-source curves (in the saturation region). Thus, to simplify circuit design, the gate may be tied to the source of a FET to provide a constant current source. (See Fig. 13.13(a, b).)

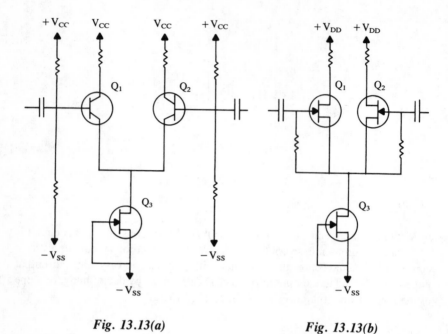

$-V_{SS}$

Fig. 13.13(a)   **Fig. 13.13(b)**

Of course, this arrangement automatically fixes the value of I_{D3} and thus fixes the currents flowing through Q_1 and Q_2 at $I_{D3}/2$. If this is an unacceptable design limitation, then a voltage divider must be used in the gate circuit. See Fig. 13.13(c).

Example (13.7). Proceeding with this design, let us compute I_{E1} $(= I_{E2})$:

$$I_{E1} = \frac{\beta + 1}{\beta} I_{C1} = \frac{101}{100} 2 \text{ mA} = 2.02 \text{ mA} (= I_{E2})$$

Also,

$$I_{C3} = 2 I_{E1} = 2(2.02) = 4.04 \text{ mA}$$
$$I_{E3} = \frac{\beta + 1}{\beta} I_{C3} = \frac{101}{100} (4.04) = 4.08 \text{ mA}$$

and

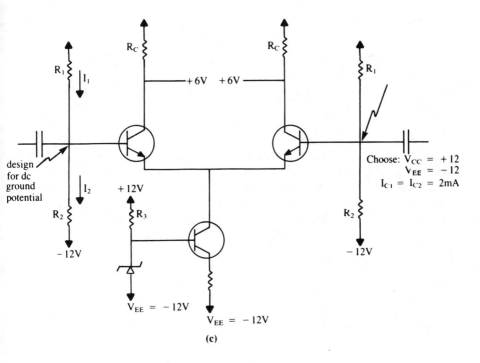

Choose: $V_{CC} = +12$
$V_{EE} = -12$
$I_{C1} = I_{C2} = 2mA$

Fig. 13.13(c)

$$I_{B1} = \frac{I_{C1}}{\beta_1} = \frac{2}{100} = 0.02 \text{ mA } (= I_{B2})$$

Also, assume $I_1 = 3I_{B1}$,

$$I_1 = 3I_{B1} = 3(0.02) = 0.06 \text{ mA}$$
$$V_{R1} = 12 \text{ V} - 0 \text{ V} = 12 \text{ V}$$
$$R_1 = \frac{V_{R1}}{I_1} = \frac{12}{0.06} = 200 \text{ k}$$
$$V_{R2} = 0 - (-12 \text{ V}) = 12 \text{ V}$$
$$\text{and } I_2 = I_1 - I_B = 0.06 - 0.02 = 0.04 \text{ mA}$$
$$R_2 = \frac{V_{R2}}{I_2} = \frac{12}{0.04} = 300 \text{ k}$$
$$V_{RC1} = 6 \text{ V}$$

therefore,

$$R_{C1} = \frac{V_{RC1}}{I_{C1}} = \frac{6 \text{ V}}{2 \text{ mA}} = 3 \text{ k}$$

$$I_{B3} = \frac{I_{C3}}{\beta} = \frac{4.04}{100} = 0.0404 \text{ mA}$$

We note that the emitters of both Q_1 and Q_2 are at -0.6 V dc potential (relative to ground). This is so because the bases have been arbitrarily designed to be at dc ground potential and, therefore, the emitter must be 0.6 V lower in dc potential. Therefore, the voltage across R_E in series with the emitter to collector of Q_3 is $-0.6 - (-12 \text{ V}) = +11.4$ V. If we divide this voltage drop equally between R_E and the emitter-to-collector potential of Q_3 we have

$$V_{RE} = V_{ECQ3} = \frac{11.4}{2} = 5.7 \text{ V}$$

therefore,

$$R_E = \frac{V_{RE}}{I_{E3}} = \frac{5.7 \text{ V}}{4.0804} = 1.4 \text{ k}\Omega$$

The voltage rating of the Zener diode must be equal to the voltage across the emitter resistor plus the emitter-to-base voltage drop:

$$V_Z = V_{RE} + V_{EB} = 5.7 + 0.6 = 6.3 \text{ V}$$

The value of the resistor R_3 must be chosen so as to ensure that the Zener remains conducting and, therefore, in regulation at all times. If the Zener current is selected as two milliamps, the base of current of .0404 mA should have negligible effect upon the Zener operation.

$$
\begin{aligned}
V_{R3} &= V_{CC} - V_Z - V_{EE} \\
&= +12 - 6.3 - (-12) \\
&= 17.7 \text{ V}
\end{aligned}
$$

$$R_3 = \frac{V_{R3}}{I_{R3}} = \frac{17.7 \text{ V}}{2 \text{ mA}} = 8.85 \text{ k}\Omega$$

Let us determine how much the dc voltages and currents shift if we should happen to replace our original transistors with new units with β values of twice the original value.

Because the Zener is assumed to remain in regulation, the voltage drop across the emitter resistor V_{RE} will remain the same. Thus, the emitter current will remain at approximately 4.08 mA. Therefore,

$$I_{C3} = \frac{\beta}{\beta + 1} I_{E3}$$

therefore,

$$I_{C3} = \frac{200}{201} (4.08) = 4.06 \text{ mA}$$

$$I_{E1} = \frac{I_{C3}}{2} = \frac{4.06}{2} = 2.03 \text{ mA}$$

$$I_{C1} = \frac{\beta}{\beta + 1} I_{E1} = \frac{200}{201} 2.03 = 2.02 \text{ mA}$$

$$V_{RC1} = I_{C1} R_{C1} = 2.02 \text{ mA (3k)} = 6.06$$

We can see that the emitter-to-collector voltage of Q_1 (and Q_2) has experienced an insignificant change. Computing I_{B1}:

$$I_{B1} = \frac{I_{C1}}{\beta} = \frac{2.02}{200} = 0.0101 \text{ mA}$$

Let us determine how much the ground-to-base voltage for Q_1 (and, similarly, Q_2) has changed due to the increased dc β for Q_1 through Q_3.

$$I_1 = I_2 + 0.0101 \text{ mA}$$
$$V_{R1} + V_{R2} = 24 \text{ V}$$

therefore,

$$200 I_1 + 300 I_2 = 24$$
$$200 (I_2 + 0.0101) + 300 I_2 = 24$$

therefore,

$$I_2 = 0.044, I_1 = 0.054 \text{ mA}$$
$$V_{R1} = I_1 R_1 = 10.8 \text{ V}$$
$$V_{R2} = I_2 R_2 = 13.2 \text{ V}$$

We note that the base dc voltage has shifted approximately 1.2 V above ground potential. Consequently, the emitter of Q_1 (and Q_2) has shifted 1.2 V above ground potential. Thus, the maximum output voltage swing has been reduced by a corresponding amount.

We note that if capacitor coupling is utilized, there is no need for two power supplies (positive and negative, 12 V each).

Since we have completed our dc design, we are now interested in the signal voltage amplification that can be anticipated. To simplify such an analysis, we will analyze the voltage gain with a signal applied to only one transistor base and determine the output at the corresponding collector. See Fig. 13.14.

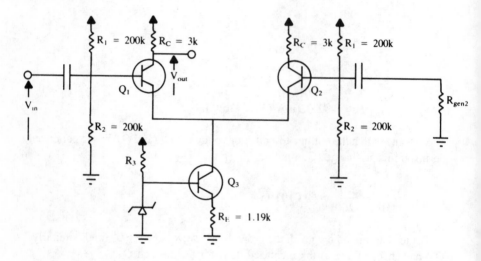

Fig. 13.14

Recalling from the small-signal analysis formulas:

$$A_V = \frac{-h_{fe} Z_L}{h_{ie} + (h_{fe} + 1) Z_E}$$

For this analysis, we will assume that

$$h_{fe} = h_{FE} = 100$$

and that

$$h_{ie} \doteq \frac{\beta\ 26\ \text{mV}}{I_E} = 1.3\ \text{k},\ Z_L = R_C,\ R_{gen} = 0.3\ \text{k}$$

The only impedance that is not an obvious value is Z_E. Recall that Z_E is the impedance "seen" by "standing" at the emitter and "looking" to ground. See Fig. 13.15.

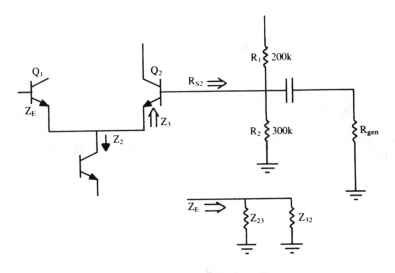

Fig. 13.15

We note that Z_E equals Z_2 of Q_3 in parallel with Z_3 of Q_2.

$$Z_E = Z_{23} \,/\!/\, Z_{32}$$

Z_3 of Q_2 may be determined by the small-signal formula:

$$Z_3 \doteq \frac{R_S + h_{ie}}{h_{fe} + 1}$$

or, for this example,

$$Z_3 \doteq \frac{R_{S2} + h_{ie2}}{h_{fe2} + 1}$$

Where R_{S2} is determined by "standing" at the base of Q_2 and "looking" outward.

We note from Fig. 13.15 that R_{S2} may be determined by

$$R_{S2} = R_1 \mathbin{/\mkern-5mu/} R_2 \mathbin{/\mkern-5mu/} R_{gen2} = 200 \text{ k} \mathbin{/\mkern-5mu/} 300 \text{ k} \mathbin{/\mkern-5mu/} 0.3 \text{ k}$$
$$R_{S2} = .299 \text{ k}$$

Recalling:

$$Z_3 \doteq \frac{R_{S2} + h_{ie2}}{h_{fe2} + 1} = \frac{.299 \text{ k} + 1.3 \text{ k}}{101} = 15.8 \ \Omega$$

Also,

$$Z_{E1} = Z_{23} \mathbin{/\mkern-5mu/} Z_{32}$$

Recalling that the impedance "seen" when "looking" into the collector of a bipolar transistor is

$$Z_2 \doteq \frac{1}{h_{oe}} \doteq 100 \text{ k}$$

we have

$$Z_{E1} \doteq Z_{23} \mathbin{/\mkern-5mu/} Z_{32} = 100 \text{ k} \mathbin{/\mkern-5mu/} .0158 \text{ k}$$

or

$$Z_{E1} \doteq Z_{32} = .0158 \text{ k} = \frac{R_{S2} + h_{ie2}}{h_{fe2} + 1}$$
thus,

$$A_{V1} \doteq \frac{-h_{fe1} Z_{L1}}{h_{ie1} + (h_{fe1} + 1) Z_{E1}} \doteq \frac{-h_{fe1} Z_{L1}}{h_{ie1} + (h_{fe1} + 1)\left(\dfrac{R_{S2} + h_{ie2}}{h_{fe2} + 1}\right)}$$

If the transistors (Q_1 and Q_2) are matched:

$$h_{fe1} = h_{fe2} = h_{fe} \text{ and } h_{ie1} = h_{ie2} = h_{ie}$$

We may write:

$$A_{V1} = \frac{-h_{fe1} Z_{L1}}{h_{ie} + (h_{fe} + 1)\left(\dfrac{R_{S2} + h_{ie}}{h_{fe} + 1}\right)} = \frac{-h_{fe} Z_{L1}}{2h_{ie} + R_{S2}} \tag{13.8}$$

Substituting:

$$A_{V1} = \frac{-100\ (3k)}{2(1.3\ k)\ +\ 0.0158\ k} = -115$$

Under what circumstances would it be desirable to use a differential amplifier? Let us consider a phonograph cartridge mounted on the end of a tone arm, as shown in Fig. 13.16.

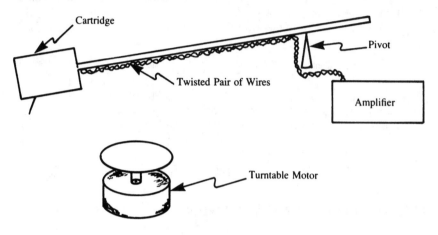

Fig. 13.16

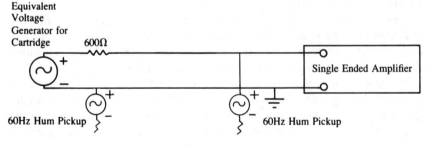

Fig. 13.17

The two wires perform as unwanted antennas and develop 60 Hz *hum* signals (see Fig. 13.17). If a single ended amplifier was used to amplify the voltage at the wire ends, two signals would be present:

1. the signal from the phono cartridge;
2. the 60 Hz *hum* signal developed in one of the twisted wire pair.

430

The output signal of the amplifier would consist of the desired signal and *hum* signal, amplified equally.

If a differential amplifier was used with neither input grounded, as shown in Fig. 13.18, we note that the 60 Hz signal will appear in phase at both amplifier input terminals and, thus, due to the large CMRR, will be greatly attenuated at the amplifier output terminal. The cartridge signal, however, is not a common-mode voltage and, therefore, will be amplified as a difference voltage. Due to impedance-matching considerations, noise generation, and ease of manufacturing, a single-ended input with twisted pairs is usually used in commercial applications.

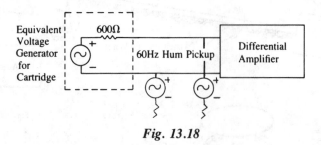

Fig. 13.18

13.2 THE OPERATIONAL AMPLIFIER

Ideally, an operational amplifier would exhibit an infinite voltage gain from either the non-inverting or inverting input to the output terminal, infinite input impedance, and zero output impedance.

Infinite input impedance would be desirable because the operational amplifier would not load the signal source. Zero output impedance is desirable because the OP AMP could then supply as much signal current to the load as required without displaying any decrease in output signal voltage levels.

At first analysis, infinite voltage gain might seem like a disadvantage because the output signal would be infinite, or, in a practical situation, the output would be limited by the available power supply voltages. For example, if a sine wave is applied to one input of an ideal operational amplifier, we will expect to see the wave form of Fig. 13.19 at the OP AMP output terminal.

Note that every point on the input waveform is amplified by the maximum possible amount, and is only limited by the output voltage power supply levels. If a commercially available operational amplifier is connected in a test circuit and a sine wave input is applied, the performance will approach that pictured in Fig. 13.19, differing only in that the minimum and maximum levels of the output square wave would not be quite $+V_{CC}$ and $-V_{EE}$, differing by one-half to one and one-half volts.

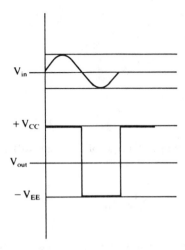

$$Fig.\ 13.19$$

By "steering" some of the output voltage back to the inverting input terminal (negative feedback), a reduced output will be observed. We can think of the output voltage, which is out of phase with the input voltage, as "cancelling out" some of the input voltage when negative feedback is employed. Thus, in most operational amplifier circuits, negative feedback is necessary to obtain useful output signals.

Table 13.1

PARAMETERS	Ideal	Commercial Unit
A_V	∞	100,000
Z_{in}	$\infty\ \Omega$	100,000 Ω
Z_0	0 Ω	100 Ω

Let us compare a typical commercially available OP AMP with an ideal OP AMP, using Table 13.1. Recall that the voltage gain for a differential amp was the output voltage divided by the *difference between* the two input voltages. Let us determine the input voltage *between* the two input terminals in a practical situation:

$$A_d = \frac{V_{out\ d}}{V_{in\ d}}$$

therefore,

$$V_{in\ d} = \frac{V_{out\ d}}{A_d}$$

If we choose a signal output voltage of 5 V and differential voltage gain of 100,000, then we have

$$V_{in\ d} = \frac{5}{100,000} = 5(10^{-5}) = 50\ \mu V$$

For most circuit designs and analyses, the potential between the two input terminals is near the value calculated. Consequently, the potential that exists between the input terminals is referred to as a *virtual short circuit*.

If one of the input terminals is connected to circuit ground, the other terminal is referred to as being at *virtual ground*. See Fig. 13.20.

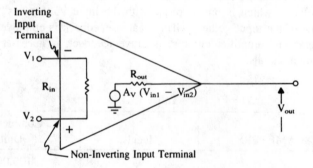

Fig. 13.20 OP AMP Equivalent Circuit

The input current can be easily calculated:

$$I_{in\ d} = \frac{V_{in\ d}}{R_{in}} = \frac{5(10^{-5})}{10^5} = 0.5\ nA$$

With this preliminary discussion of operational amplifiers, we are now in a position to discuss their use in practical circuit applications. The following are some OP AMP circuits that are commonly used in industrial applications.

13.2.1 The Inverting Amplifier

The inverting amplifier is an easily constructed circuit requiring only two resistors and an operational amplifier. Schematically it would be portrayed as shown in Figs. 13.21 and 13.22.

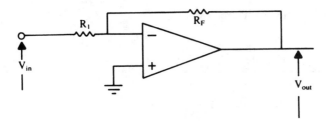

Fig. 13.21

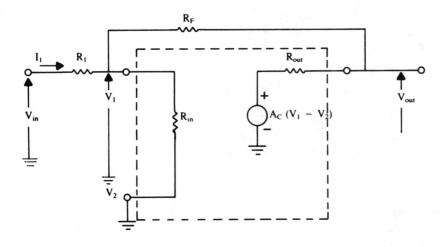

Fig. 13.22

Computing the voltage gain:

$$A_V = \frac{V_{out}}{V_{in}}$$

at input terminal 1,

434

$\Sigma I = 0$ (currents out are positive)

$$\frac{V_1 - V_{in}}{R_1} + \frac{V_1}{R_{in}} + \frac{V_1 - V_{out}}{R_f} = 0 \tag{13.9}$$

Also, at the output terminal,

$\Sigma I = 0$ (currents out are positive)

$$\frac{V_{out} - A_d\,(V_1 - V_2)}{R_{out}} + \frac{V_{out} - V_1}{R_f} = 0, \text{ but } V_2 = 0$$

therefore,

$$\frac{V_{out} - A_d\,V_1}{R_{out}} + \frac{V_{out} - V_1}{R_f} = 0$$

$$\frac{V_{out}}{R_{out}} - \frac{A_d\,V_1}{R_{out}} + \frac{V_{out}}{R_f} - \frac{V_1}{R_f} = 0$$

$$V_1\left(-\frac{A_d}{R_{out}} - \frac{1}{R_f}\right) + V_{out}\left(\frac{1}{R_{out}} + \frac{1}{R_f}\right) = 0$$

$$V_1 = \frac{V_{out}\left(\dfrac{1}{R_{out}} + \dfrac{1}{R_f}\right)}{\left(-\dfrac{A_d}{R_{out}} - \dfrac{1}{R_f}\right)} = \frac{V_{out}\dfrac{R_f + R_{out}}{R_{out}R_f}}{\dfrac{A_d R_f + R_{out}}{R_{out}R_f}} = V_{out}\left(\frac{R_f + R_{out}}{A_d R_f + R_{out}}\right) \tag{13.10}$$

Restating (13.9):

$$\frac{V_1 - V_{in}}{R_1} + \frac{V_1}{R_{in}} + \frac{V_1 - V_{out}}{R_f} = 0$$

$$\frac{V_1}{R_1} - \frac{V_{in}}{R_1} + \frac{V_1}{R_{in}} + \frac{V_1}{R_f} - \frac{V_{out}}{R_f} = 0$$

$$V_1\left(\frac{1}{R_1} + \frac{1}{R_{in}} + \frac{1}{R_f}\right) - \frac{V_{in}}{R_1} - \frac{V_{out}}{R_f} = 0$$

From (13.9) and (13.10):

$$V_{out} \left(\frac{R_f + R_{out}}{A_d R_f + R_{out}} \right) \left(\frac{1}{R_1} + \frac{1}{R_{in}} + \frac{1}{R_f} \right) - \frac{V_{in}}{R_1} - \frac{V_{out}}{R_f} = 0$$

$$V_{out} \left[\left(\frac{R_f + R_{out}}{A_d R_f + R_{out}} \right) \left(\frac{1}{R_1} + \frac{1}{R_{in}} + \frac{1}{R_f} \right) - \frac{1}{R_f} \right] = \frac{V_{in}}{R_1}$$

$$\frac{V_{out}}{V_{in}} = \frac{1}{R_1} \frac{1}{\left(\dfrac{R_f + R_{out}}{A_d R_f + R_{out}} \right) \left(\dfrac{1}{R_1} + \dfrac{1}{R_{in}} + \dfrac{1}{R_f} \right) - \dfrac{1}{R_f}}$$

$$= \frac{1}{\left(\dfrac{R_f + R_{out}}{A_d R_f + R_{out}} \right) \left(1 + \dfrac{R_1}{R_{in}} + \dfrac{R_1}{R_f} \right) - \dfrac{R_1}{R_f}} \tag{13.11}$$

$$\frac{V_{out}}{V_{in}} = \frac{1}{\left(\dfrac{1 + R_{out}/R_f}{A_d + R_{out}/R_f} \right) \left(1 + \dfrac{R_1}{R_{in}} + \dfrac{R_1}{R_f} \right) - \dfrac{R_1}{R_f}}$$

We note in the first denominator term

$$\frac{R_{out}}{R_f} \ll 1$$

therefore, $1/A_d$ remains. Since

$$A_d \gg 1$$

the entire first term tends to zero, since

$$\frac{1}{A_d} \to 0$$

Thus,

$$\frac{V_{out}}{V_{in}} = A_V \doteq -\frac{R_f}{R_1} \tag{13.12}$$

If we had initially assumed that $V_1 \doteq 0$, the derivation would have been significantly reduced.

Starting with Equation (13.9):

$$\frac{V_1 - V_{in}}{R_1} + \frac{V_1}{R_{in}} + \frac{V_1 - V_{out}}{R_f} = 0$$

$$\frac{V_1}{R_1} - \frac{V_{in}}{R_1} + \frac{V_1}{R_{in}} + \frac{V_1}{R_f} - \frac{V_{out}}{R_f} = 0$$

therefore:

$$\frac{V_{out}}{V_{in}} = A_V \doteq -\frac{R_f}{R_1}$$

Example (13.8). The calculation of the percentage error resulting from the simplified equation will be a function of the component values. Let us see what a typical error would be if

$$R_{in} = 100 \text{ k}, R_1 = 10 \text{ k}, R_f = 100 \text{ k}$$
$$A_{Vd} = 10^5, R_{out} = 0.1 \text{ k}$$

$$A_V = \cfrac{1}{\left(\dfrac{R_f + R_{out}}{A_d R_f + R_{out}}\right)\left(1 + \dfrac{R_1}{R_{in}} + \dfrac{R_1}{R_f}\right) - \dfrac{R_1}{R_f}}$$

$$= \cfrac{1}{\left(\dfrac{100 + 0.1}{-10^5(100) + 0.1}\right)\left(1 + \dfrac{10}{100} + \dfrac{10}{100}\right) - \dfrac{10}{100}}$$

$$A_V = -9.9987$$

With the approximate formula,

$$A_V = -10$$

therefore, % error is 0.012%.
   Computing the input impedance:

$$Z_{in} = \frac{V_{in}}{I_1}$$

$$I_1 = \frac{V_{in} - V_1}{R_1} = \frac{V_{in}}{R_1} - \frac{V_1}{R_1} = \frac{V_{in}}{R_1} - \frac{1}{R_1}(V_1)$$

From equation (13.10):

$$I_1 = \frac{V_{in}}{R_1} - \frac{1}{R_1}\left(V_{out}\frac{R_f + R_{out}}{A_d R_f + R_{out}}\right) = \frac{V_{in}}{R_1} - V_{out}\left(\frac{1}{R_1} \cdot \frac{R_f + R_{out}}{A_d R_f + R_{out}}\right)$$

From equation (13.11):

$$I_1 = \frac{V_{in}}{R_1} - \left[\frac{V_{in}}{\left(\dfrac{R_f + R_{out}}{A_d R_f + R_{out}}\right)\left(1 + \dfrac{R_1}{R_{in}} + \dfrac{R_1}{R_f} - \dfrac{R_1}{R_f}\right)}\right]\left(\frac{1}{R_1} \cdot \frac{R_f + R_{out}}{A_d R_f + R_{out}}\right)$$

$$\frac{I_1}{V_{in}} = \frac{1}{R_1} - \frac{1}{R_1}\left[\frac{\dfrac{R_f + R_{out}}{A_d R_f + R_{out}}}{\left(\dfrac{R_f + R_{out}}{A_d R_f + R_{out}}\right)\left(1 + \dfrac{R_1}{R_{in}} + \dfrac{R_1}{R_f}\right) - \dfrac{R_1}{R_f}}\right]$$

$$= \frac{1}{R_1} - \frac{1}{R_1}\left[\frac{1}{1 + \dfrac{R_1}{R_{in}} + \dfrac{R_1}{R_f} - \dfrac{R_1}{R_f}\left(\dfrac{A_d R_f + R_{out}}{R_f + R_{out}}\right)}\right]$$

which may be rewritten:

$$\frac{I_1}{V_{in}} = \frac{1}{R_1} - \frac{1}{R_1}\left[\frac{1}{1 + \dfrac{R_1}{R_{in}} + \dfrac{R_1}{R_f} - \dfrac{R_1}{R_f}\left(\dfrac{A_d + R_{out}/R_f}{1 + R_{out}/R_f}\right)}\right]$$

We note that

$$\frac{R_{out}}{R_f} \ll 1$$

and that A_d is a very large number, therefore, the entire second term tends toward zero:

$$\frac{V_{in}}{I_1} = Z_{in} \doteq R_1 \tag{13.14}$$

Example (13.9). Let us calculate the percent error in utilizing the approximate formula (all resistors in kilohms, $k\Omega$):

$$\frac{I_1}{V_{in}} = \frac{1}{R_1} - \frac{1}{R_1}\left[\frac{1}{1 + \dfrac{R_1}{R_{in}} + \dfrac{R_1}{R_f} - \dfrac{R_1}{R_f}\left(\dfrac{A_d R_f + R_{out}}{R_f + R_{out}}\right)}\right]$$

$$\frac{I_1}{V_{in}} = \frac{1}{10} - \frac{1}{10}\frac{1}{1 + \dfrac{10}{100} + \dfrac{10}{100} - \dfrac{10}{100}\left[\dfrac{(10^5)(100) + 0.1}{100 + 0.1}\right]}$$

$$\frac{I_1}{V_{in}} = 0.0999$$

therefore,

$$\frac{V_{in}}{I_1} = Z_{in} = 10.001 \text{ k}\Omega$$

but

$$Z_{in} \doteq R_1 = 10 \text{ k}$$

Thus,

$$\% \text{ error} = 0.01\%$$

Compute the output impedance from Fig. 13.23. We know that to determine the output impedance, we must short the input voltage source and apply a test generator to the output terminals and force current into the output terminals:

$$I_{Test} = \frac{E_{Test} - A_d(V_1 - V_2)}{R_{out}} + \frac{E_{Test}}{R_f + R_1 /\!/ R_{in}}$$

However, $V_2 = 0$. Thus,

$$I_{Test} = \frac{E_{Test} - A_d V_1}{R_{out}} + \frac{E_{Test}}{R_f + \dfrac{R_1 R_{in}}{R_1 + R_{in}}}$$

$$= \frac{E_{Test}}{R_{out}} - \frac{A_d}{R_{out}} \cdot V_1 + \frac{E_{Test}}{R_f \left(\dfrac{R_1 + R_{in}}{R_1 + R_{in}} \right) + \dfrac{R_1 R_{in}}{R_1 + R_{in}}}$$

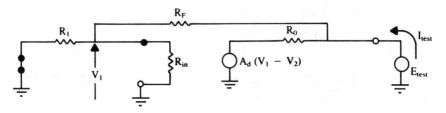

Fig. 13.23

By use of the voltage divider rule:

$$V_1 = E_{Test} \frac{R_1 /\!/ R_{in}}{R_1 /\!/ R_{in} + R_f}$$

therefore,

$$I_{Test} = \frac{E_{Test}}{R_{out}} - \frac{A_d}{R_{out}} E_{Test} \frac{\dfrac{R_1 R_{in}}{R_1 + R_{in}}}{\dfrac{R_1 R_{in}}{R_1 + R_{in}} + R_f} + \frac{E_{Test}}{R_f + \dfrac{R_1 R_{in}}{R_1 + R_{in}}}$$

therefore,

$$\frac{I_{Test}}{E_{Test}} = \frac{1}{R_{out}} - \frac{A_d}{R_{out}} \frac{\dfrac{R_1 R_{in}}{R_1 + R_{in}}}{\dfrac{R_1 R_{in}}{R_1 + R_{in}} + R_f \dfrac{R_1 + R_{in}}{R_1 + R_{in}}} + \frac{1}{\dfrac{R_1 R_{in}}{R_1 + R_{in}} + R_f \dfrac{R_1 + R_{in}}{R_1 + R_{in}}}$$

$$= \frac{1}{R_{out}} - \frac{A_d}{R_{out}} \frac{R_1 R_{in}}{R_1 R_{in} + R_f(R_1 + R_{in})} + \frac{R_1 + R_{in}}{R_1 R_{in} + R_f(R_1 + R_{in})} \qquad (13.15)$$

$$\frac{I_{Test}}{E_{Test}} = \frac{1}{R_{out}} + \frac{\dfrac{-A_d R_1 R_{in}}{R_{out}} + R_1 + R_{in}}{R_1 R_{in} + R_f(R_1 + R_{in})} \qquad (13.15a)$$

Example (13.10). Computing and recalling that $A_d = -10^5$ and that our answer will be in kilohms (kΩ):

$$\frac{I_{Test}}{E_{Test}} = \frac{1}{0.1} + \frac{- \dfrac{-10^5 (10)(100)}{0.1} + 10 + 100}{10(100) + 100(10 + 100)}$$

$$\frac{E_{Test}}{I_{Test}} = 1.2\,(10^{-5})\,\text{k}\Omega$$

$$Z_0 = 1.2\,(10^{-2})\,\Omega = 0.012\,\Omega$$

We note that the first and third terms of Equation (13.15) have little effect upon the nature of Z_{out}. Therefore,

$$\frac{I_{Test}}{E_{Test}} \doteq \frac{A_d}{R_{out}} \frac{R_1 R_{in}}{R_1 R_{in} + R_f(R_1 + R_{in})}$$

$$= \frac{A_d}{R_{out}} \frac{1}{1 + R_f \dfrac{R_1 + R_{in}}{R_1 R_{in}}} \doteq \frac{A_d}{R_{out}} \frac{1}{1 + R_f/R_1}$$

$$\frac{I_{Test}}{E_{Test}} \doteq \frac{A_d R_1}{R_{out} R_f} \tag{13.16}$$

Computing:

$$Z_{out} \doteq \frac{R_{out} R_f}{A_d R_1} = \frac{10^2 (10^5)}{10^5 (10^4)} = 0.01\,\Omega \tag{13.17}$$

The error in utilizing the approximation formula may be computed as follows:

$$\% \text{ error} = \left| \frac{0.012 - 0.01}{0.012} \right| \times 100 = 16.7\%$$

The output impedance is considerably less than one ohm, therefore, unless an unusually low value load impedance is connected to the output terminals, little error would be introduced in utilizing the approximation formula. In fact, for the remainder of the derivations dealing with circuits utilizing operational amplifiers with negative feedback, the following will be assumed:

$$A_{Vd} \to \infty, \ V_1 - V_2 \to 0, \ R_{in} \to \infty \text{ and } Z_{out} \to 0$$

Example (13.11). A simple extension of the inverting amplifier is the *summing amplifier*. In certain applications it is desired to add together (sum) voltages from different sources. For example, in industrial-control applications, the outputs of various sensors are used to control the input to a system. In a chemical process, the temperature of the chemical solution in a large retort, along with pressure and pH may be determining factors as to how much heat energy is to be supplied to the process. Each of the above parameters could be monitored with a sensor that yields a voltage which is proportional to the variable. These variables could then be summed together and the output could be used to adjust a solenoid value, which would control the amount of natural gas being used to heat the retort. See Figs. 13.24 and 13.25.

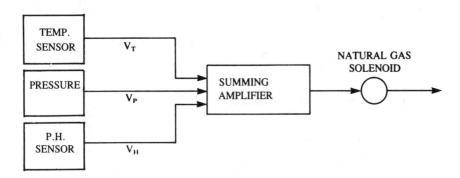

Fig. 13.24

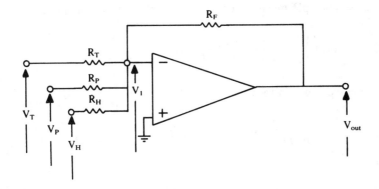

Fig. 13.25

at Node A,

$\Sigma I = 0$ (currents out are positive)

$$\frac{V_1 - V_T}{R_T} + \frac{V_1 - V_P}{R_P} + \frac{V_1 - V_H}{R_H} + \frac{V_1 - V_{out}}{R_P} = 0$$

$$\frac{V_1}{R_T} - \frac{V_T}{R_T} + \frac{V_1}{R_P} - \frac{V_P}{R_P} + \frac{V_1}{R_H} - \frac{V_H}{R_H} + \frac{V_1}{R_f} - \frac{V_{out}}{R_f} = 0$$

but $V_1 \to 0$

$$-\frac{V_T}{R_T} - \frac{V_P}{R_P} - \frac{V_H}{R_H} - \frac{V_{out}}{R_f} = 0$$

$$\frac{V_{out}}{R_F} = -\left(\frac{V_T}{R_T} + \frac{V_P}{R_P} + \frac{V_H}{R_H}\right) \Rightarrow$$

$$V_{out} = -R_f\left(\frac{V_T}{R_T} + \frac{V_P}{R_P} + \frac{V_H}{R_H}\right) \tag{13.18}$$

If all input signals are weighted equally, we have

$$V_{out} = -\frac{R_f}{R_1}(V_T + V_P + V_H) \tag{13.19}$$

Example (13.12). Suppose a chemical process required that changes in temperature be four times more effective than changes in chemical pH, and changes in pressure be twice as effective upon the output voltage than a comparable change in chemical pH. Choose a value of 120 kΩ for R_f and design the appropriate summing amplifier circuit. Given that changes in the voltage due to the chemical solution pH are amplified ten times by the summing amplifier, we have

$$V_{out} = -R_f\left(\frac{V_T}{R_T} + \frac{V_P}{R_P} + \frac{V_H}{R_H}\right)$$

Examining the equation for changes in pH only:

$$V_{out} = -R_f\left(\frac{V_H}{R_H}\right)$$

therefore,

$$R_H = -R_f\frac{V_H}{V_{out}}$$

$$R_H = -120\text{ k}\left(\frac{-1}{10}\right) = 12\text{ k}$$

For all signals:

$$V_{out} = -R_f \left(\frac{4V_T}{R_H} + \frac{2V_P}{R_H} + \frac{V_H}{R_H} \right)$$
$$= -R_f \left(\frac{V_T}{R_H/4} + \frac{V_P}{R_H/2} + \frac{V_H}{R_H} \right)$$

Therefore,

$$R_T = \frac{R_H}{4} = \frac{12 \text{ k}}{4} = 3 \text{ k}$$
$$R_P = \frac{R_H}{2} = \frac{12 \text{ k}}{2} = 6 \text{ k}$$

It should be noted that one advantage of the use of an operational amplifier compared to a discrete transistor is that involved dc bias design computations are unnecessary.

If the input and output signals are coupled through capacitors, the dc levels at the OP AMP input and output terminals will be essentially at ground potential (if one of the input terminals is referenced to ground).

13.2.2 The Non-Inverting Amplifier

If signal inversion (polarity reversal between the input and output terminals) is undesirable, the circuit of Fig. 13.26 may be used.

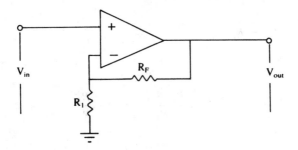

Fig. 13.26

Let us compute the voltage gain by first defining it as:

$$A_V = \frac{V_{out}}{V_{in}}$$

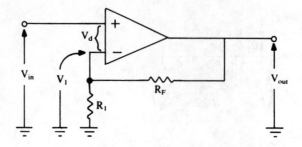

Fig. 13.27

By use of the voltage divider rule (see Fig. 13.27):

$$V_1 = V_{out} \frac{R_1}{R_1 + R_f}$$

Also,

$$V_{in} = V_1 + V_d$$

therefore, $V_1 = V_{in} - V_d$

Thus,

$$V_{in} - V_d = V_{out} \frac{R_1}{R_1 + R_f}$$

$$\frac{V_{in} - V_d}{V_{out}} = \frac{R_1}{R_1 + R_f}$$

if $V_d \rightarrow 0$

$$\frac{V_{in}}{V_{out}} = \frac{R_1}{R_1 + R_f} \quad \text{or} \quad \frac{V_{out}}{V_{in}} = \frac{R_1 + R_f}{R_1} = 1 + \frac{R_f}{R_1}$$

$$A_V = 1 + \frac{R_F}{R_1} \tag{13.20}$$

Computing the input impedance, we derive Fig. 13.28. By use of Kirchoff's voltage law, we note that

$$V_1 + V_d = V_{in}$$

therefore, $V_1 = V_{in} - V_d$ \hfill (13.21)

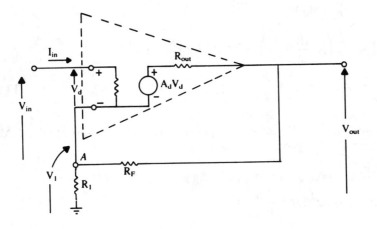

Fig. 13.28

Also, $V_{out} = V_1 + A_d V_d$ (13.22)

By use of the voltage divider rule:

$$V_1 = V_{out} \frac{R_1}{R_1 + R_f}$$

therefore,

$$V_{out} = \frac{R_1 + R_f}{R_1} \cdot V_1$$ (13.23)

Equating (13.22) and (13.23):

$$\frac{R_1 + R_f}{R_1} V_1 = V_1 + A_d V_d$$

$$\frac{R_1 + R_f}{R_1} V_1 - V_1 = A_d V_d$$

$$V_1 \left(\frac{R_1 + R_f}{R_1} - 1 \right) = A_d V_d$$

$$V_1 \frac{R_f}{R_1} = A_d V_d$$

$$V_1 = A_d V_d \frac{R_1}{R_f}$$ (13.24)

Substituting (13.21) into (13.24):

$$V_{in} - V_d = A_d V_d \frac{R_1}{R_f}$$

$$V_{in} = V_d \left(A_d \frac{R_1}{R_f} + 1 \right) = I_{in} R_{in} \left(A_d \frac{R_1}{R_f} + 1 \right)$$

$$\frac{V_{in}}{I_{in}} = Z_{in} = R_{in} \left(A_d \frac{R_1}{R_f} + 1 \right) \qquad (13.25)$$

Example (13.13). Compute the input resistance of a non-inverting OP AMP with the following parameter values:

$A_d = 200{,}000$, $R_1 = 10k$, $R_f = 100k$, and $R_{in} = 100k$

$Z_{in} = 100k(200{,}000 \frac{10k}{100k} + 1)$

neglecting the "1",

$Z_{in} = 100k(20{,}000)$
$Z_{in} = 2(10^9) \ \Omega$

We see that Equation (13.25) may be written:

$$Z_{in} \doteq \frac{A_d R_{in} R_1}{R_f} \qquad (13.26)$$

Computing the output impedance of the non-inverting amplifier we derive Fig. 13.29.

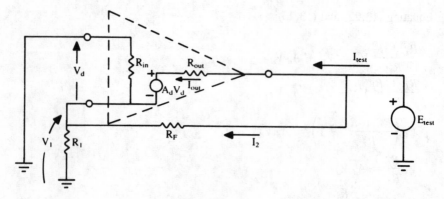

Fig. 13.29

The voltage across R_1 by the voltage divider rule is

$$V_1 = E_{Test} \frac{R_1}{R_1 + R_f} \qquad (13.27)$$

Also, by inspection:

$$V_1 = -V_d \qquad (13.28)$$

$$I_2 = \frac{E_{Test} - V_1}{R_f} \qquad (13.29)$$

$$I_{out} = \frac{E_{Test} - (A_d V_d + V_1)}{R_{out}} = \frac{E_{Test}}{R_{out}} - \frac{A_d V_d}{R_{out}} - \frac{V_1}{R_{out}}$$

$$I_{out} = \frac{E_{Test}}{R_{out}} - \frac{A_d}{R_{out}}(-V_1) - \frac{V_1}{R_{out}} = \frac{E_{Test}}{R_{out}} + V_1 \left(\frac{A_d - 1}{R_{out}} \right) \qquad (13.30)$$

By Kirchoff's current law at Node 0:

$$I_{out} + I_2 = I_{Test} \Rightarrow I_{out} = I_{Test} - I_2 \qquad (13.31)$$

from (13.29) and (13.31):

$$I_{out} = I_{Test} - \left(\frac{E_{Test} - V_1}{R_f} \right) \qquad (13.32)$$

Equating (13.32) and (13.30):

$$I_{Test} - \frac{E_{Test}}{R_f} + \frac{V_1}{R_f} = \frac{E_{Test}}{R_{out}} + V_1 \frac{(A_d - 1)}{R_{out}}$$

$$I_{Test} = \frac{E_{Test}}{R_f} + \frac{E_{Test}}{R_{out}} + V_1 \left(\frac{A_d - 1}{R_{out}} - \frac{1}{R_f} \right) \qquad (13.33)$$

Substituting (13.27) into (13.33):

$$\frac{E_{Test}}{R_f} + \frac{E_{Test}}{R_{out}} + E_{Test} \left(\frac{R_1}{R_1 + R_f} \right) \left(\frac{A_d - 1}{R_{out}} - \frac{1}{R_f} \right) = I_{Test}$$

Therefore,

$$\frac{E_{Test}}{I_{Test}} = Z_0 = \cfrac{1}{\cfrac{1}{R_f} + \cfrac{1}{R_{out}} + \cfrac{R_1}{R_1 + R_f}\left(\cfrac{A_d - 1}{R_{out}} - \cfrac{1}{R_f}\right)} \tag{13.34}$$

Example (13.14). Compute the output impedance of a non-inverting amplifier if the component values are

$R_f = 100k$, $R_1 = 10k$
$R_{out} = 0.1k$ and $A_d = 2(10^5)$

Note: For computation, all resistors are in kilohms.

$$Z_0 = \cfrac{1}{\cfrac{1}{100} + \cfrac{1}{10} + \cfrac{10}{10 + 100}\left[\cfrac{2(10^5)}{0.1} - \cfrac{1}{100}\right]}$$

$Z_0 = 5.5(10^{-6})\ k\Omega = 5.5(10^{-3})\ \Omega = 0.0055\ \Omega$

If we compare the non-inverting amplifier circuit with the inverting amplifier circuit, we note the characteristics given in Table 13.2.

Table 13.2

CONFIGURATION	A_V	Z_{in}	Z_0	Phase Reversal
Inverting	$-R_f/R_1$	R_1	$< 1\ \Omega$	yes
Non-Inverting	$1 + R_f/R_1$	$\to \infty$	$< 1\ \Omega$	no

13.2.3 The Unity Gain Voltage Follower

We recall that one of the advantages of an emitter-follower, or source-follower, stage, was that a high input impedance was transformed to a low output impedance. A similar function can be performed by an operational amplifier with the exception that higher input impedance, lower output impedance, and closer to unity voltage gain can be anticipated. We note from the schematic shown in Fig. 13.30 that V_0 is almost equal to V_{in}, differing by V_d, less than 100 microvolts (for a small load current).

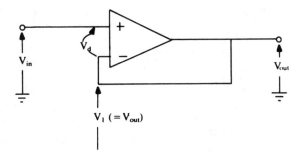

Fig. 13.30

A typical application for a unity gain voltage follower would be to isolate a high impedance source from a low impedance load. For example, a thermistor used to monitor temperature might be utilized to drive an analog meter movement. A unity gain voltage follower could be used to isolate the two devices.

13.2.4 Differentiators and Integrators

Probably the single largest use of OP AMPs when they were initially introduced was in instrumentation applications.

Early uses of analog computers entailed solving differential equations in both on-line control (real time) and off-line applications. Some early missile firings, for both military and space applications, utilized analog computers to process acceleration information from sensors and, thus, generate error-correction signals to rocket gimbals. These gimbals (swivels) could change the firing angle of the rocket motors, thereby providing a steering mechanism. The solution of differential equations requires the use of both calculus operations: *differentiation* and *integration*.

A differentiator

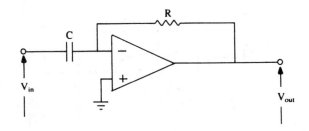

Fig. 13.31 Differentiator Circuit Without Compensation

450

Redrawing, we have the layout of Fig. 13.32. We note that $V_0 = -A_d V_d$

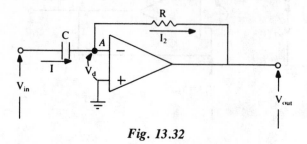

Fig. 13.32

If we assume that the current flowing into the OP AMP inverting terminal is negligible compared with other circuit currents, then we may apply Kirchoff's current law at node A:

$$\Sigma I_{in} = \Sigma I_{out}$$
$$I_1 = I_2$$
$$C \frac{d}{dt}(V_{in} - V_d) = \frac{V_d - V_{out}}{R}$$

if $V_d \to 0$

$$C \frac{dV_{in}}{dt} = -\frac{V_{out}}{R}$$

therefore,

$$V_{out} = -RC \frac{d}{dt} V_{in} \qquad (13.35)$$

Example (13.15). Determine the output waveform for the circuit and input waveform shown in Figs. 13.33 and 13.34.

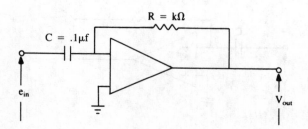

Fig. 13.33

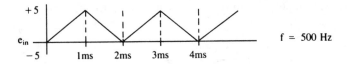

Fig. 13.34

To analyze this circuit, we must examine the input as though it were a series of discrete dc levels acting as inputs for periods of one millisecond each.

From (13.35):

$$V_{out} = -RC \frac{d}{dt} (V_{in})$$

and for a waveform composed of straight line segments: $V_{out} = -RC \frac{\Delta V_{in}}{\Delta t}$

where

$$R = 10^3, C = 10^{-7}, \text{ and } v_{in}$$

can be determined by writing the equation for a straight line.

The slope:

$$\frac{\Delta y}{\Delta x} = \frac{\Delta V}{\Delta t} = \frac{5 \text{ V}}{1 \text{ ms}}$$

therefore, $v_{in}(t) = 5000t + 0$ (y intercept), $0 < t < 1$ ms
$$= 5000 \text{ V/sec}$$

therefore,

$$v_{out}(t) = -(10^3)(10^6) \left(\frac{\Delta V}{\Delta t} \right) = -10^3 (10^6) \frac{d}{dt} (5000t)$$

$$= -5 \text{ V}$$

During the next half-cycle the equation for the straight line is

$$v_{in}(t) = -5000t + 10$$

therefore,

$$v_{out} = -(10^3)(10^{-7}) \frac{d}{dt} (-5000t + 10)$$

$$= +5 \text{ V}$$

Therefore, the output waveform is as shown in Fig. 13.35.

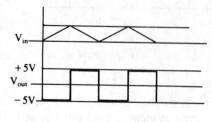

Fig. 13.35

An integrator

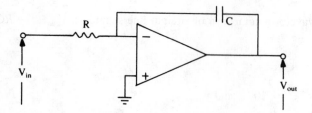

Fig. 13.36 Integrator Circuit without Compensation

Redrawing Fig. 13.36, we have the layout of Fig. 13.37.

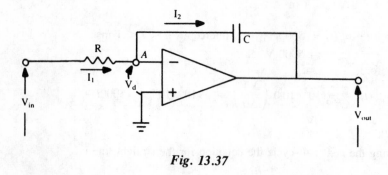

Fig. 13.37

Summing the currents at node A:

$$\Sigma I_{in} = \Sigma I_{out}$$
$$I_1 = I_2$$

$$\frac{V_{in} - V_d}{R} = C \frac{d}{dt} (V_d - V_{out})$$

if $V_d \to 0$

$$\frac{V_{in}}{R} = C \frac{d}{dt} (-V_{out})$$

therefore,

$$\frac{d}{dt} V_{out} = -\frac{1}{RC} V_{in}$$

$$dV_{out} = -\frac{1}{RC} V_{in} dt \qquad (13.36)$$

Integrating both sides:

$$V_{out} = -\frac{1}{RC} \int_0^t V_{in} \, dt$$

Example (13.16). Determine the output waveform for the circuit and input waveform shown in Fig. 13.38.

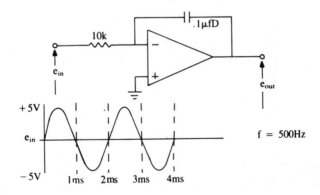

Fig. 13.38

Using Equation (13.36), we have

$$v_{out}(t) = \frac{1}{RC} \int_0^t v_{in}(t) dt$$

454

for the above input waveform, we can write

$v_{in}(t) = 5 \sin \omega t$, where $\omega = 2\pi f = 2\pi (500) = 1000\pi$
therefore,

$v_{in}(t) = 5 \sin 1000\pi t$
therefore,

$$v_{out}(t) = \frac{1}{10^4 (10^{-7})} \int 5 \sin 1000 \pi t \, dt + C$$

$$= -5000 \int \sin 1000 \pi t \, dt + C$$

$$= -\frac{5}{\pi} \cos 1000 \pi t + C$$

(where C is a dc offset level)

Plotting the output waveform, we have the curves of Fig. 13.39.

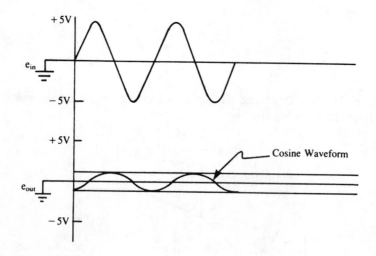

Fig. 13.39

Another common application of operational amplifiers is in *active filters*. We may recall from the section dealing with the high frequency performance of single-stage amplifiers that at a particular frequency, the cut-off frequency, the stray capacitance becomes effective and begins to "short" the signal to ground (see Fig. 13.40).

We can replace the stray capacitance with a discrete capacitor, and thereby develop a circuit that will pass low frequency signals up to a selected frequency which we have designed (see Fig. 13.41). If the frequencies that the generator applies to the lowpass filter exceed the cut-off frequency, the output signal will be reduced. The amount of reduction is 6 dB per octave (above the cut-off frequency), which means that for every doubling of the generator signal frequency, the output signal will be halved. Ideally, we would like a filter circuit that, in most cases, would completely attentuate the input signal above the cut-off frequency and cause no attenuation below the cut-off frequency. See Figs. 13.42 and 13.43.

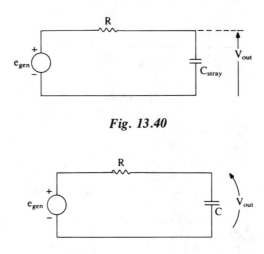

Fig. 13.40

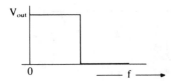

Fig. 13.41 Losspass Filter

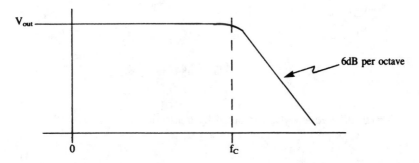

Fig. 13.42 Ideal Lowpass Filter Frequency Response

Fig. 13.43 Single-Section Filter Frequency Response

456

If a load resistance is connected to the output of a filter, the cut-off frequency is increased (see Fig. 13.44). The increase in the cut-off frequency can be seen by examining the Thevenin equivalent circuit of Fig. 13.45.

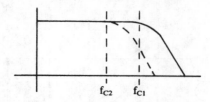

Fig. 13.44

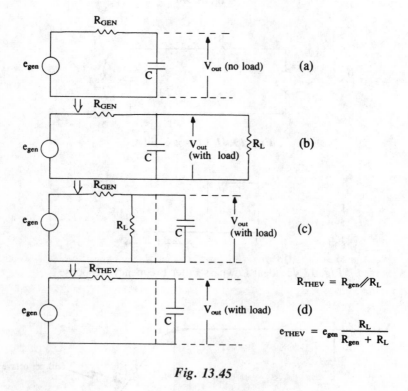

Fig. 13.45

The cut-off frequency is computed by the formula:

$$f_0 = \frac{1}{2\pi \, RC}$$

where R is the resistor in series with the capacitance. We note that R_{Thev} is a lower value resistance than R_{gen}. Therefore, the cut-off frequency will be a higher value when a load resistance is connected in parallel with the capacitor.

In a practical circuit design, the effect upon the cut-off frequency is not necessarily a troublesome problem because the capacitance value can be adjusted to yield a desired cut-off frequency. Once a design is "frozen," however, changes in the load resistor value present a problem. For example, if the load resistance represents the input resistance of a bipolar transistor, and should the transistor require field replacement, the new transistor may present a different load to the filter. Thus, a new cut-off frequency would be exhibited by the filter. How then can a filter be made independent of the amplifier circuits that "feed" a signal into it and "take" a signal from it? An operational amplifier will provide a solution to this problem.

First, if the signal source has "zero" output impedance, then the designer need not be concerned with adding the generator resistance to a fixed value resistor in his computation of the cutoff frequency (see Fig. 13.46).

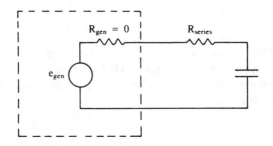

Fig. 13.46

If the output impedance of the generator was a significant value (compared with R_{series}), the designer would be concerned with being able to determine its resistance value accurately. Also, the designer would be concerned with the output resistance of the source changing with respect to time.

Second, if the input impedance to a succeeding amplifier is infinite, then the input impedance will have no effect on the cutoff frequency (see Fig. 13.47).

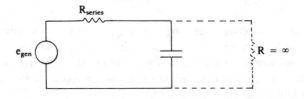

Fig. 13.47

458

An operational amplifier can be connected so as to have an output impedance of less than one ohm and an input impedance of greater than 10^9 ohms (see Fig. 13.48).

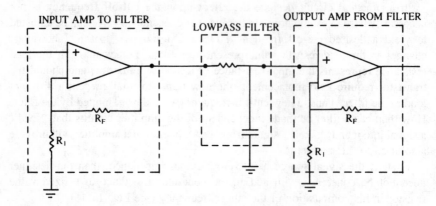

Fig. 13.48

Example (13.17).   Design a lowpass filter to have a cut-off frequency of 100 Hz. Choose the capacitance value equal to 0.1 microfarad (μF), and design the input and output amplifiers to have gains of 10. To design the amplifier, choose $R_1 = 10$ k and determine R_f

$$A_V = 1 + \frac{R_f}{R_1}$$

$$A_V - 1 = \frac{R_f}{R_1}$$

therefore,

$$R_f = R_1 (A_V - 1) = 10k (10 - 1) = 90 \text{ k}$$

For the filter,

$$f_C = \frac{1}{2\pi RC} \Rightarrow \left[R = \frac{1}{2\pi f_C C} = \frac{1}{2\pi (100)(0.1)(10^{-6})} = 15.9 \text{ k} \right]$$

If the transition between passing low frequencies and alternating high frequencies is to be very abrupt, i.e., approaching an ideal filter in performance, then additional lowpass filter sections may be added in cascade. For each additional filter section that is added in series, an additional 6 dB per octave of attenuation will occur (see Fig. 13.49).

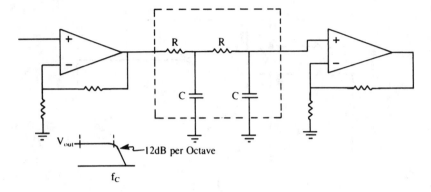

Fig. *13.49* 2-Section Lowpass R-C Filter 12 dB/Octave Attenuation

If the resistance and capacitance are interchanged, a highpass filter results.

13.2.4 The Difference Amplifier

An additional circuit that utilizes the unique characteristics of an OP AMP is the difference amplifier. A schematic for a difference amplifier is shown in Fig. 13.50.

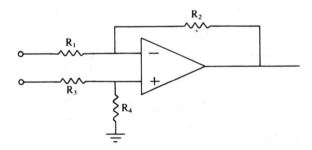

Fig. *13.50* Difference Amplifier

The difference amplifier can function in a manner similar to the single-stage differential amplifier composed of discrete bipolar transistors which was discussed earlier. A difference amplifier has the same desirable characteristics as the single-stage differential amplifier, namely, a high common-mode rejection ratio, but also does not require extensive dc bias design. Let us determine the voltage gain of a difference amplifier (see Fig. 13.51).

460

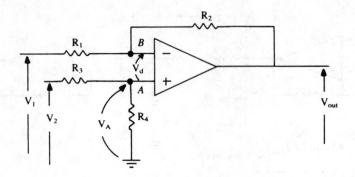

Fig. 13.51

For ease of computation, let us assume that $V_d \rightarrow 0$, and that the signal current into the inverting and non-inverting terminals is zero.

At node B,

$$\Sigma I = 0$$

$$\frac{V_1 - V_A}{R_1} + \frac{V_{out} - V_A}{R_2} = 0$$

$$\frac{V_1}{R_1} - \frac{V_A}{R_1} + \frac{V_{out}}{R_2} - \frac{V_A}{R_2} = 0$$

$$\frac{V_1}{R_1} + \frac{V_{out}}{R_2} = \frac{V_A}{R_1} + \frac{V_A}{R_2}$$

$$\frac{V_1}{R_1} + \frac{V_{out}}{R_2} = V_A \left(\frac{1}{R_1} + \frac{1}{R_2} \right)$$

$$\frac{V_1}{R_1} + \frac{V_{out}}{R_2} = V_A \left(\frac{R_1 + R_2}{R_1 R_2} \right) \tag{13.37}$$

By the voltage divider rule, we see that

$$V_A = V_2 \frac{R_4}{R_3 + R_4} \tag{13.38}$$

Substituting (13.38) into (13.37):

$$\frac{V_1}{R_1} + \frac{V_{out}}{R_2} = \left(V_2 \frac{R_4}{R_3 + R_4} \right) \left(\frac{R_1 + R_2}{R_1 R_2} \right)$$

$$\frac{V_{out}}{R_2} = V_2 \frac{R_4}{R_3 + R_4} \cdot \frac{R_1 + R_2}{R_1 R_2} - \frac{V_1}{R_1}$$

$$\frac{V_{out}}{R_2} = V_2 \frac{R_4}{R_3 + R_4} \cdot \frac{R_1 + R_2}{R_1 R_2} - \frac{V_1 R_2}{R_1 R_2}$$

$$V_{out} = V_2 \frac{R_4 (R_1 + R_2)}{R_1 (R_3 + R_4)} - V_1 \left(\frac{R_2}{R_1}\right) \qquad (13.39)$$

If we set

$$\frac{R_4 (R_1 + R_2)}{R_1 (R_3 + R_4)} = A \text{ and } \frac{R_2}{R_1} = B$$

we note that since all resistors are positive values, A and B must be a positive value.

From the above: $R_2 = BR_1$

Substituting:

$$\frac{R_4 (R_1 + BR_1)}{R_1 (R_3 + R_4)} = A$$

$$A = \frac{R_4 (1 + B)}{R_3 + R_4}$$

therefore,

$$\frac{R_3 + R_4}{R_4} = \frac{1 + B}{A}$$

or

$$1 + \frac{R_3}{R_4} = \frac{1 + B}{A}$$

therefore,

$$\frac{R_3}{R_4} = \frac{1 + B}{A} - 1 \qquad (13.40)$$

Because R_3 and R_4 represent resistor values, they must both be positive values:

$$\frac{1 + B}{A} - 1 > 0$$

$$\frac{1 + B}{A} > +1$$

or

$$1 + B > A \qquad (13.41)$$

Note that if $B = A$, the inequality of (13.41) is satisfied.

Example (13.18). Let us use the above criterion to design a phono preamplifier (Fig. 13.52).

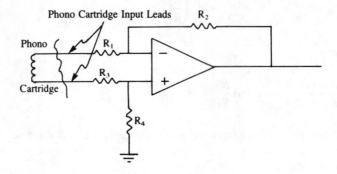

Fig. 13.52

Let us assume that the cartridge has optimum frequency characteristics when it "sees" a load of 44 k. Also, let us design for a difference voltage gain of 10. If the cartridge develops a signal voltage of 3 mV and if the *hum* pickup is 5 mV in the phono cartridge input leads, let us compute the desired signal and hum signal at the amplifier output terminal. Given that the OP AMP has a CMRR 110 dB.

Solution:

Recalling that the voltage between the inverting and non-inverting terminals approaches a short, the total resistance seen by the phono cartridge is

$$Z_{LOAD} = R_1 + R_3$$

If these resistors are made equal, they would be 22 kΩ each:

$R_1 = 22k, R_2 = 22k,$
therefore, $Z_{LOAD} = 44 \text{ k}\Omega$

Rewriting (13.39):

$$V_{out} = V_2 \frac{R_4 (R_1 + R_2)}{R_1 (R_3 + R_4)} - V_1 \frac{R_2}{R_1}$$
$$V_{out} = V_2 A - V_1 B$$

If the gain of each input to the common output is made equal,

$$V_{out} = V_2 B - V_1 B = B (V_2 - V_1) \tag{13.42}$$

However, the difference amplification was required to be 10. Thus,

$B = 10 = A$

therefore,

$B = \dfrac{R_2}{R_1} = 10$ and $R_1 = 22$ k

therefore,

$R_2 = 10R_1 = 10(22\text{k}) = 220$ k

From (13.40), if $A = B$,

$$\dfrac{R_3}{R_4} = \dfrac{1 + B}{A} - 1 = \dfrac{1 + A}{A} - 1 = \dfrac{1 + A - A}{A} = \dfrac{1}{A}$$

$$\dfrac{R_3}{R_4} = \dfrac{1}{A} \Rightarrow \dfrac{R_4}{R_3} = A \qquad\qquad (13.43)$$

$$R_4 = AR_3 = 10(22\text{k}) = 220 \text{ k}$$

If the signal voltage developed by the cartridge is 3 mV, then the output voltage would be from (13.42):

$$V_{out} = B\,(V_2 - V_1) = BV_d$$
$$= 10(3\text{mV}) = 30 \text{ mV}$$

The common mode voltage would be calculated:

$$A_C = \dfrac{V_{oC}}{V_{inC}} \Rightarrow V_{oC} = A_C\, V_{inC}$$

However,

$$CMRR = \dfrac{A_d}{A_C}$$

therefore,

$$A_C = \dfrac{A_d}{CMRR}$$

therefore,

$$V_{oC} = \dfrac{A_d}{CMRR} \cdot V_{inC}$$

Given: $A_d = 10$

$\qquad\quad CMRR = 110$ dB $= 316,000$

$\qquad\quad V_{inC} = 5$ mV

$$V_{oC} = \frac{10}{316,000} (5 \text{ mV})$$
$$V_{oC} = 0.000158 \text{ mV}$$

We see that the signal voltage has been increased significantly and the hum voltage is a negligible value.

PROBLEMS

1. If a differential amplifier has a differential gain of ten, and the input to the inverting terminal is a 10 peak sinewave and the amplitude of the non-inverting terminal is 3 mV peak, sketch both inputs and the output waveform underneath one another. The input waveforms are in phase with one another.

2. The amplifier of problem number one has a CMRR of 80 dB. If 5 mV is simultaneously applied to both input terminals, what output will result?

3. For the circuit shown below, design for the given dc condition:

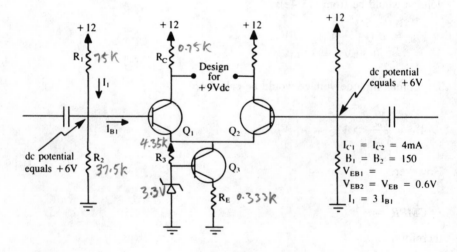

Fig. 13.53

4. If the source impedances connected to Q_1 and Q_2 were 600 Ω each and if the ac β equals the dc β of each transistor, what is the stage gain?

5. Design an inverting amplifier utilizing an OP AMP to have a gain of 15. The input impedance must be 15 kΩ.

6. If the OP AMP used in the design of Problem 5 has an open loop gain of 150,000, an input resistance of 100 kΩ and output resistance of 150 Ω, what is the output impedance of the circuit of the circuit designed in Problem 5?

7. Design a summing amplifier to satisfy the following requirements. An automobile fuel injection system is designed such that gas pedal depression position is to be weighted 10 times more important than engine operating temperature and the relative humidity of the induction air is to be weighted three times more important than the engine operating temperature. Design for a gain of 5 for the engine operating temperature and use a feedback resistor of 100 kΩ.

8. Design a non-inverting amplifier utilizing an OP AMP to have a gain of 15. Choose $R_f = 120$ kΩ.

9. If the OP AMP utilized in Problem 8 has the characteristics of an open loop gain of 150,000, an input resistance of 100 kΩ and an output resistance of 150 Ω, what is the input and output impedance of the amplifier designed in Problem 8?

10. The input waveform shown below is applied to the circuit shown below. Sketch and label the output waveform.

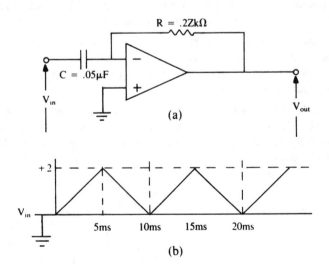

Fig. 13.54

11. The input waveform shown below is applied to the circuit shown below. Sketch and label the output waveform.

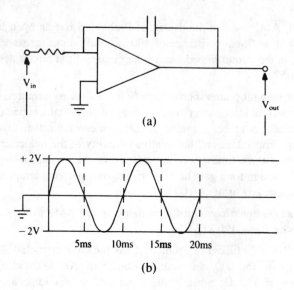

(a)

(b)

Fig. 13.55

12. Design an active lowpass filter to have the following characteristics: An input and an output amplifier with a voltage gain of 15, a cut-off frequency of 60 Hz. Choose the value of the capacitor to be 0.2 μF.

13. Design a difference amplifier to amplify the output signal from a remotely mounted carbon microphone that produces a 50 mV difference signal. The voltage amplification should be 40 and the microphone should see an input impedance of 1000 Ω.

14. If the OP AMP used in Problem 12 had a manufacturers' published CMRR of 100 dB and the two 30 feet long wires from the microphone to the amplifier each developed 100 mV of *hum* signal, what would the output *hum* voltage be?

Chapter 14

Oscillators

14.1   OSCILLATOR THEORY AND APPLICATIONS

There are sections of electronic systems that require a repetitive input wave-form as an input signal, such as a square wave (Fig. 14.1(a)) or a sine wave (Fig. 14.1(b)).

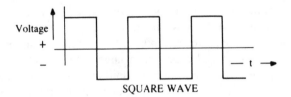

SQUARE WAVE

Fig. 14.1(a)

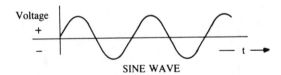

SINE WAVE

Fig. 14.1(b)

A repetitive square wave is used for many applications, such as to trigger a magnetron in a radar system, to provide the system clock for a microprocessor chip, and, after some modification in its waveshape, to drive the yoke in a video display (CRT, VDT, TV, *et cetera*).

A sine wave oscillator is used in communications to develop the radio frequency carrier, which will ultimately carry the information from the transmitting antenna to the receiving antenna.

467

Generally, a square wave is developed in a *relaxation oscillator* and a sine wave is developed in a *sine wave oscillator*. However, the oscillator roles may be interchanged.

A sine wave oscillator may be used to develop a square wave by using a clipper circuit, and a square wave oscillator may be used to develop a sine wave by passing the square output through a lowpass filter.

14.1.1 Sine Wave Oscillators

Generally, sine wave oscillators rely upon two basic electrical phenomena for their operation: positive feedback and resonant circuits.

We know that it is possible to take the output signal from the collector or drain of a common-emitter bipolar or common-source field-effect transistor, and there will be a 180-degree phase shift between the base to collector (bipolar) or gate to drain (FET). If we detect the signal from the collector and connect it back to the base through a resistor, the gain of the amplifier would be reduced.

If there was a method of increasing the phase shift an additional 180°, then the total phase shift would be 180° because of the transtor and 180° because of a phase shift network, or a total of 360° (see Fig. 14.2). Let us examine the effect of such a connection. Suppose that there was a small positive-going pulse at the base caused by noise (or any other phenomenon). This positive-going pulse would be amplified and appear as a larger negative-going pulse at the collector. After "going through" the phase shift network, it would appear back at the base as a positive-going pulse, thus reinforcing the original positive-going pulse.

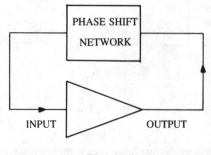

Fig. 14.2

The pulse would grow in amplitude until the transistor collector-to-emitter voltage decreased to almost zero volts (saturation). See Fig. 14.3. If reactive elements were used in the feedback path, L-C or R-C networks, the amplification process would reverse because a charge on a capacitor, or flux around an inductor, cannot be maintained in the presence of an energy dissipating element (resistance) without continuous energy input. Recall that the transistor output is in saturation and thus the collector potential can no longer decrease.

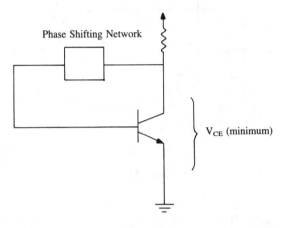

Fig. 14.3

Once saturation is reached, the collector-to-emitter potential begins to increase. Thus, a sine wave is developed. Two questions arise:

1. How does the phase shift network achieve a 180° phase shift?
2. How is the frequency of oscillation determined?

To answer the first question, if we consider the effect of a transformer, we know that, depending upon how we connect the secondary, one of the leads may be increasingly positive or negative with respect to ground (see Fig. 14.4(a,b)).

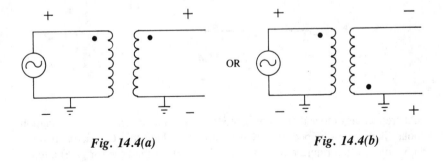

Fig. 14.4(a)   **Fig. 14.4(b)**

Thus, if we use a transformer to "feed" the output back to the input, we can "force" the amplifier stage to oscillate (see Fig. 14.5).

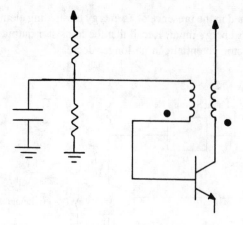

Fig. 14.5

Recalling from ac theory, we know that for a series resonant circuit, a low impedance will occur at the resonant frequency. For a series-parallel resonant circuit, at the resonant frequency, the maximum impedance will be "seen" when "looking into" the L-C parallel combination (see Fig. 14.6).

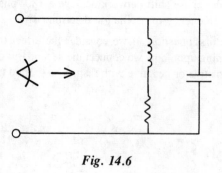

Fig. 14.6

At frequencies other than resonance, the impedance of the L-C combination "looks like" a low impedance. If we were to include this L-C series-parallel combination, often referred to as a "tank" circuit, in the base (or gate) circuit, the only frequency where the signal would *not* be shorted to ground would be at the tank resonant frequency. See Fig. 14.7(a) and Fig. 14.7(b,c) with the bias resistor.

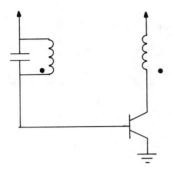

Fig. 14.7(a)

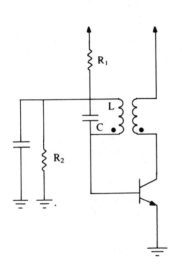

Fig. 14.7(b)

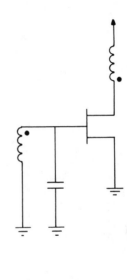

Fig. 14.7(c)

Recalling from ac theory for a series-parallel resonant circuit:

$$f_0 = \frac{1}{2\pi \sqrt{LC}} \sqrt{1 - \frac{1}{Q^2}}$$

and if the circuit Q is sufficiently high:

$$f_0 \doteq \frac{1}{2\pi \sqrt{LC}} \qquad (14.1)$$

472

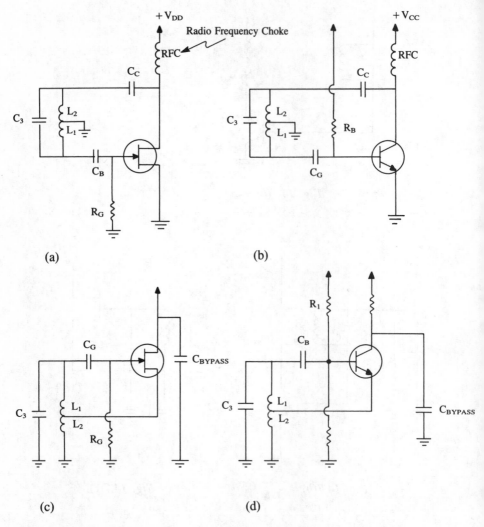

Fig. 14.8 Hartley Oscillators

In most design applications, the frequency is normally a desired or known quantity, and the inductance of the transformer is also a known value. Thus, the capacitor value must be adjusted to yield the desired frequency. Solving (14.1) for C we have

$$C = \frac{1}{4\pi^2 f_0^2 L}$$

(14.2)

It should be pointed out that the L term in Equation (14.2) must also include the inductance reflected from the primary into the secondary. The above oscillator circuit is referred to as an "Armstrong oscillator," named after its developer. The Armstrong oscillator is not generally used in low power oscillator circuits because two distinct coils are required. Separation between the coils will affect the inductance and, consequently, the oscillator frequency. Also, two coils represent added éxpense. Thus, utilization of an oscillator circuit that could function with a single coil would be preferred.

Various sine wave oscillator circuits have been developed to overcome the need to use the distinct coils. The "Hartley oscillator" circuit requires a single tapped coil, as shown in Fig. 14.8(a–d).

Referring to Figs. 14.8(c) and 14.9(a), if we assume that at a particular instant in time the drain potential is decreasing with respect to ground, then the potential developed across the coil would be as shown in Fig. 14.9(b).

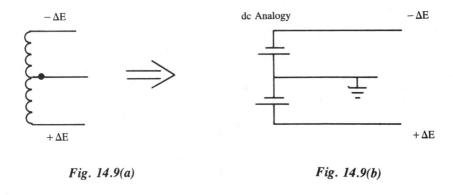

Fig. 14.9(a)　　　　　Fig. 14.9(b)

We see that the gate of the FET is experiencing an incrementally positive voltage, reinforcing the original positive voltage, which must have been present.

Thus, the amplifier stage is subjected to positive feedback and will oscillate. The capacitor value can be varied to adjust the frequency of oscillation. One potential problem to which the oscillator could be subjected is being "over-driven." If the amplitude of the sine wave appearing at the gate is sufficiently large, the transistor can be driven into saturation or cut-off. A simple method for deterring the transistor oscillator from over-driving itself is to employ some scheme whereby the output signal amplitude (and, thus, the input signal amplitude) is regulated to a fixed value. A simple technique to accomplish this goal is the inclusion of an R-C network, which will provide self-bias. Referring to Fig. 14.10, R_G and C_G accomplish this goal. When the input sine wave is positive-going, current will flow into the gate circuit (which is normally a reversed-biased diode). The current flow will leave the capacitor with a charge across its plates,

474

as shown in Fig. 14.11. However, the left side of the capacitor is connected to dc ground through the center tapped coil, as shown in Fig. 14.12.

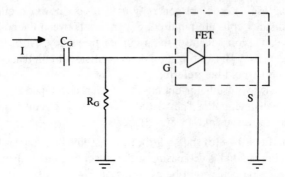

Fig. 14.10

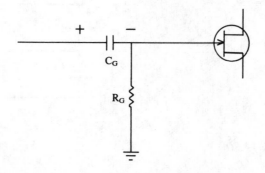

Fig. 14.11

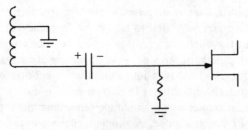

Fig. 14.12

Thus, the gate "sees" a negative self-bias, which tends to *reduce* the *average drain current*. As a transistor ages and its drain current tends to decrease, less signal is generated. Less signal results in less negative bias and, therefore, *more drain current*. Thus, the R-C network tends to self-regulate the transistor oscillator output signal and bias Q point.

Interestingly, a large negative bias will cut off drain current during the majority of the oscillator cycle as shown in Fig. 14.13. Thus, the efficiency of the oscillator stage will be high:

$$\eta = \frac{P_{out}}{P_{in}} = \frac{\text{ac Power Out}}{\text{dc Power In}}$$

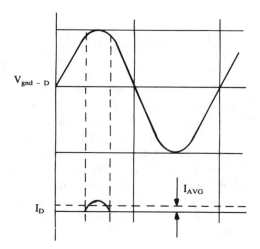

Fig. 14.13

The high efficiency is a result of the fact that the dc power is

$$P_{dc} = E_{dc}\,I_{dc} = E_{dc}\,I_{avg}$$

and since current only flows for a brief period each cycle, I_{avg} is quite low.

We may wonder why the output voltage is a sinusoid when the signal current is a short pulse. If we recall that a tank circuit is a parallel resonant circuit, which has a high impedance at one frequency, and at higher (or lower) frequencies the tank impedance approaches a short; therefore, the drain circuit tank will oscillate at only one frequency, and the function of the current pulses is to supply energy lost in the tank as a result of resistance and radiation.

We return to the question of how the value of R_G and C_G are computed? A generally accepted rule of thumb is to make the $R_G\,C_G$ time constant equal to ten times the oscillator period, or

476

$$R_G C_G = 10T = \frac{10}{f_0}$$

If we choose a convenient value for the gate resistor, then we have

$$C_G = \frac{10}{f_o R_G}$$

Example (14.1). Design a gate self-bias network for an oscillator operating at 1 MHz. Choose

$$R_G = 100 \text{ k}$$
$$C_G = \frac{10}{f_o R_G} = \frac{10}{10^6(10^5)} = 10^{-10} = 100(10^{-12}) = 100 \text{ pF}$$

The tapped coil used in a Hartley oscillator circuit may be an undesirable component in a production run where large quantities of identical components must be constructed. It may be difficult to locate the tap at exactly the same number of turns on each coil, or the wire may be very small and, therefore, tapping the winding may be difficult or expensive. A circuit where a tapped coil is unnecessary is the Collpitts oscillator. Referring to the circuit shown in Fig. 14.14, we note that instead of the tapped coil, two series capacitors form a signal voltage divider (see Fig. 14.15). In this circuit, as in the Hartley circuit, the gate will receive a positive feedback signal (Fig. 14.16), thereby stimulating the amplifier stage into oscillation.

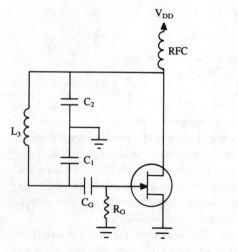

Fig. 14.14(a)

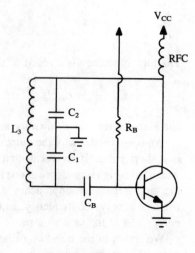

Fig. 14.14(b)

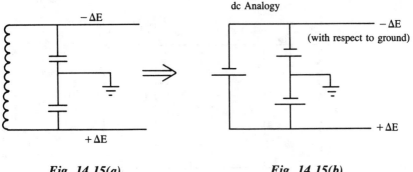

Fig. 14.15(a)   Fig. 14.15(b)

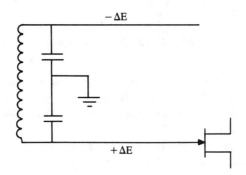

Fig. 14.16

The coil inductance in this circuit can be adjusted to vary the frequency of oscillation. C_1 or C_2 could also be adjusted, but their ratio is fixed by feedback requirements (as we will investigate in the next section). Thus, if we vary C_1 to adjust the oscillator frequency, we will also be forced to adjust C_2 to yield the proper feedback ratio. Inductances are not usually as readily adjusted as capacitances, particularly for air-cored inductances used at high frequencies. Thus, we are faced with a dilemma: the Collpitts oscillator is desirable because it utilizes a single untapped coil, but frequency adjustment is less desirably accomplished than for the Hartley. A way around this problem is by using the Clapp oscillator circuit, a variant of the Collpitts oscillator (see Fig. 14.17).

We note the inclusion of C_4 in addition to the other components. The reactances of C_1 and C_2 are deliberately chosen to be small compared with C_4. Therefore, C_1 and C_2 form the required voltage divider for feedback and oscillation, and C_4 is adjusted to achieve the desired resonant frequency.

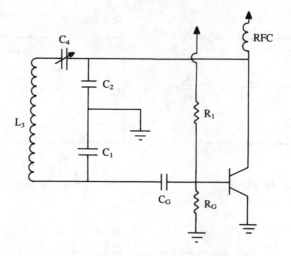

Fig. 14.17 Clapp Oscillator

We have discussed various oscillator circuits and the determination of the self-bias network, $R_G C_G$. However, we have not discussed the quantitative relationship between L_1, L_2, and C_3 for a Hartley oscillator and C_1, C_2, and L_3 for a Collpitts oscillator.

In general, we know that the voltage gain is defined as the ratio of the output voltage to the input voltage. Thus, for the circuit shown in Fig. 14.18, the voltage gain of the amplifier by itself (*not* the entire stage) is

$$A_V = \frac{e_{out}}{e'_{in}}$$

If

$$e'_{in} = e_{in} + \beta\, e_{in}$$

due to the summing network, then the voltage gain *with* feedback is

$$A_{Vf} = \frac{e_{out}}{e_{in}} = \frac{e_{out}}{e'_{in} - \beta\, e_{out}}$$

If we divide the numerator and denominator by e'_{in}, we have

$$A_{Vf} = \frac{e_{out}/e'_{in}}{e'_{in}/e'_{in} - \beta\, e_{out}/e'_{in}} = \frac{A}{1 - \beta\, A}$$

where A is the amplification without feedback and A_{Vf} is the overall voltage gain with feedback. We recall that the feedback network must create a 180° phase shift for oscillation (a transformer was used in our earlier example) and we know that the amplifier voltage gain A has a 180° phase shift associated with it. Therefore, the term βA is a positive term. If βA is adjusted equal to unity, then

$$A_{Vf} = \frac{A}{1 - 1}$$

which tends toward infinity.

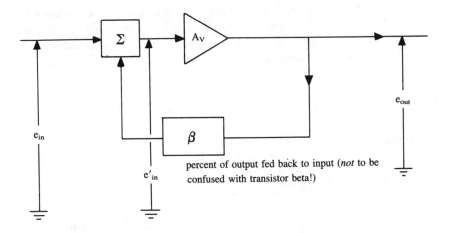

Fig. 14.18

A transistor amplifier with a voltage gain of -100 and a feedback network with a magnitude of $-1/100$ ($= -0.01$) would yield $\beta A = 1$. It is easy to see why an amplifier design often turns out to be an excellent oscillator design.

If the feedback factor β is a positive value, then the overall feedback is negative and the voltage gain of the stage is reduced (instead of being increased, as with an oscillator).

Negative feedback in amplifiers has certain desirable characteristics. Among these amplifier changes are:

1. Lower amplifier stage distortion;
2. Wider amplifier stage bandwidth.

Example (14.2). Of course, we do not "get something for nothing." The "loss" is a reduced amplifier stage gain. If we use the above numbers for voltage gain and feedback factor β for an amplifier stage with negative feedback, β becomes +0.01

$$A_{Vf} = \frac{A_V}{1 - \beta A_V} = \frac{-100}{1 - (+0.01)(-100)}$$

Thus,

$$A_{Vf} = \frac{-100}{1 - (-1)} = \frac{-100}{2} = -50$$

Suppose that over a period of time the transistor β (not to be confused with a feedback factor) had decreased and, therefore, the voltage gain had decreased from 100 to 90. This is a 10% reduction in amplification. Let us see what happens to our stage gain with feedback:

$$A_{Vf} = \frac{A}{1 - \beta A} = \frac{-90}{1 - (0.01)(-90)} = \frac{-90}{1 + 0.9} = -47.4$$

The percent decrease may be computed:

$$\% \text{ decrease} = \left| \frac{50 - 47.4}{50} \right| \times 100 = 5.26\%$$

Thus, *with negative feedback, less percent change in the voltage gain will be experienced due to changes in the non-feedback voltage gain.*

Returning to our sine wave oscillator design investigation, if we arrange the amplifier such that three reactances are connected as shown in the diagram of Fig. 14.19, and also assume that the amplifier input impedance does not appreciably load the network, then we can write the following, for a FET (or electron tube):

$$A = -g_m (r_d \;/\!/\; Z_L)$$

If the gate resistor is a sufficiently large value, then

$$Z_L = X_2 \;/\!/\; (X_1 + X_3)$$
$$= \frac{X_2 (X_1 + X_3)}{X_1 + X_2 + X_3}$$

or

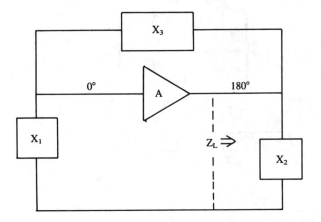

Fig. 14.19

$$A = -g_m \frac{r_d Z_L}{r_d + Z_L}$$

However, we know that

$$A_{vf} = \frac{A}{1 - \beta A}$$

and for oscillation

$$\beta A = +1$$

Thus,

$$\beta A = \beta \left[-g_m \frac{r_d Z_L}{r_d + Z_L} \right] = +1$$

where, according to the voltage divider rule, we have

$$\beta = \left(\frac{X_1}{X_1 + X_3} \right) \left(-g_m \frac{r_d Z_L}{r_d + Z_L} \right) = 1$$

therefore,

$$\left(\frac{X_1}{X_1 + X_3}\right)(-g_m)\left\{\frac{r_d\left[\dfrac{X_2\,(X_1 + X_3)}{X_1 + X_2 + X_3}\right]}{r_d + \dfrac{X_2\,(X_1 + X_3)}{X_1 + X_2 + X_3}}\right\} = 1$$

$$X_1\,(-g_m)\,\frac{r_d\left(\dfrac{X_2}{X_1 + X_2 + X_3}\right)}{\dfrac{r_d + (X_1 + X_2 + X_3) + X_2\,(X_1 + X_3)}{X_1 + X_2 + X_3}} = 1$$

hence,

$$\frac{-g_m\,V_d\,X_1\,X_2}{r_d\,(X_1 + X_2 + X_3) + X_2\,(X_1 + X_3)} = 1$$

We know that $r_d\,(X_1 + X_2 + X_3)$ is an imaginary number, since we are multiplying a real term, r_d, times an imaginary term, $(X_1 + X_2 + X_3)$. However, the right-hand side of

$$\frac{-g_m\,r_d\,X_1\,X_2}{r_d\,(X_1 + X_2 + X_3) + X_2\,(X_1 + X_3)} = 1$$

is a real number (1 is real). Therefore, $r_d\,(X_1 + X_2 + X_3)$ must be zero, or the left-hand side of the above formula would be a complex number, and a complex number cannot equal a real number (unity). Therefore,

$$-g_m\,\frac{r_d\,X_1\,X_2}{X_2\,(X_1 + X_3)} = 1$$

or $-g_m\,\dfrac{r_d\,X_1}{X_1 + X_3} = 1$

or $-g_m\,r_d\,X_1 = X_1 + X_3$ (14.3a)

but $X_1 + X_2 + X_3 = 0$ (since $r_d \neq 0$) (14.3b)
$X_1 + X_3 = -X_2$

and $-g_m\,r_d\,X_1 = -X_2$ (from (14.3a) and (14.3b))
$g_m\,r_d\,X_1 = X_2$

or $g_m\,r_d\,\dfrac{X_1}{X_2} = +1$ (14.3c)

We can conclude then that X_1 and X_2 must be the same type of reactances, i.e., either both inductances or both capacitances. If this were not true, then the left side of (14.3) would be a negative value.

We may also conclude that reactance X_3 must be the opposite of X_1 and X_2 since

$$X_1 + X_2 + X_3 = 0$$

or

$$X_1 + X_2 = -X_3$$

Example (14.3). Let us design a Hartley oscillator with a FET that must operate at 1 MHz (see Fig. 14.20).

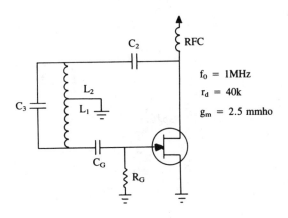

$f_0 = 1\text{MHz}$

$r_d = 40\text{k}$

$g_m = 2.5 \text{ mmho}$

Fig. 14.20

If we employ a tapped coil for L_1 and L_2 we know that

$$X_{LT} = X_{L1} + X_{L2} \pm 2X_M$$

However, if we employ an air-cored coil with a low coupling coefficient, we may write

$$X_{LT} \doteq X_{L1} + X_{L2}$$

or

$$L_T \doteq L_1 + L_2$$

Let us choose $R_G = 500$ k and $L_1 = 10$ μH. To ensure that the oscillator will continue to oscillate, even if the transconductance should decrease, let us choose $\beta A \geq 1$, thereby ensuring positive feedback:

Let $g_m r_d \dfrac{X_1}{X_2} = 2$

therefore,

$$g_m r_d \frac{j \omega L_1}{j \omega L_2} = 2$$
$$g_m r_d L_1 = 2L_2$$
$$2.5(10^{-3}) \, 40(10^3) \, 10(10^{-6}) = 2L_2$$

therefore,

$$L_2 = \frac{2.5(40) \, 10(10^{-6})}{2}$$
$$L_2 = 500 \ \mu H$$

and

$$C_3 = \frac{1}{4\pi^2 f_0^2 L_T} = \frac{1}{4\pi^2 \, (10^6)^2 \, 510(10^{-6})}$$
$$C_3 = 49.7 \text{ pF}$$

Also, for R_G chosen equal to 500 k, we have

$$C_G = \frac{10}{R_G f_0} = \frac{10}{500(10^3)(10^6)}$$
$$C_G = 20 \text{ pF}$$

If we had chosen a one millihenry (mH) RFC, its reactance would be

$$X_{L \, (RFC)} = 2\pi f_0 L = 2\pi \, (10^6)(10^{-3}) = 6.3 \text{ k}\Omega$$

Choosing the coupling capacitor reactance to be one-twentieth of the RFC, we have

$$X_{C \, C} = \frac{1}{20} X_{L \, (RFC)} = \frac{6.3(10^3)}{20} = 310 \ \Omega$$

and

$$X_{CC} = \frac{1}{2\pi f C_C}$$

therefore,

$$C_C = \frac{1}{2\pi f X_{CC}} = \frac{1}{2\pi (10^6)(310)}$$
$$C_C = 500 \text{ pF}$$

and our design is complete!

Example (14.4). Design a Clapp oscillator to oscillate at 1 MHz for the circuit shown in Fig. 14.21. We note that if $C_4 << C_1$ and $C_4 << C_2$:

$$C_T = \frac{1}{\dfrac{1}{C_1} + \dfrac{1}{C_2} + \dfrac{1}{C_4}} = \frac{C_4}{\dfrac{C_4}{C_1} + \dfrac{C_4}{C_2} + \dfrac{C_4}{C_4}} \doteq C_4$$

therefore,

$$C_T = \frac{1}{4\pi^2 f_0^2 L} = C_4$$

Thus,

$$f_0 = \frac{1}{2\pi \sqrt{LC_T}}$$

Also, choose $C_1 = 10C_4$, therefore, C_1 and C_2 will not affect the tank resonant frequency.

Choose $L = 500 \text{ }\mu H$, therefore,

$$C_4 = \frac{1}{4\pi^2 f_0^2 L} = \frac{1}{4\pi^2 (10^6)^2 500(10^{-6})}$$
$$C_4 = 50.7 \text{ pF}$$

Choose a variable capacitor 10–75 pF:

$$C_1 = 10C_4 = 10(50.7) = 507 \text{ pF}$$

$$g_m r_d \frac{X_1}{X_2} = 2 \rightarrow g_m r_d \frac{\left(\dfrac{1}{j\omega C_1} \right)}{\left(\dfrac{1}{j\omega C_2} \right)} = 2$$

$$g_m r_d C_2 = 2 C_1$$

therefore,

$$C_2 = \frac{2C_1}{g_m\, r_d} = \frac{2\,(507)\,(10^{-12})}{2.5(10^{-3})(40)\,(10^3)}$$
$$C_2 = 1013 \text{ pF}$$

Choose .001 μF:

$$C_G = \frac{10}{R_G\, f_0} = \frac{10}{500(10^3)(10^6)} = 20 \text{ pF}$$

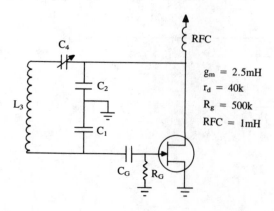

Fig. 14.21

The value of C_4 may be selected as a 5–100 pF variable capacitor in order to allow the capacitor to be adjusted over a wide range from its midpoint (50 pF). C_1 can be selected as 500 pF—a standard value. It should be noted that oscillators designed for very high frequency applications must function with inter-electrode and wiring capacitance. Therefore, these stray capacitances must be included in the design equations.

Another variation of the sine wave oscillator is the *crystal controlled oscillator*. Certain materials, when under pressure, will develop a potential difference between parallel faces of a crystalline mass. This is referred to as the *piezo electric effect*. The opposite transducer effect can also be demonstrated, whereby the application of a voltage to parallel faces of a crystal will cause a change in the mechanical shape of the crystal. Sonar, a commonly employed underwater detection method, relies on the stressing of a piezo electric crystal to force it to emit sound waves through water. The application of most interest to us in this chapter is the use of crystals to control sine wave oscillators. Why use a crystal? Regular L-C oscillators will change frequency as a result of changes in temperature, or other factors that affect the component values. A crystal can be used instead of the tuned circuit (tank circuit). It will vibrate at one of two natural resonant frequencies. The resonant frequencies may be as low as 10 kHz or as high as 100 MHz. A typical low frequency oscillator application which has become popular recently is the use of crystals in electronic wrist watches ($\doteq$ 32 kHz). The most common long-term crystal application is in the generation of the local oscillator frequency in radio (also TV and radar) transmitters.

Both the long-term and short-term stability of a crystal controlled oscillator is far superior to an L-C oscillator. Short-term stability is due, in part, to the crystal being a rigid mass whose expansion coefficient is relatively low, and also to the fact that the equivalent electrical resonant circuit Q is on the order of 20,000. With an extremely high Q, the bandwidth is very narrow and, therefore, the cyclical drift in the oscillator frequency is minimal. Long-term stability can be anticipated because the most common crystal material is quartz, which is relatively impervious to degeneration with aging. To further ensure that the crystal will not be affected by environmental conditions, crystals used in oscillators are packaged, or "housed," in hermetically sealed metal containers, referred to as "crystal cans." In cases where extreme oscillator frequency stability is required, as in commercial radio and TV broadcasting stations, the crystal is also operated in a small thermostatically controlled oven.

The electrical equivalent of a crystal is shown in Fig. 14.22, where L and C are the electrical equivalents of the mass and the elasticity of the crystal, while R represents the vibrational energy losses. C_p is the capacitance caused by the parallel metal plates used to input electrical energy into the crystal. The combination LRC forms an equivalent series resonant circuit, but the use of the L and C_p will allow the crystal to be operated in a high impedance series-parallel resonant circuit mode. A sketch of the equivalent electrical impedance with respect to a crystal frequency is shown in Fig. 14.23.

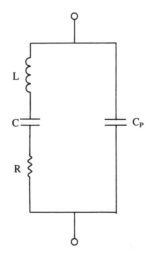

Fig. 14.22

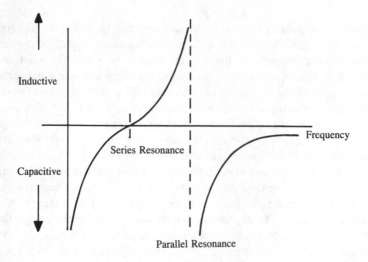

Inductive

Capacitive

Series Resonance

Frequency

Parallel Resonance

Fig. 14.23

The difference between the series and parallel resonant frequency operation of a crystal is a small value. The mode of operation is dependent upon whether the designer decides to utilize the low impedance of the series resonant equivalent circuit or the high impedance of the series-parallel resonant circuit. Typical crystal oscillator circuits utilizing the series resonant mode of operation will provide a low impedance path from the amplifier output to the input, as shown in Fig. 14.24(a,b).

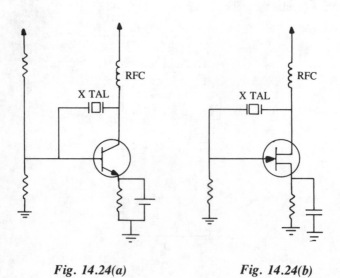

RFC

X TAL

RFC

X TAL

Fig. 14.24(a)   *Fig. 14.24(b)*

For operation in the parallel resonant mode, we need a circuit in which a low impedance is presented by the crystal at all frequencies except at the desired frequency of operation. In the circuits shown in Fig. 14.25(a,b) a low impedance path to ground is provided by the crystal for all frequencies except the parallel resonant frequency of the crystal.

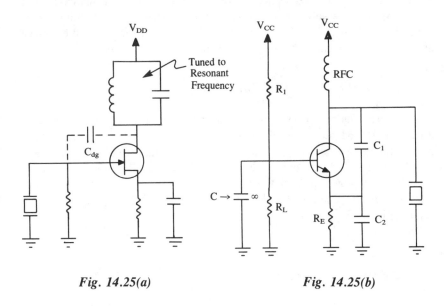

$$\textbf{Fig. 14.25(a)} \qquad\qquad \textbf{Fig. 14.25(b)}$$

One advantage of employing the crystal in a parallel resonant mode is that a small trimmer capacitor may be placed across the crystal to allow a small amount of trimming of the oscillator frequency. Crystals are somewhat fragile and may be damaged by a mechanical shock or by applying too large a dc bias voltage or too large a signal voltage.

14.2 LOW FREQUENCY SINUSOIDAL OSCILLATORS

There are certain applications where a low frequency sinusoidal source is required. For example, the simultation of the 60 Hz power-line signal may be required for a "back-up" emergency power supply. One problem that arises with the previously discussed oscillator circuits can be appreciated by examining the formula for resonant frequency:

$$f_0 = \frac{1}{2\pi \sqrt{LC}}$$

Solving for L, we have

$$L = \frac{1}{4\pi^2 f_0^2 C}$$

and we note that as the values f_0 and C decrease, the value of L increases. The use of large inductor values is undesirable, from the standpoints of both physical size and cost. Thus, if oscillator circuits could be devised that would eliminate the use of an inductance, the above problems would be eliminated.

14.3 RC PHASE SHIFT OSCILLATORS

In the previous design examples, a tuned L-C resonant circuit was used to provide the 180-degree phase shift, which was necessary to stimulate an amplifier into oscillation. We know, however, that an R-C voltage divider can provide a phase shift approaching, but not equaling, 90 degrees. See Fig. 14.26(a,b). Thus, if we employ a minimum of three R-C networks, each with a phase shift of 60 degrees, the necessary 180-degree phase shift for oscillation will be achieved (see Fig. 14.27).

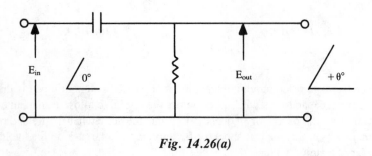

Fig. 14.26(a)

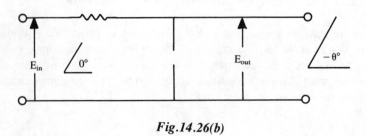

Fig.14.26(b)

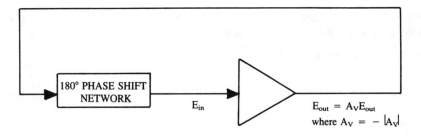

Fig. 14.27

If we employ the network of Fig. 14.28, the last resistance shown would include the input resistance of the transistor. Any losses in the successive R-C voltage divider networks must be compensated by the voltage amplification of the active device. If we use *mesh analysis* to determine the characteristics of the network, we have the configuration of Fig. 14.29.

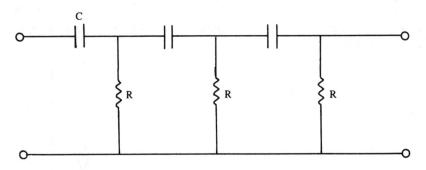

Fig. 14.28

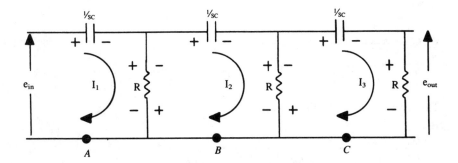

Fig. 14.29

492

Note: $1/SC$ is the shorthand notation that is used in place of $1/j\omega c$ when performing a Laplace transform circuit analysis.

at A summing voltages ccw:

$$I_1 R - I_2 R + I_1 \frac{1}{SC} - e_{in} = 0$$

at B ccw:

$$I_2 R - I_3 R + I_2 \frac{1}{SC} + I_2 R - I_1 R = 0$$

at C ccw:

$$I_3 R + I_3 \frac{1}{SC} + I_3 R - I_2 R = 0$$

If we collect like terms and solve the simultaneous equations, we would arrive at the following Δ determinant:

$$\Delta = \begin{vmatrix} R + \dfrac{1}{SC} & -R & 0 \\ -R & 2R + \dfrac{1}{SC} & -R \\ 0 & -R & 2R + \dfrac{1}{SC} \end{vmatrix}$$

We also note from the schematic of Fig. 14.29 that $e_{out} = I_3 R$. Solving for I_3 we have

$$I_3 = \frac{e_{in} R^2}{R^3 - \dfrac{5R}{\omega^2 C^2} + j\left[\dfrac{1}{\omega^3 C^3} - \dfrac{6R^2}{\omega C}\right]}$$

However, $e_{out} = I_3 R$, therefore,

$$\frac{e_{out}}{e_{in}} = \frac{I_3 R}{e_{in}} = \frac{R^3}{R^3 - \dfrac{5R}{\omega^2 C^2} + j\left[\dfrac{1}{\omega^3 C^3} - \dfrac{6R^2}{\omega C}\right]}$$

The imaginary term in the denominator must be equal to zero because the voltage fed back to the input has a 180-degree phase shift. (Minus one is a real number, not a complex number). Therefore,

$$\frac{1}{\omega^3 C^3} - \frac{6R^2}{\omega C} = 0$$

$$\frac{1}{\omega^3 C^3} = \frac{6R^2}{\omega C} \quad \text{or} \quad \frac{1}{\omega^2 C^2} = 6R^2$$

$$\omega^2 = \frac{1}{6R^2 C^2}$$

$$(2\pi f)^2 = \frac{1}{6R^2 C^2}$$

$$f^2 = \frac{1}{24\pi^2 R^2 C^2}$$

therefore, $f = \dfrac{1}{2\sqrt{6}\pi\, RC}$

$$f_0 = \frac{0.065}{RC} \tag{14.4}$$

Also,

$$\frac{e_{out}}{e_{in}} = \frac{R^3}{R^3 - \dfrac{5R}{\omega^2 C^2}} = \frac{R^2}{R^2 - \dfrac{5}{\omega^2 C^2}}$$

but $\dfrac{1}{\omega^2 C^2} = 6R^2$

Thus, $\dfrac{e_{out}}{e_{in}} = \dfrac{R^2}{R^2 - 5(6R^2)} = \dfrac{R^2}{R^2\, 30R^2} = -\dfrac{1}{29}$ \qquad (14.5)

Therefore, the signal appearing at the output of the R-C network is 1/29 of the input signal. Therefore, the amplifier must have a voltage gain of at least 29 in order to overcome the network loss.

There are other combinations of R-C networks, which will also develop a 180-degree phase shift, such as those shown in Fig. 14.30(a–d).

We note that by using extra R-C sections, the network voltage loss decreases. Thus, the amplification required of the active devicer (bipolar or field-effect transistor, or operational amplifier) is lessened.

494

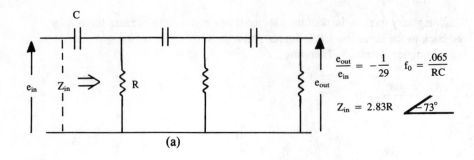

$$\frac{e_{out}}{e_{in}} = -\frac{1}{29} \qquad f_0 = \frac{.065}{RC}$$

$$Z_{in} = 2.83R \quad \angle -73°$$

(a)

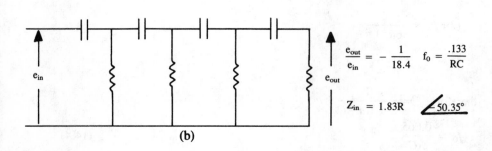

$$\frac{e_{out}}{e_{in}} = -\frac{1}{18.4} \qquad f_0 = \frac{.133}{RC}$$

$$Z_{in} = 1.83R \quad \angle -50.35°$$

(b)

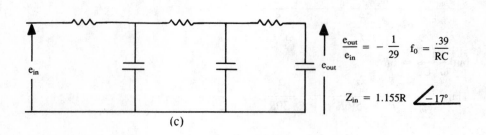

$$\frac{e_{out}}{e_{in}} = -\frac{1}{29} \qquad f_0 = \frac{.39}{RC}$$

$$Z_{in} = 1.155R \quad \angle -17°$$

(c)

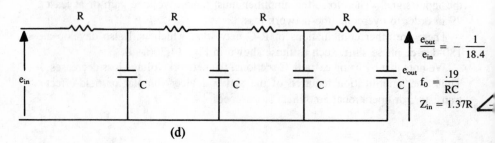

$$\frac{e_{out}}{e_{in}} = -\frac{1}{18.4}$$

$$f_0 = \frac{.19}{RC}$$

$$Z_{in} = 1.37R \quad \angle$$

(d)

Fig. 14.30

Example (14.5). Let us examine a design for a phase shift oscillator employing a three section R-C network, as shown in Fig. 14.31(a).

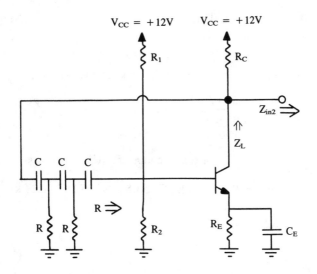

Fig. 14.31(a)

Given: $V_{CC} = +12$ V
$h_{FE} = h_{fe} = 100$
$I_{CQ} = 2$ mA
$I_1 = 2 I_B$
$V_{RE} = 10\% V_{CC}$
$f_0 = 60$ Hz
$Z_{in\ 2} = 28.9$ k (due to next stage)

Thus,

$$V_{RE} = 0.1\ (12) = 1.2\ \text{V and}\ I_E = I_C \frac{\beta + 1}{\beta} = 2.02\ \text{mA}$$

$$V_{RC} = \frac{V_{CC} - V_{RE}}{2} = \frac{12 - 1.2}{2} = 5.4\ \text{V}, \qquad R_C = \frac{V_{RC}}{I_C} = 2.7\ \text{k}\Omega$$

$$R_E = \frac{V_{RE}}{I_E} = 0.594\ \text{k, choose 560 }\Omega$$

$$V_{R2} = V_{RE} + V_{EB\ Q} = 1.2 + 0.7 = 1.9\ \text{V}$$

$$R_2 = \frac{V_{R2}}{I_2} = \frac{1.9V}{0.02 \text{ mA}} = 95 \text{ k, choose } 100 \text{ k}$$

$$I_1 = 2(0.02) = 0.04$$

$$I_2 = I_1 - I_B$$

$$I_2 = 0.04 - 0.02 = 0.02 \text{ mA}$$

$$R_1 = \frac{V_{CC} - V_{R2}}{I_1} = \frac{12 - 1.9}{0.04} = 252.5 \text{ k, choose } 270 \text{ k}$$

$$h_{ie} = \frac{0.026 \text{ V}}{I_B} = \frac{26 \text{ mV}}{0.02 \text{ mA}} = 1.3 \text{ k}$$

$$X_{CE} = 0.1 \, R_E = 0.0594 \text{ k}$$

$$C_E = \frac{1}{2\pi \, (60) \, (59.4)} = 44.6 \text{ }\mu\text{F, choose } 50 \text{ }\mu\text{F}$$

$$Z_{in1} = h_{ie} \parallel R_1 \parallel R_2 = 1.3 \text{ k} \parallel 252.5 \text{ k} \parallel 95 \text{ k} \doteq 1.28 \text{ k}$$

$$f = \frac{1}{2\pi \, \sqrt{6} \, RC}$$

therefore, $C = \dfrac{1}{2\pi f \, \sqrt{6} \, R} = \dfrac{1}{2\pi \, (60) \, \sqrt{6} \, (1.28) \, (10^3)}$

$$C = 0.849 \text{ }\mu\text{F}$$

Choose 0.5 μF, then R may be computed:

$$R = \frac{1}{2\pi f \sqrt{6} \, C} - \frac{10^6}{2\pi \, 60 \, \sqrt{6} \, (0.5)} = 2.17 \text{ k}$$

Add a resistor in series with base and we have

$$R_{add} = 2.17 \text{ k} - 1.28 \text{ k} = 0.89 \text{ k} = 890 \text{ }\Omega$$

For an oscillator to oscillate, gain must be greater than 29. Assume that an emitter follower is the load ($= 28.9$ k). Therefore,

$$A_V = \frac{-h_{fe} \, Z_L}{h_{ie}}$$

where

$$Z_L = R_{C1} \mathbin{/\!/} Z_{in2} \mathbin{/\!/} Z_N$$

and

$$Z_N = 2.83R \; \underline{/-73°}$$
$$= 2.83\,(2.17) \; \underline{/-73°}$$
$$= 6.14 \text{ k} \; \underline{/-73°}$$

$$Z_L = \cfrac{1}{\cfrac{1}{2.7} + \cfrac{1}{28.9} + \cfrac{1}{6.14 \; \underline{/-73°}}}$$

$$Z_L = 2.27 \text{ k} \; \underline{/-20.7°}$$

Thus,

$$A_V = \frac{-h_{fe}\,Z_L}{h_{ie}} = \frac{-100\,(2.27 \; \underline{/-20.7°})}{1.3} = -174.5 \; \underline{/-20.7°}$$

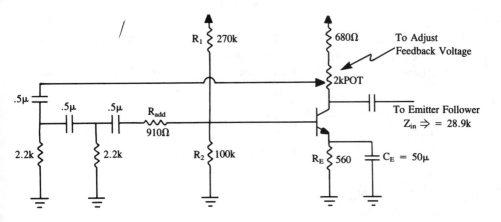

Fig. 14.31(b)

Our final schematic is given in Fig. 14.31(b). In completing this section on phase shift oscillators, it is worthwhile to note that if an R-C network is used with the capacitance to ground, the transistor input impedance should be a high value (FET would be appropriate). If the impedance is a low value, the last R-C element will not be symmetrical to the preceding elements.

The Wein bridge oscillator is another R-C sinusoidal oscillator configuration. The use of an operational amplifier in the design minimizes the number of components that are necessary. Referring to the schematic shown in Fig. 14.32, we note that both the inverting and non-inverting inputs to the OP AMP are used. The inverting terminal signal input replaces signal voltage that is lost in the phase shifting network which is connected to the non-inverting terminal.

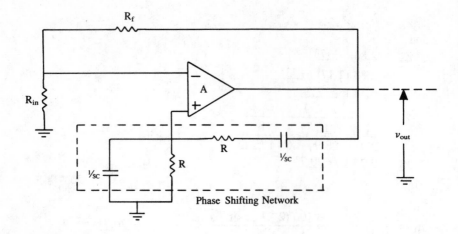

Fig. 14.32

The R-C networks, along with the gain determining network R_f/R_i, provide the necessary phase shift to ensure oscillation.

The voltage at the non-inverting terminal $v_{in(+)}$ by the voltage divider rule is

$$v_{in(+)} = v_{out} \frac{R \mathbin{/\mkern-5mu/} \dfrac{1}{SC}}{R \mathbin{/\mkern-5mu/} \dfrac{1}{SC} + R + \dfrac{1}{SC}}$$

Solving, we have

$$\frac{v_{in(+)}}{v_{out}} = \frac{1}{3 + j\left(R\omega C - \dfrac{1}{R\omega C}\right)}$$

and for oscillation, the feedback voltage is in phase with the input voltage at the non-inverting terminal, i.e., there is no phase shift. Thus, the ratio of $v_{in(+)}/v_{out}$ must be a real number, i.e.,

$$\left(R\omega C - \frac{1}{R\omega C}\right) = 0$$

therefore,

$$\frac{v_{in(+)}}{v_{out}} = \frac{1}{3} \tag{14.6}$$

and if $R\omega C - \dfrac{1}{R\omega C} = 0$

or $v_{in(+)} = \dfrac{v_{out}}{3}$,

then $R\omega C = \dfrac{1}{R\omega C}$

$$\omega^2 = \frac{1}{R^2 C^2} \tag{14.7}$$

$$f = \frac{1}{2\pi RC}$$

Thus, the voltage fed back to the inverting terminal must compensate for the loss in the voltage fed back to the non-inverting terminal. For safety sake, let us provide a bit more input voltage at the inverting terminal than is lost at the non-inverting terminal, i.e.,

$$v_{in(-)} \geqslant v_{in(+)}$$

By use of the voltage divider rule:

$$v_{in(-)} = v_{out} \frac{R_{in}}{R_{in} + R_f}$$

However, we stated that

$$v_{in(-)} \geqslant v_{in(+)}$$

Thus,

$$v_{out} \frac{R_{in}}{R_{in} + R_f} \geqslant \frac{v_{out}}{3}$$

$$\frac{R_{in}}{R_{in} + R_f} \geqslant \frac{1}{3}$$

therefore,

$$3R_{in} \geqslant R_{in} + R_f$$

$$2R_{in} \geqslant R_f \tag{14.8}$$

Example (14.6). Design a Wein bridge oscillator circuit to oscillate at 60 Hz. Use a 741 OP AMP. Let us start the design by choosing a large value for R, let us say, for example, 100 k. (Since we are designing a low frequency oscillator, a long time constant is required.)

$$f = \frac{1}{2\pi\,RC}$$

therefore, $C = \dfrac{1}{2\pi f R} = \dfrac{1}{2\pi\ 60\ (10^5)}$

$$C = 2.65\ (10^{-8})$$
$$C = 0.0265\ \mu F$$

Let us choose a standard value for C. Let

$$C = 0.02\ \mu F$$

Thus,

$$R = \frac{1}{2\pi f C} = \frac{1}{2\pi\ (60)(0.02)(10^{-6})} = 133\ k\Omega$$

Choose $R = 130$ k, 5%   (color code value);

For R_{in}, choose 56 k and $R_f = 100$ k:
$2R_{in} \geqslant R_f$?
$2(56\ k) \geqslant 100\ k$

Since $2R_{in} \geqslant R_f$, then our design will work.

14.4 NON-SINUSOIDAL OSCILLATORS

Non-sinusoidal oscillators are generally referred to as relaxation oscillators because at least one of the active devices is turned off during part of the oscillator cycle, i.e., no current is flowing through the device or the input control terminal has a negative bias applied to it.

At one time, the most commonly used type of relaxation oscillator was the *astable multivibrator* (Fig. 14.33) constructed of two discrete transistors. Oscillator operation requires one of the transistors to be in saturation while the other is cut off.

To analyze the operation of the astable multivibrator, we assume that the transistors have just changed state. Therefore, assume that Q_2 has just been driven into saturation and Q_1 has just been cut-off.

While making the transition from saturation to cut-off, the ground-to-collector voltage of Q_1 changes from $+0.2$ V to $+12$ V. This positive voltage is coupled

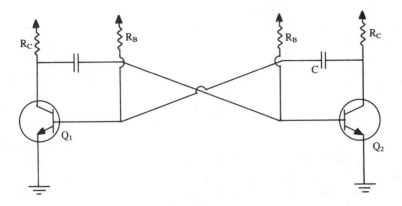

Fig. 14.33 Astable Multivibrator

to the base of Q_2, thereby driving it into saturation. The ground-to-collector potential at Q_2 changes from $+12$ to $+0.2$ Vdc. Because the voltage across a capacitor cannot be changed instantaneously, this change in potential (11.8 V) appears at the base of Q_1. Thus, the base of Q_1 changes from $+0.7$ V ($V_{EB\ SAT}$) to -11.1 V. If planar transistors are used, caution must be exercised to ensure that the maximum base-to-emitter reverse voltage rating is not exceeded, i.e., use protection diodes. The resistor R_B and the capacitor C form an RC charging network. The base potential increases until base current conductions again take place and the cycle is continuously repeated. In recent years, all of the components necessary to form an astable multivibrator have been incorporated into integrated circuit packages, such as the 555.

The cost of a 555 is considerably less than the cost of assembling the components for a discrete component astable multivibrator. Therefore, a mathematical analysis of the astable multivibrator will not be undertaken. The output waveform at a transistor collector would appear as shown in Fig. 14.34.

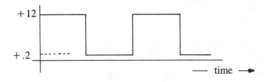

Fig. 14.34

If the R-C networks were asymmetrical, the output would appear as shown in Fig. 14.35.

502

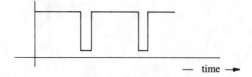

— time →

Fig. 14.35

A very simple—in terms of component count—relaxation oscillator is one that is constructed with a unijunction transistor. The UJT derives its name from the fact that a single *p-n* junction is employed in its construction (Fig. 14.36(a)). We note that the unijunction transistor symbol (Fig. 14.36(b)) is similar to a FET symbol, except that the arrow is pointing downwards.

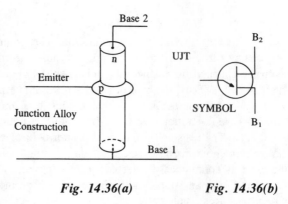

Fig. 14.36(a)   *Fig. 14.36(b)*

The operation of a UJT is as follows: the bulk resistance of the *n* material is anywhere from 4 k to 10 k. If a 10 Vdc power supply is applied across the transistor, by Ohm's law,

$$I = \frac{V}{R} = \frac{10 \text{ V}}{5 \text{ k}} = 2 \text{ mA}$$

If the *p* material is placed halfway between B_1 and B_2, the potential from ground to emitter would be 5 V (by the voltage divider rule). The bulk resistance from B_1 to B_2 may be thought of as two distinct resistances, R_{B1} and R_{B2}, with a lead brought out for the emitter connection (see Fig. 14.37(a,b)). Because the emitter connection is a *p-n* junction, diode action takes place between the *p* material and the *n* cylinder. A more correct equivalent circuit is shown in Fig. 14.38.

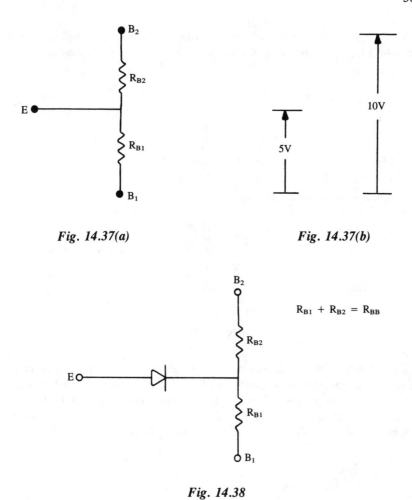

Fig. 14.37(a)

Fig. 14.37(b)

$R_{B1} + R_{B2} = R_{BB}$

Fig. 14.38

As long as the voltage source from ground to the emitter terminal in our hypothetical transistor is less than 5.7 V (5 + 0.7 for the diode), the diode is back-biased and no emitter current flows. At the point where the emitter terminal rises in potential above the voltage across R_{B1}, emitter current will begin to flow because the diode has become forward-biased. At this point, holes are injected into the *p-n* junction and the resistance of R_{B1} begins to decrease. With the decrease in resistance, more current flows and *avalanche* action takes place with a final equivalent resistance for R_{B1} of between 5 and 25 ohms.

Before proceeding with a description of the usage of a UJT, let us discuss the electrical characteristics in a little more depth. First, a more accurate portrayal of the equivalent circuit of a UJT is as given in Fig. 14.39. We note that R_{B1}

504

is shown as a variable resistor where its resistance varies from its bulk resistance value of 2 k to 6 k, and decreases to 5 to 25 ohms in the avalanche state.

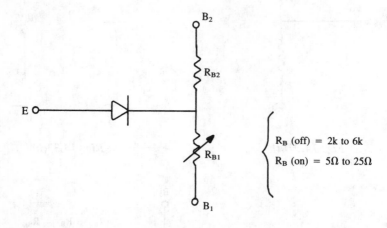

$$R_B \text{ (off)} = 2k \text{ to } 6k$$
$$R_B \text{ (on)} = 5\Omega \text{ to } 25\Omega$$

Fig. 14.39

The location of the p material may be thought of as a certain percentage of the vertical height from the bottom of the cylindrically shaped n material. Modern UJTs are made by a different process than the junction alloy method depicted herein. The location of the p material determines the ratio of R_{B1} to R_{BB} and is assigned a special symbol, the Greek letter eta, η. Thus,

$$\eta = \frac{R_{B1}}{R_{B1} + R_{B2}} = \frac{R_{B1}}{R_{BB}}$$

A typical value for η is 0.6, but a range of values is usually specified in the data contained in a manufacturer's specification sheet.

Some other key electrical characteristics of a UJT are depicted in the voltage *versus* current curve shown in Fig. 14.40. We note the negative resistance portion of the curve, which enables a UJT to be used as an oscillator without a positive feedback network.

In Fig. 14.40, we note I_P and V_P. $V_{peak}(V_P)$ represents the peak voltage at the emitter, at the point where avalanche action is about to begin. Therefore, from Fig. 14.41, we have

$$V_P = \eta \, V_{BB} + V_{diode}$$
$$V_E = V_P \text{ (at switching)}$$

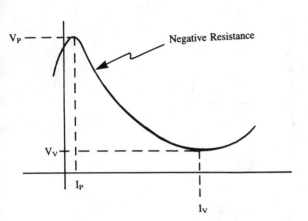

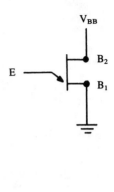

Fig. 14.40

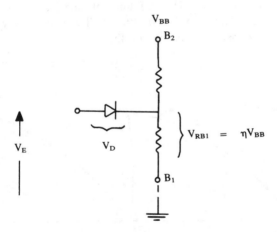

Fig. 14.41

I_{peak} (I_P) is the small amount of forward diode current that flows just before avalanche action takes place. Typically, I_{peak} can vary from 1 to 100 μA, depending upon the UJT construction.

After switching, the voltage from emitter to base 1, is referred to as the valley voltage, V_V (typically, $V_V = 2$ V) and I_V (typically, $I_V = 5$ mA).

Quite often, a UJT is used with an R-C charging network in the emitter circuit as shown in Fig. 14.42.

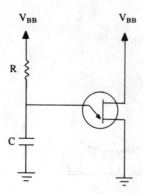

Fig. 14.42

When the voltage across the capacitor charges up to V_P, the UJT fires and discharges the capacitor to V_V. The UJT turns off and the process is repeated. Thus, a timer or oscillator may be constructed. The voltage waveform at the emitter would appear as shown in Fig. 14.43.

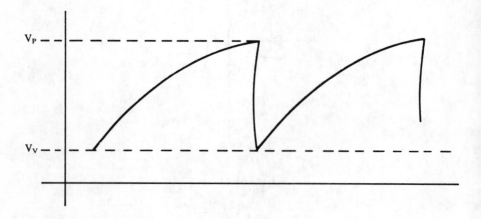

Fig. 14.43

Instead of sensing the voltage at the emitter, a simple method of detecting the oscillator output is to include a low value resistance from the base 1 terminal to ground (see Fig. 14.44).

By sensing the output from the base 1 terminal, the oscillator has a Thevenin equivalent resistance of approximately 20 ohms and the output pulse train would appear as shown in Fig. 14.45.

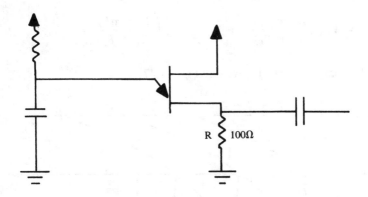

Fig. 14.44

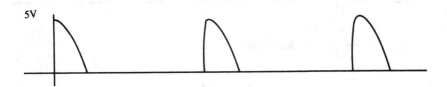

Fig. 14.45

At this point, let us derive the oscillator design equation, with reference to Fig. 14.46:

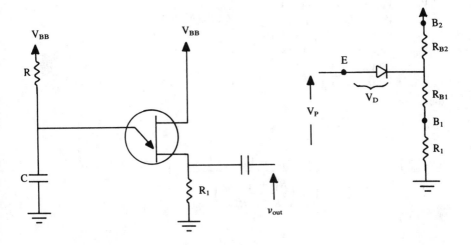

Fig. 14.46

$$V_P = V_{BB} \left(\frac{R_1 + R_{B1}}{R_1 + R_{BB}} \right) + V_D = V_{BB} \left(\frac{R_1 + \eta R_{BB}}{R_1 + R_{BB}} \right) + V_D$$

and the changing curve is given by Fig. 14.47.

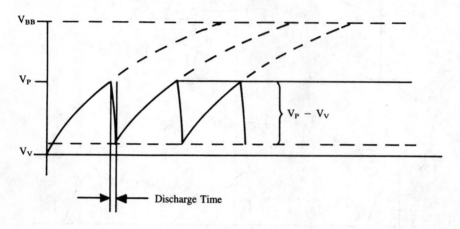

Fig. 14.47

If we assume that the discharge time is negligible compared to the charge time, we have, for *any* exponentially increasing curve,

$$v_{out} = v_{total} (1 - \epsilon^{-t/RC})$$

or for our design analysis:

$$V_P - V_V = (V_{BB} - V_V)(1 - \epsilon^{-t/RC})$$
$$V_P - V_V = V_{BB} - V_V - (V_{BB} - V_V)\,\epsilon^{-t/RC}$$
$$V_P - V_{BB} = -(V_{BB} - V_V)\,\epsilon^{-t/RC}$$
$$V_{BB} - V_P = (V_{BB} - V_V)\,\epsilon^{-t/RC}$$
$$\epsilon^{-t/RC} = \frac{V_{BB} - V_V}{V_{BB} - V_P}$$

therefore,

$$\epsilon^{+t/RC} = \frac{V_{BB} - V_V}{V_{BB} - V_P}$$

$$t = RC \ln \left[\frac{V_{BB} - V_V}{V_{BB} - V_{BB} \left(\dfrac{R_1 + \eta R_{BB}}{R_1 + R_{BB}} \right) - V_D} \right]$$

If

$t = T$ (period),

then

$$T = \frac{1}{f}$$

or

$$f = \frac{1}{RC \ln \left[\dfrac{V_{BB} - V_V}{V_{BB} - V_{BB} \left(\dfrac{R_1 + \eta R_{BB}}{R_1 + R_{BB}} \right) - V_D} \right]}$$

If this equation is rearranged slightly, we have

$$f = \frac{1}{RC \ln \left[\dfrac{1 - V_V/V_{BB}}{1 - \dfrac{R_1 + \eta R_{BB}}{R_1 + R_{BB}} - \dfrac{V_D}{R_{BB}}} \right]} \qquad (14.9)$$

Rewriting:

$$f = \frac{1}{RC \ln \left[\dfrac{1 - \dfrac{V_V}{V_{BB}}}{1 - \dfrac{\dfrac{R_1}{R_{BB}} + \eta}{\dfrac{R_1}{R_{BB}} + 1} - \dfrac{V_D}{V_{BB}}} \right]}$$

If

$R_{BB} \gg R_1$ $(5000 \gg 100)$
$V_{BB} \gg V_V$ $(15 \gg 2)$
$V_{BB} \gg V_D$ $(15 \gg 0.7)$

then

$$f \doteq \frac{1}{RC \ln \dfrac{1}{1 - \eta}} \qquad (14.10)$$

Example (14.7). Let us design a 60 Hz pulse source.
Let

$$\eta = 0.6, \ V_{BB} = +15 \text{ V}, \ V_D = 0.7, \ V_V = 2 \text{ V}, \ R_1 = 100 \ \Omega,$$
$$R_{BB} = 10 \text{ k, choose } C = 0.5 \ \mu\text{F}$$

By the approximate formula (14.10), we have

$$R \ \frac{1}{fC \ln\left(\dfrac{1}{1 - \eta}\right)}$$

$$R = \frac{1}{60(0.5)(10^{-6}) \ln\left(\dfrac{1}{1 - 0.6}\right)} = 36.4 \text{ k}$$

By formula (14.9), we calculate

$$R = \frac{1}{fC \ln\left[\dfrac{\dfrac{1 - V_V}{V_{BB}}}{1 - \dfrac{R_1 + \eta R_{BB}}{R_1 + R_{BB}} - \dfrac{V_D}{V_{BB}}}\right]}$$

thus,

$$R = \frac{1}{60(0.5)(10^{-6}) \ln\left[\dfrac{1 - \dfrac{2}{15}}{1 - \dfrac{0.1 + 0.6(10)}{0.1 + 10} - \dfrac{0.7}{15}}\right]}$$

therefore,
$$R = 36.7 \text{ k}$$

and our design is complete!

A semiconductor device that performs a function similar to that of a UJT is a programmable *unijunction* transistor (PUT). This is something of a misnomer because a PUT is really a four-layer device, similar to a SCR, SCS, or other *three-junction* devices. A detailed explanation of four-layer devices will not be presented here, but a brief discussion is necessary to facilitate an understanding of the use of PUTs.

If two transistors, a *pnp* and a *npn* are connected in series as shown in Fig. 14.48, we note that a small current ΔI_{B1} into the base of Q_1 develops a larger collector current, ΔI_{C1}. ΔI_{C1}, is really ΔI_{B2}. With ΔI_{B2} flowing, an amplified current results in the collector of Q_2, ΔI_{C2}. However, ΔI_{C2} becomes the input current to the base of Q_1, thus, reinforcing the current already flowing into the base of Q_1. This positive feedback action causes one transistor to feed the other, and both units quickly "snap" into saturation with approximately one-volt drop across the series pair. It should be emphasized that discrete transistors are not used in a four-layer device. The current flowing through the equivalent transistor pair is limited only by series external resistances and, therefore, a limiting resistance must be included in all designs. Depending upon the individual transistor β values, breakdown voltage, *et cetera*, the four-layer device may be used for different applications, e.g., SCRs, SCSs, DIACs, TRIACs, and PUTs. To use a four-layer device as a PUT, if we connect the equivalent transistor pair as shown in the schematic in Fig. 14.49 (a,b), action similar to UJT operation will take place.

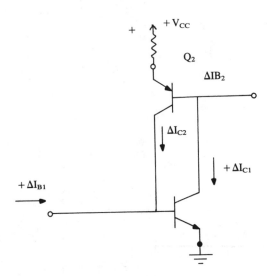

Fig. 14.48

Referring to Fig. 14.49 (a), R_1 and R_2 form a voltage divider. Let us assume R_1 and R_2 are each 10 k and that $V_{BB} = 20$ V. The voltage at the "top" of R_2 is $+10$ V with respect to ground. If the capacitor is discharged, then the base-to-emitter junction of Q_2 is back-biased. When the capacitor voltage has increased

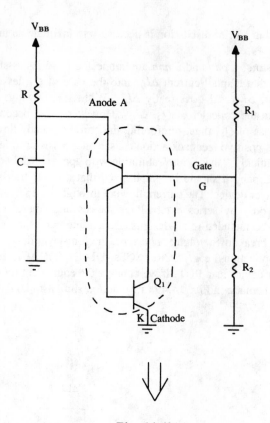

Fig. 14.49(a)

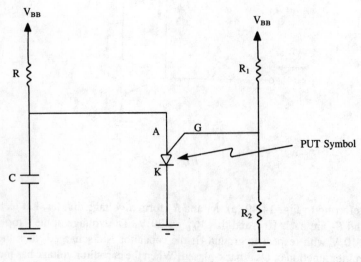

Fig. 14.49(b)

to the divider voltage plus the forward diode drop of V_{BEQ2}, current will flow and regenerative action will take place. The capacitor C will be discharged through the path shown in Fig. 14.50.

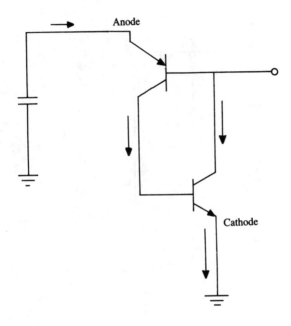

Fig. 14.50

The four-layer device will extinguish and the process will repeat continuously. Let us develop the formula for the frequency of an oscillator (Fig. 14.51) based upon a PUT:

$$V_{gnd - G} = V_{R2} = V_{BB} \frac{R_2}{R_1 + R_2}$$

When V_{gnd-A} increases to $V_{gnd-G} + V_{BE2}$, the PUT fires:

$$V_{gnd - A} = V_{BB} \frac{R_2}{R_1 + R_2} + V_{BE2}$$

In general, the capacitor will charge and discharge between the limits shown in Fig. 14.52.

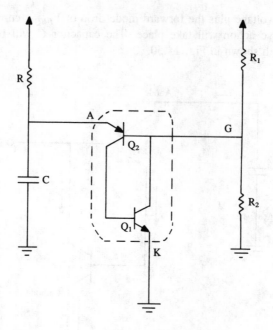

Fig. 14.51

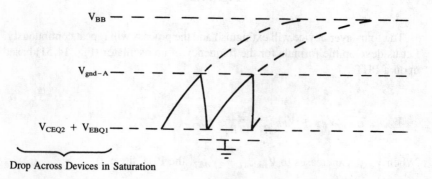

V_BB

V_{gnd-A}

$V_{CEQ2} + V_{EBQ1}$

Drop Across Devices in Saturation

Fig. 14.52

In general:

$$v(t) = V_{max}(1 - \epsilon^{-t/RC})$$

$$V_A - (V_{CE2} + V_{EB1}) = [V_{BB} - (V_{CE2} + V_{EB1})](1 - \epsilon^{-t/RC})$$

$$V_A - (V_{CE2} + V_{EB1}) = V_{BB} - (V_{CE2} + V_{EB1})$$
$$- [V_{BB} - (V_{CE2} + V_{EB1})]\epsilon^{-t/RC}$$

$$V_{BB} - V_A = V_{BB} - (V_{CE2} + V_{EB1})\epsilon^{-t/RC}$$

$$\epsilon^{-t/RC} = \frac{V_{BB} - V_{BB}\dfrac{R_2}{R_1 + R_2} - V_{EB2}}{V_{BB} - V_{CE2} - V_{EB1}}$$

$$t = RC \ln \frac{V_{BB} - V_{CE2} - V_{EB1}}{V_{BB}\dfrac{R_1}{R_1 + R_2} - V_{EB2}}$$

$$t = RC \ln \frac{V_{BB} - V_{CE2} - V_{EB1}}{V_{BB} - V_{BB}\dfrac{R_2}{R_1 + R_2} - V_{EB2}}$$

$$t = RC \ln \frac{V_{BB} - V_{CE2} - V_{EB1}}{V_{BB}\dfrac{R_1}{R_1 + R_2} - V_{EB2}}$$

Therefore, assuming the discharge time is negligible, the oscillator frequency is

$$f = \frac{1}{T} = \frac{1}{RC \ln \left[\dfrac{V_{BB} - V_{CE2} - V_{EB}}{V_{BB}\dfrac{R_1}{R_1 + R_2} - V_{EB2}}\right]} \tag{14.11}$$

If $V_{EB1} = V_{EB2} = 0.6$ V, and if $V_{CE2\,SAT} = 0.4$ V, we have

$$f = \frac{1}{RC \ln \left[\dfrac{V_{BB} - 1}{V_{BB}\dfrac{R_1}{R_1 + R_2} - 0.6}\right]} \tag{14.12}$$

Let us examine a typical example.

516

Example (14.8). Design a 60 Hz oscillator using a PUT, if $V_{BB} = +15$, $R_1 = R_2 = 10$ k and $C = 0.5$ μF. Thus,

$$R = \cfrac{1}{fC \ln\left[\cfrac{V_{BB} - 1}{V_{BB}\cfrac{R_1}{R_1 + R_2} - 0.6}\right]} = \cfrac{1}{60(0.5)(10^{-6}) \ln\left[\cfrac{15 - 1}{15(0.5) - 0.6}\right]}$$

$R = 47.1$ k

PROBLEMS

1. A relaxation oscillator is so named because _____.

2. Sine wave oscillators employ an active device to provide a 180-degree phase shift. An additional 180-degree phase shift is accomplished in a phase shifting network. State an additional function of the active device.

3. In the construction of L-C oscillators, the L-C network determines the oscillator frequency of oscillation. Why is this so?

4. A disadvantage of the Armstrong type of oscillator *versus* the Hartley or Collpitts type is _____.

5. The resonant frequency of a series-parallel resonant circuit, as shown in Fig. 14.53, which is used in oscillators, is given approximately by the formula:

$$f = \frac{1}{2\pi \sqrt{LC}}$$

Why is this approximation valid?

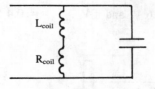

Fig. 14.53

6. In connecting an Armstrong circuit, if the transformer secondary leads are not connected properly, the circuit will not oscillate. Why is this so?

7. In a Hartley oscillator, it is said that autotransformer action takes place in the inductor. Describe the meaning of this statement.

8. The purpose of the R-C network at the gate or base lead of a Hartley or Collpitts oscillator is to perform what function?

9. If the gate current is a non-sinusoidal pulse train, why are the Hartley or Collpitts oscillators referred to as sinusoidal oscillators?

10. The Clapp oscillator is a variation of what other basic oscillator design?

11. Compare the advantages of the Hartley *versus* the Collpitts oscillator and *vice versa*.

12. Why is the Clapp a desirable oscillator to use?

13. We set the following:

$$g_m r_d \frac{x_1}{x_2} > 1$$

Why?

14. Design a Hartley oscillator to operate at 2 MHz. Use a FET with

$$g_m = 4 \text{ mmho}, \; r_d = 25 \text{ k}, \; R_G = 500 \text{ k}$$

15. Repeat Problem 14 for a Collpitts oscillator. Use $L = 500 \; \mu\text{H}$.

16. The short-term stability of a crystal oscillator is ensured because _____.

17. The most commonly used material in oscillator crystals is _____.

18. A crystal may be made to oscillate at either of two frequencies. These frequencies are determined by operating the crystal in what modes?

19. R-C phase shift oscillators are desirable for the generation of a low frequency sine wave because _____.

20. The phase shift network must accomplish what electrical function?

21. The active device used in a phase shift oscillator must have a gain at least equal to _____.

22. Design a four-section (R to ground) R-C phase shift oscillator that employs a bipolar transistor with the following characteristics:

$$h_{FE} = h_{fe} = 75, I_{CQ} = 2.5 \text{ mA} \quad I_1 = 2I_B, V_{RE} = 10\% \; V_{CC}$$
$$V_{CC} = +15 \text{ V}, f_0 = 1000 \text{ Hz}, Z_{in2} = 100 \text{ k}$$

23. Design a Wein bridge oscillator that employs a 741 operational amplifier and oscillates at 1000 Hz. Choose a standard value for each of the components used.

24. Design a UJT oscillator to operate at 1000 Hz with

$V_{BB} = +12$ V, $V_D = 0.7$ V
$V_V = 2$ V, $R_1 = 100$ Ω, $R_{BB} = 8$ k, choose $C = 0.05$ μF

25. Design a PUT oscillator to operate at 1000 Hz with

$V_{BB} = +12$ V, $R_1 = R_2 = 10$ kΩ and $C = 0.05$ μF

Chapter 15

Tuned Amplifier Stages

15.1 TUNED AMPLIFIERS

When active devices are used at very high frequencies, the amount of amplification that will occur will decrease with increasing frequency. At a sufficiently high frequency, no voltage amplification will occur. If this particular frequency is exceeded, the output voltage will be *less* than the input voltage, so that we no longer experience voltage gain, but instead we notice a voltage *loss*. The cause of this loss in amplification may be attributed to many factors, such as increased charge transit flow time in the active device, as well as losses resulting from distributed capacitance to ground.

Because capacitance can be minimized but not eliminated, the solution is to make the capacitance to ground appear to be a very high impedance. This can be accomplished by forming a parallel resonant circuit to ground. Implicit in the construction of parallel resonant circuits is a resonant frequency passband. (See Fig. 15.1).

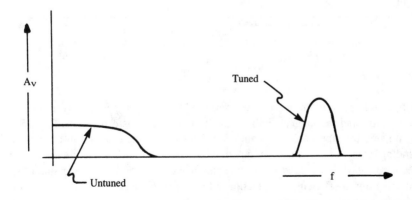

Fig. 15.1

519

The resonant frequency passband will only enable the amplification of a narrow band of frequencies. The wider the band of frequencies to be amplified, the less amplification that will occur. The fact that only a narrow band of frequencies is amplified, at first thought, might be considered as a shortcoming. However, the requirements for amplifying high frequencies are usually such that we only wish to amplify a selected band of frequencies. This selectivity requirement is particularly pertinent for communications equipment. Both transmitters and receivers have major portions of the equipment designed around tuned circuits.

Most active devices would not amplify signals if used in *untuned amplifier circuits* and at communications frequencies, but will amplify quite well if the appropriate tuned circuit is employed. In communications equipment, interconnections between amplifier stages employing tuned circuits may consist of a simple parallel resonant circuit, or the connection between active devices may be performed with a transformer. One or both of the transformer windings may be tuned to a particular frequency. The primary and secondary windings of the transformer may be tuned to the frequency that we wish to amplify (Fig. 15.2) or to a frequency slightly greater or less than the desired frequency. The purpose of tuning the resonant circuits (primary and secondary) to a frequency slightly greater or less than the desired frequency is to increase the band of frequencies (bandwidth) to be amplified.

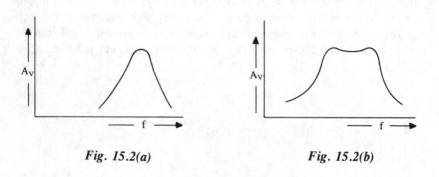

Fig. 15.2(a)   *Fig. 15.2(b)*

If a wide bandwidth is required, a second method used is to "over-couple" the primary and secondary windings of an interstage transformer. See Fig. 15.3.

In using this approach to obtain wider bandwidth, the primary and secondary windings may or may not be tuned to the desired *middle frequency* (center frequency, f_0) of the band of frequencies to be amplified. To understand how the coupling coefficient can be adjusted in a practical circuit, a description of interstage transformers must first be undertaken.

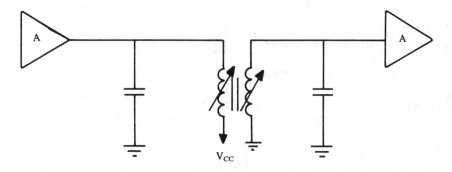

Fig. 15.3

15.2 INTERSTAGE TRANSFORMER

An interstage transformer at audio frequencies is simply a laminated steel-cored transformer. At higher frequencies, which are usually employed in inter-mediate-stage amplifier (IF amps) used in communications receivers, the trans-former normally consisted of two coils wound around a cardboard or plastic hollow tube. The tube is typically between 1/8 inch to 1/4 inch in diameter and from 3/4 inch to 2 inches long. Inside the threaded tube are two threaded molded ferite slugs, as shown in Fig. 15.4.

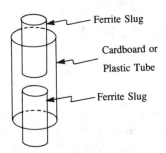

Fig. 15.4

A ferrite slug is a cylindrically shaped, 1/8 inch to 1/4 inch diameter threaded rod. It is molded from powdered ferrite with a binder. The ferrite material (or powdered iron and iron oxide) is used so that eddy current losses are minimized as compared with a solid metal core. On the outside of the hollow tube, which is used to hold the ferrite slugs, the secondary and primary coil windings are wound. See Fig. 15.5.

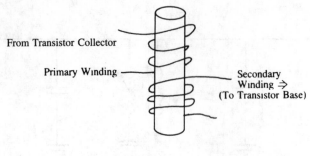

Fig. 15.5

The final assembly is then sometimes mounted inside of a rectangular-shaped aluminum can. The can has all the terminals needed for connection to active devices located on the bottom periphery. See Fig. 15.6.

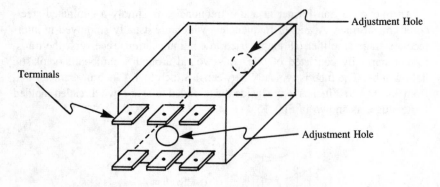

Fig. 15.6

Adjustment access holes are located at the top and bottom so that the two slugs may be driven inward or outward (for adjustment of the resonant frequencies and bandwidth). The inward movement of a slug has the effect of increasing the inductance of the coil wound directly over the slug being adjusted. However, as a result of the magnetic lines of flux coupling from one slug to the other, the movement of one slug changes the resonant frequency of both coils and affects the bandwidth of the entire interstage network.

At some operating frequency, depending upon the application, perhaps above 100 MHz, the losses in ferrite core materials become unacceptably high and at this point air-cored transformers are usually utilized. The inductance of each winding of an air-cored transformer cannot be varied as readily as ferrite slug tuned transformers. For example, at several hundred megahertz, few turns of heavy gauge wire may be used. These windings may or may not be wound on a coil form. If a coil form is not used (to minimize losses), then the primary winding of the interstage transformer will be supported by the terminals of the preceding active device and the secondary winding, shown in Fig. 15.7, will be mounted and supported by the terminals of the succeeding active device. Increasing or decreasing the spacing of the windings will determine the inductance, and the proximity of the primary to secondary windings will affect the coupling coefficient and, thus, affect both the resonant frequency of each winding and the passband bandwidth.

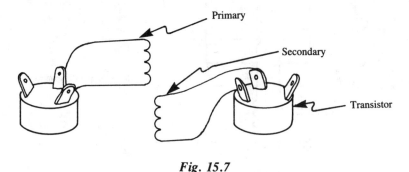

Fig. 15.7

If the coupling is varied, we would expect to see variations in the passband as shown in Fig. 15.8.

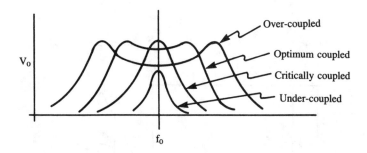

Fig. 15.8 Frequency Response Curve for a Double-Tuned Inductively Coupled RF Transformer with Various Coupling Coefficients.

A typical connection for a double-tuned transformer in the intermediate frequency amplifier of a commercial broadcast receiver might appear as shown in Fig. 15.9. We note that the secondary of the transformer is tapped part way in order to facilitate impedance matching for the output of Q_1 into the low impedance of the base circuit of Q_2. The two capacitances C_B and resistances R_1 and R_2 shown in Fig. 15.9 are included in the circuit for decoupling the radio frequency signal, which may be generated in one high frequency amplifier stage, and ensuring that the signal does not enter the input of another amplifier through the power supply.

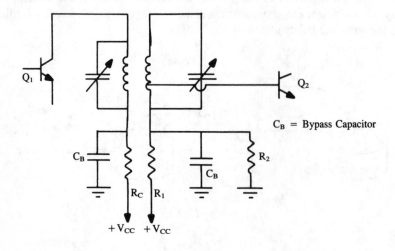

Fig. 15.9

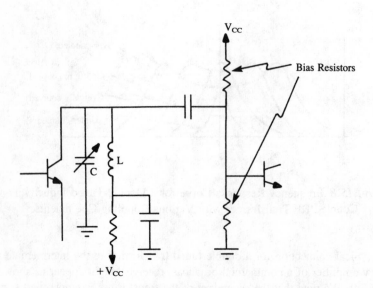

Fig. 15.10 Single-Tuned Interstage Coupling Network

Where wide bandwidths are not required, a single-tuned circuit (Fig. 15.10) is sometimes employed. Single-tuned circuit design is less involved, but the performance is inferior to a double-tuned design.

15.3 INDUCTANCE, CAPACITANCE, AND RESISTANCE

For impedance-matching purposes, the inductance may be tapped (auto-transformer action) so that the low input impedance of the second amplifier will receive a maximum power transfer.

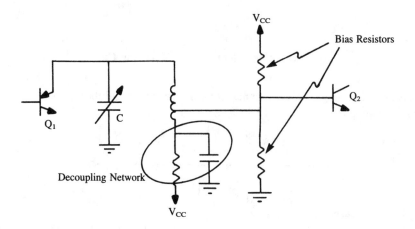

Fig. 15.11

Some of the capacitance, which determines the resonant frequency of the tuned circuit, is contributed by the inter-electrode capacitance of the active devices and some capacitance is contributed by the interconnecting wiring. The discrete capacitor C, shown in Fig. 15.11, is usually a variable capacitor because variations in component placement and inter-electrode capacitance would vary the total capacitance of the interstage resonant circuit (tank circuit). Therefore, by employing a variable capacitor, the exact resonant frequency of the tank may be adjusted.

We recall from our study of ac electricity that the resonant frequency of a series or parallel resonant circuit is given by the formula:

$$f_0 = \frac{1}{2\pi\sqrt{LC}}$$

Reviewing further, we also know that no matter how well designed and constructed an inductor may be, there is always some resistance in its windings. See Fig. 15.12.

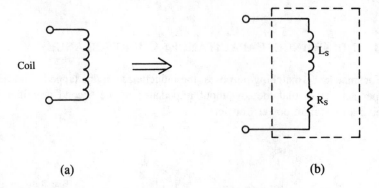

Coil

L_S

R_S

(a)   (b)

Fig. 15.12 Equivalent Circuit of Tuning Coil

A complete single-tuned interstage tuning network would then consist of the components pictured in Fig. 15.13.

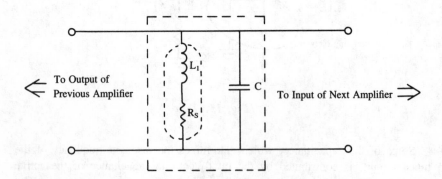

To Output of Previous Amplifier

L_1

R_S

C

To Input of Next Amplifier

Fig. 15.13 Actual Tank Circuit Components

For analysis purposes, it is easier to work with a ''pure'' parallel resonant circuit, as shown in Fig. 15.14. We know that any network consisting of real and reactive components may be translated into an R, L, and C, connected either in series or parallel.

The translation process is accomplished by computing the real and reactive circuit values for the circuit under consideration, and setting these results equal to the real and reactive values for the ''pure'' parallel (or series) resonant circuit.

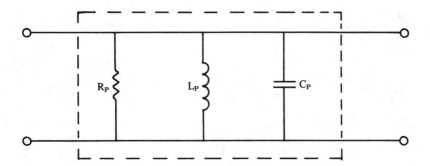

Fig. 15.14

For the tank circuit consisting of a coil and capacitor connected in parallel, we need only concern ourselves about changing the coil with its series-connected equivalent inductance and resistance with a parallel-connected inductor and resistor. The capacitor value remains the same for either the actual tank or its "pure" parallel equivalent circuit. Therefore, we need not concern ourselves about including the capacitor in the circuit translation (see Fig. 15.15). Thus, we have

$$Y_{inS} = \frac{1}{Z_{in}} = \frac{1}{R_S + j\,\omega\,L_S} = \frac{1}{R_S + j\,\omega\,L_S}\left(\frac{R_S - j\,\omega\,L_S}{R_S - j\,\omega\,L_S}\right)$$

$$= \frac{R_S - j\,\omega\,L_S}{R_S^2 + \omega^2\,L_S^2} = \frac{R_S}{R_S^2 + \omega^2\,L_S^2} - j\,\frac{\omega\,L_S}{R_S^2 + \omega^2\,L_S^2}$$

(15.1)

and the input admittance for the parallel network is simply

$$Y_{inP} = \frac{1}{R_P} + \frac{1}{j\,\omega\,L_P} = \frac{1}{R_P} - j\,\frac{1}{\omega\,L_P}$$

Since the two circuits are equivalent when their admittances are equal,

$$Y_{inP} = Y_{inS}$$
$$\frac{1}{R_P} - j\,\frac{1}{\omega\,L_P} = \frac{R_S}{R_S^2 + \omega^2\,L_S^2} - j\,\frac{\omega\,L_S}{R_S^2 + \omega^2\,L_S^2}$$

(15.2)

The real part of the expression on the left side of the equals sign must equal the real part of the expression on the right side:

528

$$\frac{1}{R_P} = \frac{R_S}{R_S^2 + \omega^2 L_S^2}$$

$$R_P = \frac{R_S^2 + \omega^2 L_S^2}{R_S} = R_S \left(\frac{R_S^2 + \omega^2 L_S^2}{R_S^2} \right)$$

$$R_P = R_S (1 + Q_S^2) \tag{15.3}$$

It is generally accepted that if $Q_S \geqslant 10$,

$$R_P \doteq Q_S^2 R_S \tag{15.4}$$

Also, the imaginary terms of (15.2) are equal:

$$\frac{1}{\omega L_P} = \frac{\omega L_S}{R_S^2 + \omega^2 L_S^2} \Rightarrow \omega L_P = \frac{R_S^2 + \omega^2 L_S^2}{\omega L_S}$$

$$L_P = \frac{R_S^2 + \omega^2 L_S^2}{\omega^2 L_S} = L_S \left(\frac{R_S^2 + \omega^2 L_S^2}{\omega^2 L_S^2} \right) = L_S \left(1 + \frac{R_S^2}{\omega^2 L_S^2} \right)$$

$$L_P = L_S \left(1 + \frac{1}{Q_S^2} \right) \tag{15.5}$$

It is generally accepted that if $Q_S \geqslant 10$,

$$L_P \doteq L_S \tag{15.6}$$

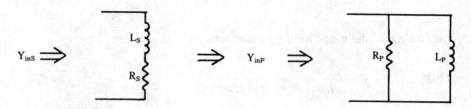

Fig. 15.15

The resonant frequency occurs when the equivalent parallel inductive reactance value equals the capacitance reactance. For our equivalent "pure" parallel resonant circuit, we have

$$X_{LP} = X_{CP} \Rightarrow \omega_0 L_P \Rightarrow \frac{1}{\omega_0 C}$$

where $\omega_0 =$ resonant radial frequency.

From equation (15.5), substituting for L_P, we have

$$\omega_0 \left[L_S \left(1 + \frac{1}{Q_S^2} \right) \right] = \frac{1}{\omega_0 C}$$

$$\omega_0^2 = \frac{1}{L_S C (1 + 1/Q_S^2)}$$

$$(2\pi f_0)^2 = \frac{1}{L_S C (1 + 1/Q_S^2)}$$

$$f_0^2 = \frac{1}{4\pi^2 L_S C (1 + 1/Q_S^2)}$$

$$f_0 = \frac{1}{2\pi \sqrt{L_S C (1 + 1/Q_S^2)}} = \frac{1}{2\pi \sqrt{L_S C} \sqrt{1 + 1/Q_S^2}} \qquad (15.7)$$

$$f_0 = \frac{1}{2\pi \sqrt{L_S C}} \frac{Q_S}{\sqrt{1 + Q_S^2}}$$

Again, we note that if $Q_S \geq 10$,

$$f_0 \doteq \frac{1}{2\pi \sqrt{L_S C}} \qquad (15.8)$$

We note that if the value of the quality factor of the inductance decreases below a value of 10, the resonant frequency of a series-parallel resonant circuit, will begin to decrease.

Example (15.1). Compute the resonant frequency of the series-parallel resonant circuit shown in Fig. 15.16.

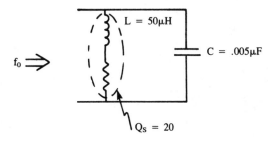

$$f_0 \Rightarrow \qquad L = 50\mu H$$
$$C = .005\mu F$$
$$Q_S = 20$$

Fig. 15.16

$$f_0 = \frac{1}{2\pi \sqrt{L_S C}} = \frac{1}{2\pi \sqrt{50(10^{-6})(.005)(10^{-6})}} = .318 \text{ MHz}$$

530

Example (15.2). Compute the resonant frequency of the series-parallel resonant circuit shown in Fig. 15.17.

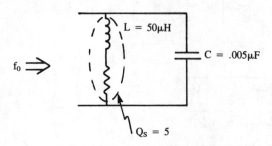

Fig. 15.17

$$f_0 = \frac{1}{2\pi \sqrt{L_S C}} \frac{1}{\sqrt{1/Q_S^2}} = .318 \text{ MHz} \frac{1}{\sqrt{1 + 1/5^2}} = .312 \text{ MHz}$$

We also know that for our equivalent "pure" parallel resonant circuit, from Equation (15.1):

$$Y_{inS} = \frac{R_S}{R_S^2 + \omega^2 L_S^2} - j \frac{\omega L_S}{R_S^2 + \omega^2 L_S^2}$$

which may be rewritten:

$$Y_{inS} = G - jB$$

where

$$G = \frac{R_S}{R_S^2 + \omega^2 L_S^2}, \, B = \frac{\omega L_S}{R_S^2 + \omega^2 L_S^2}$$

or

$$R_P = \frac{1}{G}$$

$$R_P = \frac{R_S^2 + \omega^2 L_S^2}{R_S}, \qquad X_{LP} = \frac{R_S^2 + \omega^2 L_S^2}{\omega L_S}$$

By definition at resonance, $\omega = \omega_0$ and $Z_{in} = R_p$ (i.e., no reactance is present in the expression for the tank impedance):

$$Z_{in} = R_P = \frac{R_S^2 + \omega_0^2 L_S^2}{R_S}$$

Also, $X_{LP} = X_C$

$$\frac{R_0^2 + \omega_0^2 L_S^2}{\omega_0 L_S} = \frac{1}{\omega_0 C}$$

$$R_S^2 + \omega_S^2 L_S^2 = \frac{L_S}{C}$$

Substituting:

$$Z_{in} = R_P = \frac{L_S}{R_S C} \tag{15.9}$$

15.4 BANDWIDTH

Let us determine the factors that control the bandwidth and Q of a single tuned interstage network (Fig. 15.18).

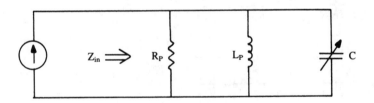

Fig. 15.18

At the -3 dB point, the admittance has increased by $\sqrt{2}$ and, therefore the voltage across the tank, if fed by a constant current source, will decrease by $\sqrt{2}$ (in contrast to the value at resonance, $1/R_P$).

In general,

$$Y_{in} = \frac{1}{R_P} + \frac{1}{jX_{LP}} - \frac{1}{jX_C} = \frac{1}{R_P} + j\left(\frac{1}{X_C} - \frac{1}{X_L}\right)$$

at the half power point:

if $X_L > X_C$, then $f_2 > f_0$

$$\frac{\sqrt{2}}{R_P} = \sqrt{\frac{1}{R_P^2} + \left(\frac{X_{LP} - X_C}{X_{LP} X_C}\right)^2}$$

squaring both sides:

$$\frac{2}{R_P^2} = \frac{1}{R_P^2} + \left(\frac{X_{LP} - X_C}{X_{LP} X_C}\right)^2 \Rightarrow \left(\frac{X_{LP} - X_C}{X_{LP} - X_C}\right) = \frac{1}{R_P^2}$$

$$\frac{X_{LP} - X_C}{X_{LP} X_C} + R_P \Rightarrow \frac{\omega_2 L_P - \dfrac{1}{\omega_2 C}}{\omega_2 L_P \left(\dfrac{1}{\omega_2 C}\right)} = \frac{1}{R_P^2}$$

$$\omega_2 L_P - \frac{1}{\omega_2 C} = \frac{L_P}{R_P C} \Rightarrow \omega_2 L_P - \frac{L_P}{R_P C} - \frac{1}{\omega_2 C} = 0$$

$$\omega_2^2 + \omega_2 \left(-\frac{1}{R_P C}\right) - \frac{1}{L_P C} = 0$$

$$\omega_2 = \frac{1}{2R_P C} \pm \frac{1}{2}\sqrt{\frac{1}{R_P^2 C^2} - 4\left(-\frac{1}{L_P C}\right)}$$

$$= \frac{1}{2R_P C} \pm \frac{1}{2}\sqrt{\frac{1}{R_P^2 C^2} + \frac{4}{L_P C}}$$

$$\omega_2 = \frac{1}{2R_P C} + \frac{1}{2}\sqrt{\frac{1}{R_P^2 C^2} + \frac{4}{L_P C}}$$

The negative sign is disregarded because ω_2 is a positive angular frequency and the term in radical form is larger than the other term, therefore the radical cannot be subtracted. Thus,

$$f_2 = \frac{1}{2\pi}\left(\frac{1}{2R_P C} + \frac{1}{2}\sqrt{\frac{1}{R_P^2 C^2} + \frac{4}{L_P C}}\right) \tag{15.10}$$

If the frequency applied to the tank circuit is less than the resonant frequency:

$X_C > X_{LP}$

$$\frac{\sqrt{2}}{R_P} = \sqrt{\frac{1}{R_P^2} + \left(\frac{X_C - X_{LP}}{X_L X_{LP}}\right)^2}$$

$$\frac{2}{R_P^2} = \frac{1}{R_P^2} + \left(\frac{X_C - X_{LP}}{X_C - X_{LP}}\right)^2 \Rightarrow \left(\frac{X_C - X_{LP}}{X_L - X_{LP}}\right)^2 = \frac{1}{R_P^2}$$

$$\frac{X_L - X_{LP}}{X_C X_{LP}} = \frac{1}{R_P} \Rightarrow X_C - X_{LP} = \frac{X_C X_{LP}}{R_P}$$

$$\frac{1}{\omega_1 C} - \omega_1 L_P = \frac{\dfrac{1}{\omega_1 C} - \omega_1 L_P}{R_P} = \frac{L_P}{R_P C}$$

$$\omega_1 L_P + \frac{L_P}{R_P C} - \frac{1}{\omega_1 C} = 0$$

$$\omega_1^2 + \omega_1 \frac{1}{R_P C} - \frac{L_P}{C} = 0$$

$$\omega_1 = -\frac{1}{2R_P C} \pm \frac{1}{2} \sqrt{\frac{1}{R_P^2 C^2} - 4\left(-\frac{L_P}{C}\right)}$$

$$= -\frac{1}{2R_P C} \pm \frac{1}{2} \sqrt{\frac{1}{R_P^2 C^2} + \frac{4L_P}{C}}$$

Disregarding the negative sign because $\omega_1 > 0$:

$$\omega_1 = -\frac{1}{2R_P C} + \frac{1}{2} \sqrt{\frac{1}{R_P^2 C^2} + \frac{4L_P}{C}}$$

Thus,

$$f_1 = \frac{1}{2\pi} \left(-\frac{1}{2R_P C} + \frac{1}{2} \sqrt{\frac{1}{R_P^2 C^2} + \frac{4L_P}{C}}\right) \tag{15.11}$$

The bandwidth is by definition:

$$BW = f_2 - f_1$$

$$BW = \frac{1}{2\pi} \left(\frac{1}{2R_P C} + \frac{1}{2} \sqrt{\frac{1}{R_P^2 C^2} + \frac{4}{L_P C}}\right)$$

$$-\frac{1}{2\pi}\left(-\frac{1}{2R_PC} + \frac{1}{2}\sqrt{\frac{1}{R_P^2C^2} + \frac{4}{L_PC}}\right)$$

$$BW = \frac{1}{2\pi}\left(\frac{2}{2R_PC}\right) = \frac{1}{2\pi\,R_PC} \tag{15.12}$$

However,

$$R_P = R_S\,(1 + Q_S^2)$$

therefore, $R_S = R_P\,\dfrac{1}{1 + Q_S^2}$

and $L_P = L_S\left(1 + \dfrac{1}{Q_S^2}\right) = L_S\left(\dfrac{Q_S^2 + 1}{Q_S^2}\right)$

Therefore, $L_S = L_P\left(\dfrac{Q_S^2}{1 + Q_S^2}\right)$

$$\omega L_S = \omega L_P\left(\frac{Q_S^2}{1 + Q_S^2}\right)$$

By definition:

$$Q_S = \frac{\omega L_S}{R_S} = \frac{\omega L_P\left(\dfrac{Q_S^2}{1 + Q_S^2}\right)}{R_P\left(\dfrac{1}{1 + Q_S^2}\right)} = \frac{\omega L_P\,Q_S^2}{R_P}$$

Thus,

$$Q_S = \frac{R_P}{\omega L_P} = \frac{R_P}{X_{LP}}$$

In a previously used equation, $Q_S = X_{LS}/R_S$. Note that the roles of X_L and R_S are interchanged.

Employing equation (15.12), we may write,

$$BW = \frac{1}{2\pi R_P C}$$

However,

$$Q = \frac{X_{LS}}{R_S} = \frac{\omega L_S}{R_S}$$

Also, $\omega = 2\pi f$. Thus,

$$\frac{R_S}{L_S} = \frac{\omega}{Q} = \frac{2\pi f}{Q}$$

Recalling that

$$R_P = \frac{L_S}{R_S C}$$

we may write:

$$BW = \frac{1}{2\pi \left(\dfrac{L_S}{R_S C}\right)} = \frac{1}{2\pi}\frac{R_S}{L_S} = \frac{1}{2\pi}\frac{2\pi f}{Q}$$

$$BW = \frac{f}{Q} \qquad\qquad (15.13)$$

We note from Equation (15.12) that the bandwidth is inversely proportional to the equivalent parallel resistance of the tank and the capacitance:

$$BW = \frac{1}{2\pi R_P C}$$

However, if the tank is located between the output of one transistor amplifier and the input of another, the equation (15.12) would be written as follows:

$$BW = \frac{L}{2\pi R_T C}$$

where $R_T = Z_0 \ /\!/ \ R_P \ /\!/ \ Z_{in2}$;

where R_T is the total resistance between stages (Figs. 15.19 and 15.20).

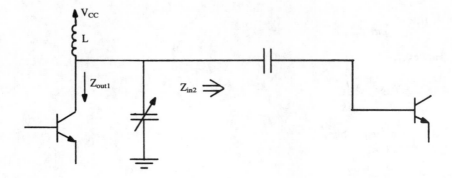

Fig. 15.19

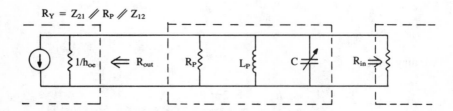

$$R_Y = Z_{21} \parallel R_P \parallel Z_{12}$$

Fig. 15.20 Interstage Equivalent Circuit

Example (15.3). Determine the bandwidth and voltage given for the interstage network and transistor shown in Fig. 15.21.

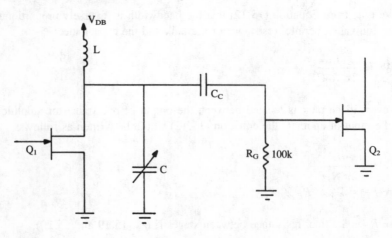

Fig. 15.21

Computing the equivalent parallel resistance of the tank, we have

$$R_P = \frac{L}{R_S C} = \frac{100(10^{-6})}{10(200)(10^{-12})} = 5(10^4) \ \Omega$$

The total impedance that Q_1 must develop an output voltage across is

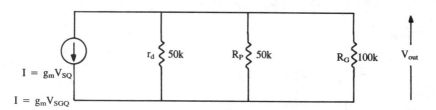

Fig. 15.22

$$Z_L = R_D \ /\!/ \ R_P \ /\!/ \ R_G = 50 \text{ k} \ /\!/ \ 50 \text{ k} \ /\!/ \ 100 \text{ k} = 20 \text{ k}$$
$$A_v = -g_m Z_L = -10,000 \ (10^{-6}) \ 20(10^3) = 200$$

If the input voltage was 40 microvolts to the gate of Q_1, the input voltage to the gate of Q_2 would be

$$V_0 = A_V V_{in} = 200(40) = 8000 \ \mu V = 8 \text{ mV}$$

The unloaded bandwidth of the tank is computed by equation (15.12):

$$BW = \frac{1}{2\pi \, R_P C}$$
$$BW_U = \frac{1}{2\pi \, (50k)(200)(10^{-12})} = 15.9 \text{ kHz}$$

and the resonant frequency is

$$f_0 = \frac{1}{\sqrt{LC}} = \frac{1}{2\pi \, \sqrt{100(10^{-6})200(10^{-12})}} = 1.125 \text{ MHz}$$

The unloaded Q is

$$Q_U = \frac{f_0}{BW} = \frac{1.125(10^6)}{15.9(10^3)} = 70.7$$

The loaded Q is

$$Q_L = \frac{Z_L}{X_{LP}} \doteq \frac{Z_L}{X_{LS}}$$

where

$$X_{LS} = 2\pi f_0 L = 2\pi \,(1.125)(10^6)(100)(10^{-6}) = 707 \ \Omega$$

Thus,

$$Q_L = \frac{20(10^3) \ \Omega}{707 \ \Omega} = 28.3$$

We see that the use of the approximation formulas are valid since the loaded Q is greater than 10. The loaded bandwidth is

$$BW_L = \frac{f_0}{Q_L} = \frac{1.125(10^6)}{28.3} = 39.8 \ \text{kHz}$$

For the example just covered, the expression for the voltage gain was

$$A_V = g_m \, Z_L$$

15.5 OTHER PARAMETERS

Unfortunately, at very high frequencies the internal inter-electrode capacitances and the time that it takes for the majority carriers to travel from one electrode to another (*transit time*), require that all the small-signal parameters be treated as complex numbers. At high frequencies, various parameters other than the h parameter are used to describe the equivalent circuit for active devices. Some of these may be expressed as y, z, s, r, or g parameters.

Example (15.4).   Let us examine a different case where one stage is coupled to another by using a single tuned coil. This time we will use bipolar transistors as the active devices, but also use a coil with the same characteristics as in the previous example (see Fig. 15.23).
For the coil,

$$L_{coil} = 100 \ \mu\text{H}, \ R_{coil} = 10 \ \Omega, \ C = 200 \ \text{pF}$$

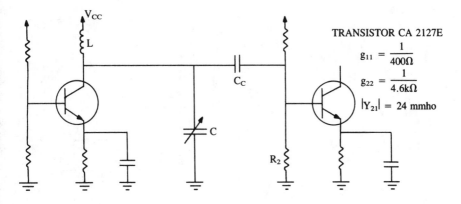

Fig. 15.23

for the transistors RE (y_{21}), the real part of the admittance,

$$y_{21} = 24 \text{ mmho}, \; g_{22} = \frac{1}{4.6 \text{ k}\Omega}, \; g_{11} = \frac{1}{400 \text{ }\Omega}$$

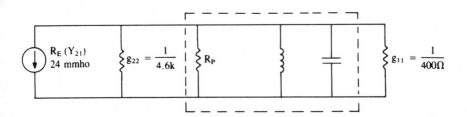

Fig. 15.24

The unloaded impedance of the tank from the previous example was 50 kΩ. If the output impedance $1/g_{22}$ and input impedance of the next stage are connected to the coil (Fig. 15.24), the loaded Q of the coil would be extremely low in value, yielding a wide bandwidth and low gain. If, however, g_{22} and g_{11} can be made to "look like" a high impedance, a better Q_L would be obtained, resulting in better selectivity and voltage gain.

If a ferrite-cored coil is used, we may assume that the coupling coefficient approaches unity, in which case the transformer relationship:

$$\frac{Z_1}{Z_2} = \left(\frac{N_1}{N_2}\right)^2 = a^2$$

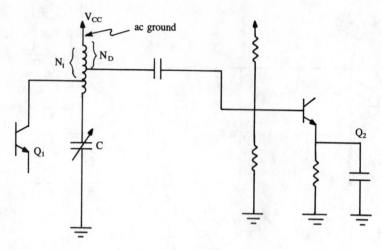

Fig. 15.25

If we wish to make the output of Q_1 "look like" 50 kΩ, then

$$\frac{Z_1}{Z_0} = a^2$$

or

$$\frac{4.6 \text{ k}\Omega}{50 \text{ k}\Omega} = a^2$$

Thus,

$$a^2 = .092$$

Therefore,

$$a = .303$$

The coil would be tapped 30% of the way from the coil end connected to V_{CC} (ac ground). For the coil connection to the next stage,

$$\frac{.4 \text{ k}}{50 \text{ k}} = a^2$$

Thus,

$a^2 = .008$

Therefore,

$a = .0894$

Thus, the coil would be tapped 9% of the way from the coil end connected to V_{CC} (ac ground). The unloaded Q would still be 70.7 (see Example (15.3)) and $BW_U = 15.9$ kHz.

The loaded Q may be computed:

$$Q_L = \frac{Z_L}{X_{LP}} \doteq \frac{Z_L}{X_{LS}} = \frac{50k \parallel 50k \parallel 50k}{707\ \Omega} = \frac{16.67\ k\Omega}{.707\ k\Omega}$$

$$Q_L = 23.6$$

and $BW_L = \dfrac{f_0}{Q_L} = \dfrac{1.125(10^6)}{23.6} = 47.7$ kHz

The voltage gain may be computed:

$$A_V = g_m Z_L$$
$$A_V = RE(y_{21})\, Z_L = 24(10^{-3})(16.67)(10^3)$$
$$= 400$$

Thus, if an input signal to the first stage (Q_1) was 40 microvolts, we would expect to see

$$V_0 = A_V V_{in} = 400(40)(10^{-6}) = 16\ \text{mV}$$

With large gain values and a high Q coil, care must be taken to ensure that the amplifier stage does not oscillate. Stability criteria have been developed, which permit the designer to "check" for instability.

PROBLEMS

1. At high frequencies an active device would not provide amplification without the incorporation of tuned circuits. This is so because _____ .

2. An advantage of a double-tuned circuit as compared with a single-tuned circuit is _____ .

3. Wide bandwidth may be achieved in a double-tuned circuit by two techniques. They are _____ .

4. Why are ferrite material is employed in tuning coils?

5. In a double-tuned transformer, adjusting one tuning slug affects three variables. They are _____ .

6. At very high frequencies, air-cored transformers are employed because _____ .

7. Sketch the frequency response curves for a double-tuned circuit that is a.) under-coupled, b.) critically coupled, c.) optimally coupled, and d.) over-coupled.

8. Why are interstage tuning coils tapped?

9. Under what conditions is there sufficient accuracy in using the formula

$$f_0 = \frac{1}{2\pi \sqrt{LC}}$$

to determine the resonant frequency of a tank circuit?

10. The equivalent parallel resistance of a tank circuit is a considerably larger value than the series resistance of the coil winding. How are these two resistances related?

11. Determine the resonant frequency for the circuit shown in Fig. 15.26.

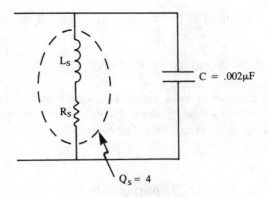

Fig. 15.26

12. For Problem 11, determine the resistance of the coil winding.

13. Determine the value of the tank impedance at resonance. *Hint:* Recall that at resonance, $Z_{in} = R_P$.

14. Determine the bandwidth for the circuit of Problem 11.

15. Determine the bandwidth and voltage gain for the interstage network and transistors shown in Fig. 15.27.

Given:

$L_{coil} = 40 \ \mu\text{H}$, $R_{coil} = 4 \ \Omega$, $C = 75$ pF, $g_m = g_{fs} = 12,000 \ \mu\text{mho}$, $r_d = 50$ k

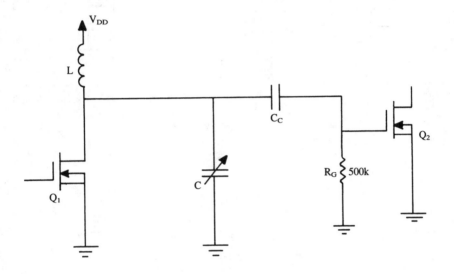

Fig. 15.27

16. Determine the bandwidth and voltage gain for the interstage network and transistors shown in Fig. 15.28. Determine the coil tap locations such that the source and load impedances equal the impedance of the unloaded coil. Given

$$g_{11} = \frac{1}{1000 \ \Omega}$$

$$g_{22} = \frac{1}{20 \ \text{k}\Omega}$$

$$|Y_{21}| = 15 \ \text{mmho}$$

where

$L_{coil} = 120 \ \mu\text{H}$, $R_{coil} = 15 \ \Omega$, $C = 100$ pF

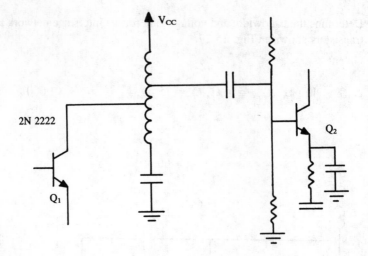

Fig. 15.28

APPENDIX A-1
Computer Program IPL91

```
      LIST    IPL91    CDT
00010 PRINT"THIS PROGRAM COMPUTES THE COMPONENT
VALUES FOR"
00020 PRINT"GIVEN CURRENTS & VOLTAGES FOR A COMMON
EMITTER"
00030 PRINT"STAGE WITH UNIVERSAL BIAS."
00040 REM I1 = DELTA*IB, VRE = GAMMA*VCC, IC = BETA*IB,
VCE = VRE
00050 PRINT"INPUT THE NOMINAL VALUE OF BETA, THE VALUES
OF:"
00060 PRINT"DELTA, GAMMA, VCC, ICQ, VBEQ"
00070 INPUTB, D,G,V9,I8,V8
00080 R3 = V9*B/(I8*(B + 1))
00090 R4 = G*V9*B/(I8*(B + 1))
00100 R1 = (B*(V9*(1-G)-V8))/(D*I8)
00110 IFD = 1THEN140
00120 R2 = (B*(G*V9 + V8))/(I8*(D-1))
00130 GOTO150
00140 R2 = 9.9E20
00150 PRINTUSING160,R1
00160 :THE VALUES WERE COMPUTED AS FOLLOWS: R1###.##K,
00170 PRINTUSING180,R2,R3,R4
00180 :R2 = ##.##!!!!K, R3 = ###.###K, R4 = ##.#####K
00190 END
EDIT
```

```
RUN IPL91 CDT
DELTA = 5, GAMMA = 1
```
THIS PROGRAM COMPUTES THE COMPONENT VALUES FOR
GIVEN CURRENTS & VOLTAGES FOR A COMMON EMITTER
STAGE WITH UNIVERSAL BIAS. GIVEN: I1 = DELTA*IB
VRE=GAMMA*VCC, IC=BETA*IB, VCE=VRC.
INPUT THE NOMINAL VALUE OF BETA, THE VALUES OF:
DELTA, GAMMA, VCC, ICQ, VBEQ
```
    ?    100,5,.1,10,5,.6
```
THE VALUES WERE COMPUTED AS FOLLOWS: R1 = 33.60K,
R2 = 8.00K, R3 = 0.90K, R4 = 0.19802K
EDIT

Table A.1
Common Emitter

$V_{CC} = +10V$, $V_{EB} = 0.6$, $V_{RC} = V_{EC}$, I_{CQ} (nominal) $= 5$ mA,
β (nominal) $= 100$

Results of Running Computer Program IPL91, CDT

Universal Stabilized Common Emitter			Delta (δ)			
		(kΩ)	1	2	4	8
	$R_C = 1$ k	R_1	188k	94k	47k	23.5
$\gamma = 0$	$R_E = 0$	R_2	∞	12k	4k	1.71k
	$R_C = 0.95$k	R_1	178k	89k	44.5k	22.25k
$\gamma = 0.05$	$R_E = 0.9901$k	R_2	∞	22k	7.33k	3.14k
	$R_C = 0.9$k	R_1	168k	84k	42k	21k
$\gamma = 0.1$	$R_E = 0.19802$k	R_2	∞	32k	10.67k	4.57k
	$R_C = 0.8$k	R_1	148k	74k	37k	18.5k
$\gamma = 0.2$	$R_E = 0.39604$k	R_2	∞	52k	17.33k	7.43k

```
       LIST     IPL91     STB
00010 PRINT"THIS PROGRAM COMPUTES THE CHANGE IN DVRC/
DB,"
00020 PRINT"VRC, VRE, & VCE FOR VARIOUS VALUES OF BETA
FOR A"
00030 PRINT"COMMON EMITTER STAGE WITH UNIVERSAL
STABILIZATION"
00040 PRINT
00050 REM I5 = IB, V9 = VCC, V8 = VBE, R3 = RC, V(N) = VRC,
U(N) = VRE
00060 REM R4 = RE, W(N) = VCE
00070 PRINT"INPUT THE VALUES OF: VCC, VBE, R1, R2, Rc, RE"
00080 INPUTV9,V8,R1,R2,R3,R4
00090 PRINT"INPUT THE NUMBER OF DIFFERENT VALUES OF
BETA"
00100 PRINT"THAT ARE TO BE COMPUTED"
00110 INPUTM
00120 PRINT"INPUT THE VALUE(S) OF BETA"
00130 FORN = 1TOM
00140 INPUTB(N)
00150 NEXTN
00160 FORN = 1TOM
00170 I(N) = (V9-V8*(1 + R1/R2))/(R1 + (B(N) + 1)*R4*(1 + R1/R2))
00180 V(N) = B(N)*I(N)*R3
00190 U(N) = (B(N) + 1)*I(N)*R4
00200 W(N) = V9-V(N)-U(N)
00210 T = R3*(V9-V8*(1 + R1/R2))*(R1 + R4*(1 + R1/R2))
00220 R(N) = T/((B(N) + 1)*R4*(1 + R1/R2) + R1)**2
00230 NEXTN
00240 PRINT
00250 PRINTUSING260
00260 : BETA     DVRC/DB     VRC     VRE     VCE
00270 PRINT
00280 PRINT
00290 FORN = 1TOM
00300 PRINTUSING310,B(N),R(N),V(N),U(N),W(N)
00310 :#####     #.##!!!!     ##.##     ##.##     ##.##
00320 NEXTN
00330 PRINT
00340 END
EDIT
```

```
      RUN    IPL91    STB
THIS PROGRAM COMPUTES THE CHANGE IN DVRC/DB, VRC, VRE,
& VCE FOR VARIOUS VALUES OF BETA FOR A. COMMON EMITTER
STAGE WITH UNIVERSAL STABILIZATION

INPUT THE VALUES OF: VCC, VBE, R1, R2, RC, RE
?  10,.6,33.6,8,.9,.19802
INPUT THE NUMBER OF DIFFERENT VALUES OF BETA
THAT ARE TO BE COMPUTED
?  10
INPUT THE VALUE(S) OF BETA (DELTA = 5, GAMMA = 0.1)
?  10
?  25
?  50
?  75
?  100
?  150
?  200
?  500
?  1000
?  10000
```

BETA	DVRC/DB	VRC	VRE	VCE
10	1.06E-01	1.38	0.33	8.29
25	5.88E-02	2.56	0.59	6.85
50	2.89E-02	3.60	0.81	5.60
75	1.71E-02	4.15	0.93	4.92
100	1.13E-02	4.50	1.00	4.50
150	6.00E-03	4.91	1.09	4.00
200	3.71E-03	5.15	1.14	3.71
500	7.10E-04	5.63	1.24	3.12
1000	1.89E-04	5.82	1.28	2.90
10000	2.01E-06	5.99	1.32	2.69

```
EDIT
```

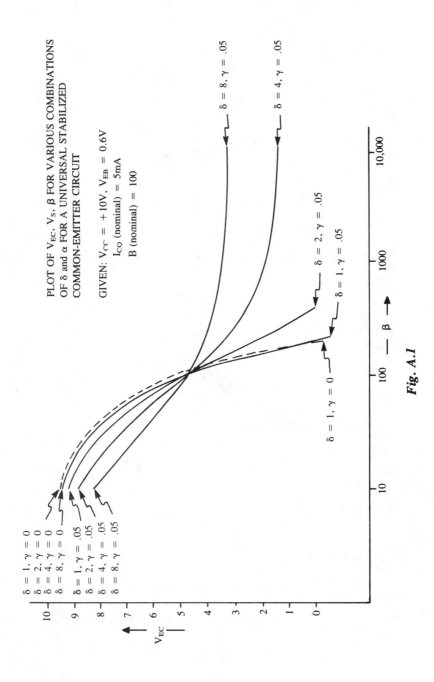

PLOT OF V_{EC}, V_S, β FOR VARIOUS COMBINATIONS
OF δ and α FOR A UNIVERSAL STABILIZED
COMMON-EMITTER CIRCUIT

GIVEN: $V_{CC} = +10V$, $V_{EB} = 0.6V$
I_{CO} (nominal) $= 5mA$
B (nominal) $= 100$

Fig. A.1

550

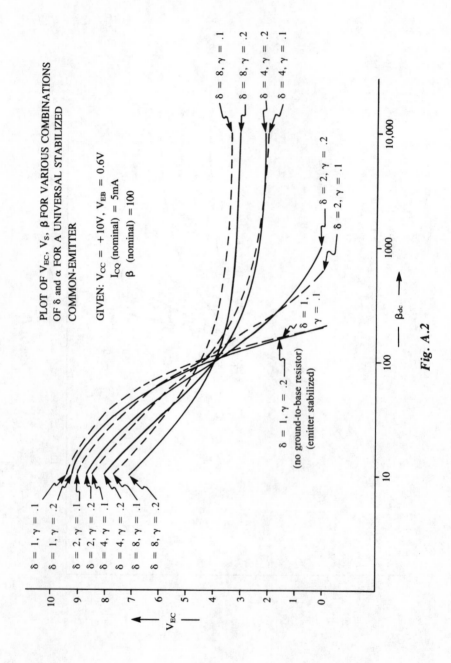

PLOT OF V_{EC}, V_S, β FOR VARIOUS COMBINATIONS
OF δ and α FOR A UNIVERSAL STABILIZED
COMMON-EMITTER

GIVEN: $V_{CC} = +10V$, $V_{EB} = 0.6V$
       I_{CQ} (nominal) $= 5mA$
       β (nominal) $= 100$

Fig. A.2

```
        IPL91    CDF
00010  PRINT"THIS PROGRAM COMPUTES THE COMPONENT VALUES
FOR GIVEN"
00020  PRINT"CURRENTS AND VOLTAGES FOR A COMMON EMITTER
STAGE WITH"
00030  PRINT"COLLECTOR FEEDBACK AND AN EMITTER RESISTOR"
00040  PRINT
00050  REM    I1 = DELTA*IB,    VRE = GAMMA*VCC,    IC = BETA*IB,
VCE = VRC
00060  PRINT
00070  PRINT"INPUT THE NOMINAL VALUE OF BETA, THE VALUES
OF: DELTA,"
00080  PRINT"GAMMA, VCC, ICQ, VBEQ."
00090  INPUTB,D,G,V9,I8,V8
00100  PRINT
00110  R3 = (B*V9*(1-G))/(2*I8*(B + D))
00120  R4 = G*V9*B/(I3*(B + 1))
00130  IFD  = 1THEN160
00140  R2 = (B*(G*V9 + V*))/(I8*(D-1))
00150  GOTO170
00160  R2 = 9.9E20
00170  R6 = (B*(V9*(1-G)-2*V8))/(2*D*I3)
00180  PRINTUSING190,R6
00190  :THE VALUES WERE COMPUTED AS FOLLOWS: RF = ###.##K,
00200  PRINTUSING210,R2,R3,R4
00210  "R2 = ##.##!!!!K, R3 = ###.###K, R4 = ##.#####K
00220  PRINT
00230  END
EDIT
```

```
EDIT SAVE
EDIT RUN    IPL91    CDF
THIS PROGRAM COMPUTES THE COMPONENT VALUES FOR GIVEN
CURRENTS AND VOLTAGES FOR A COMMON EMITTER STAGE WITH
COLLECTOR FEEDBACK AND AN EMITTER RESISTOR

INPUT THE NOMINAL VALUE OF BETA, THE VALUES OF: DELTA,
GAMMA, VCC, ICQ, VBEQ.
?    100,1,0,10,5,.6
```

DELTA = 1, GAMMA = 0
THE VALUES WERE COMPUTED AS FOLLOWS: RF = 88.00K,
R2 = 990.00E + 18K, R3 = 00.99K, R4 = 0.00000K

Table A.2
Common Emitter

$V_{CC} + +10$ V, $V_{EB} = 0.6$ V, $V_{RC} = V_{EC}$, I_{EQ} (nominal) = 5 mA,
β (nominal) = 100

Results of Running Computer Program IPL91, CDF

with Collector Feedback	(kΩ)	1	2	4	8
	R_G	88k	44k	22k	11k
	R_2	∞	12k	4k	1.71k
$\gamma = 0$	R_C		0.9804k	0.9615k	0.9259k
	R_E	0	0	0	0
	R_G	83k	41.5k	20.75k	10.38k
	R_2	∞	22k	7.33k	3.14k
$\gamma = 0.05$	R_C	0.94k	0.9314k	0.9135k	0.8796k
	R_E	0.9901k	0.09901k	0.09901k	0.09901k
	R_G	78k	39k	19.5k	9.75k
	R_2	∞	32k	10.67k	4.57k
$\gamma = 0.1$	R_C	0.89k	0.8824k	0.8654k	0.8333k
	R_E	0.19802k	0.19802k	0.19802k	0.19802k
	R_G	68k	34k	17k	8.5k
	R_2	∞	52k	17.33k	7.43k
$\gamma = 0.2$	R_C	0.79k	0.7843k	0.7692k	0.7407k
	R_E	0.39604k	0.39604k	0.39604k	0.39604k

```
      LIST   IPL91   STF
00010 PRINT
00020 PRINT"THIS PROGRAM COMPUTES DVRC/DB, VRC, VRE &
VCE"
00030 PRINT"FOR VARIOUS VALUES OF BETA FOR A COMMON
EMITTER"
00040 PRINT"STAGE WITH COLLECTOR FEEDBACK"
00050 REM I5 = IB, V9 = VCC, V8 = VBE, R3 = RC, V(N) = VRC,
U(N) = VRE,
00060 REM R4 = RE, W(N) + VCE
00070 PRINT"INPUT THE VALUES OF: VCC,VBE,RF,R2,RC & RE"
00080 INPUTV9,V8,R6,R2,R3,R4
00090 PRINT"INPUT THE NUMBER OF DIFFERENT VALUES OF
BETA"
00100 PRINT"THAT ARE TO BE COMPUTED"
00110 INPUTM
00120 PRINT"INPUT THE VALUE(S) OF BETA"
00130 FORN = 1TOM
00140 INPUTB(N)
00150 NEXTN
00160 L = 1 + (R6 + R3)/R2
00170 T = R3*(V9-V8*L)*(1 + R6-D*(R3 + R4*L))
00180 FORN = 1TOM
00190 I(N) = (V9-V8*L)/(R6 + (B(N) + 1)*(R3 + R4*L))
00200 U(N) = B(N) + 1)*I(N)*R4
00210 V(N) = (B(N) + 1*I(N)*R3*(1 + R4/R2) + V8*R3/R2
00220 W(N) = V9-V(N)-U(N)
00230 R(N) = T/((1 + R6 + B(N)*(R3 + R4*L))**2)
00240 NEXTN
00250 PRINT
00260 PRINTUSING270
00270 : BETA          DVRC/DB        VRC        VRE        VCE
00280 PRINT
00290 FORN = 1TOM
00300 PRINTUSING310,B(N),R(N),V(N),U(N),W(N)
00310 :#####        #.##!!!!   ##.##        ##.##        ##.##
00320 NEXTN
00330 PRINT
00340 END
EDIT
```

554

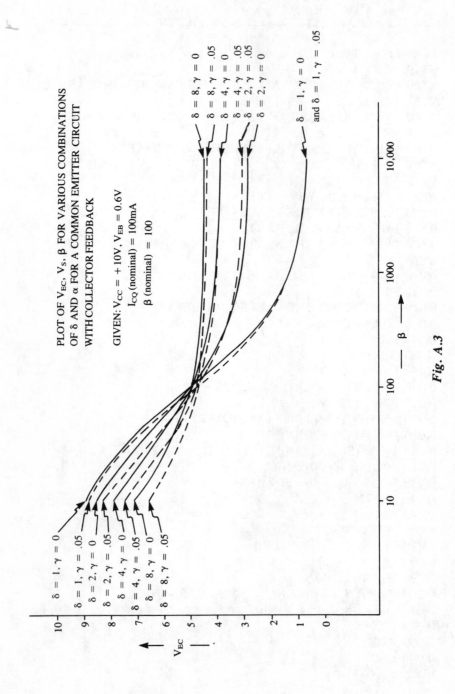

PLOT OF V_{EC}, V_S, β FOR VARIOUS COMBINATIONS
OF δ AND α FOR A COMMON EMITTER CIRCUIT
WITH COLLECTOR FEEDBACK

GIVEN: $V_{CC} = +10V$, $V_{EB} = 0.6V$
I_{CQ} (nominal) = 100mA
β (nominal) = 100

$\delta = 8, \gamma = 0$
$\delta = 8, \gamma = .05$
$\delta = 4, \gamma = 0$
$\delta = 4, \gamma = .05$
$\delta = 2, \gamma = .05$
$\delta = 2, \gamma = 0$

$\delta = 1, \gamma = 0$

and $\delta = 1, \gamma = .05$

10,000

1000

100

10

$\delta = 1, \gamma = 0$
$\delta = 1, \gamma = .05$
$\delta = 2, \gamma = 0$
$\delta = 2, \gamma = .05$
$\delta = 4, \gamma = 0$
$\delta = 4, \gamma = .05$
$\delta = 8, \gamma = 0$
$\delta = 8, \gamma = .05$

10 9 8 7 6 5 4 3 2 1 0

V_{EC}

$\beta \longrightarrow$

Fig. A.3

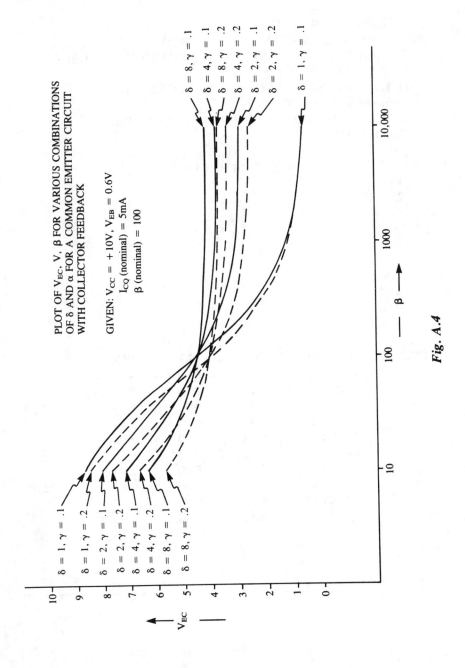

PLOT OF V_{EC}, V, β FOR VARIOUS COMBINATIONS
OF δ AND α FOR A COMMON EMITTER CIRCUIT
WITH COLLECTOR FEEDBACK

GIVEN: V_{CC} = +10V, V_{EB} = 0.6V
I_{CQ} (nominal) = 5mA
β (nominal) = 100

Fig. A.4

```
      RUN    IPL91    STF
THIS PROGRAM COMPUTES DVRC/DB,VRC,VRE & VCE
FOR VARIOUS VALUES OF BETA FOR A COMMON EMITTER
STAGE WITH COLLECTOR FEEDBACK
INPUT THE VALUES OF: VCC, VBE, RF, R2, RC & RE
   ?  10,.6,88,9.9E20,.99,0
INPUT THE NUMBER OF DIFFERENT VALUES OF BETA
THAT ARE TO BE COMPUTED
   ?  10
INPUT THE VALUE(S) OF BETA (DELTA= 1, GAMMA= 0)
   ?   10
   ?   25
   ?   50
   ?   75
   ?   100
   ?   150
   ?   200
   ?   500
   ?   1000
   ?   10000
```

BETA	DVRC/DB	VRC	VRE	VCE	
10	8.47E-02	1.04	0.00	8.96	
25	6.40E-02	2.13	0.00	7.87	
50	4.32E-02	3.43	0.00	6.57	+ 79.2%
75	3.11E-02	4.33	0.00	5.67	
100	2.34E-02	5.00	0.00	5.00	
150	1.47E-02	5.92	0.00		
200	1.01E-02	6.52	0.00	3.48	
500	2.43E-03	7.98	0.00	2.02	− 72.6%
1000	7.11E-04	8.63	0.00	1.37	
10000	8.30E-06	9.32	0.00	0.68	

```
EDIT
```

```
        LIST    DCANAL
00010 PRINT
00020 PRINT"THIS PROGRAM, DCANAL, ENABLES THE USER TO
INPUT THE RESISTOR"
00030 PRINT"VALUES FOR A COMMON EMITTER CONNECTED
BIPOLAR TRANSISTOR LINEAR
00040 PRINT"AMPLIFIER AND DETERMINE THE RESULTING DC
CURRENTS AND VOLTAGES."
00050 PRINT
00060 PRINT"THE PROGAM WILL ENABLE THE USER TO
COMPUTE"
00070 PRINT"VALUES FOR A UNIVERSAL STABILIZED, EMITTER
STABILIZED OR"
00080 PRINT"UNSTABILIZED COMMON EMITTER STAGE."
00090 PRINT"WHEN COMPUTATIONS ARE MADE FOR AN EMITTER
STABILIZED AMP,"
00100 PRINT"INPUT THE VALUE: 5E20 FOR R2. FOR AN
UNSTABILIZED AMP,"
00110 PRINT"INPUT THE VALUE ZERO FOR RE."
00120 PRINT"WHEN THE TERMINAL DISPLAYS A QUESTION
MARK, INPUT THE"
00130 PRINT"FOLLOWING: R1, R2, RC, RE, VCC, BETA(D.C.),
VBE..."
00140 PRINT"(INPUT ALL RESISTANCE VALUES IN KILOHMS)"
00150 PRINT
00160 PRINT
00170 INPUTR1,R2,R3,R4,V,B,V5
00180 I2 = (V-V5*(1 + R1/R2))/(R1 + (B + 1)*R4*(1 + R1/R2))
00190 I3 = B*I2
00200 I4 = I3 + I2
00210 V3 = I3*R3
00220 V4 = I4*R4
00230 V6 = V-V3-V4
00240 PRINT"---------------------------------------------"
00250 PRINT"FOR R1 = ";R1;"K, R2 = ";R2; "K, RC = ";R3; "K,
RE = ";R4; "K, VCC = ";V;"
00260 PRINT"BETA = ";B;" VBE = ' ;V5
00270 PRINT"THE RESULTING CURRENTS AND VOLTAGES ARE:"
00280 PRINT
00290 PRINTUSING300,I2,I3,I4
00300 ! IB = ###.######(M.A.)    IC = ####.#####(M.A.)
```

IE = ####.#####(M.A.)
00310 PRINT
00320 PRINTUSING330,V3,V4,V6
00330 !VRC = ###.####(VOLTS)   VRE = ###.####(VOLTS)
VCE = ###.####(VOLTS)
00340 PRINT
00350 PRINT"IF THE DISPLAYED VOLTAGE VALUES EXCEED THE
APPLIED VCC"
00360 PRINT"VALUE, THEN THE RESISTANCE VALUES ARE
INCOMPATABLE WITH"
00370 PRINT"THE OTHER CIRCUIT PARAMETERS..."
00380 END

     RUN   DCANAL
THIS PROGRAM, DCANAL, ENABLES THE USER TO INPUT THE
RESISTOR VALUES FOR A COMMON EMITTER CONNECTED
BIPOLAR TRANSISTOR LINEAR AMPLIFIER AND DETERMINE THE
RESULTING DC CURRENTS AND VOLTAGES.
THE PROGRAM WILL ENABLE THE USER TO COMPUTE VALUES
FOR A UNIVERSAL STABILIZED, EMITTER STABILIZED OR
UNSTABILIZED COMMON EMITTER STAGE.
WHEN COMPUTATIONS ARE MADE FOR AN EMITTER STABILIZED
AMP, INPUT THE VALUE: 5E20 FOR R2. FOR AN UNSTABILIZED
AMP, INPUT THE VALUE ZERO FOR RE.
WHEN THE TERMINAL DISPLAYS A QUESTION MARK, INPUT THE
FOLLOWING: R1, R2, RC, RE, VCC, BETA (D.C.), VBE... (INPUT ALL
RESISTANCE VALUES IN KILOHMS)

FOR R1 = 100 K, R2 = 10 K, RC = 5K, RE = .2K, VCC = 12
BETA = 100 VBE = .7
THE RESULTING CURRENTS AND VOLTAGES ARE:

IB =   0.013346(M.A.)   IC =   1.33457(M.A.)
IE =   1.34792 (M.A.)   VRC =   6.6729(VOLTS)
VRE +   0.2696(VOLTS)   VCE =   5.0575 (VOLTS)
IF THE DISPLAYED VOLTAGE VALUES EXCEED THE APPLIED VCC
VALUE, THEN THE RESISTANCE VALUES ARE INCOMPATIBLE
WITH THE OTHER CIRCUIT PARAMETERS...

APPENDIX A-2

Derivation of Resonant Frequency and Voltage Loss In a Three-element R-C Network of a Phase Shift Oscillator

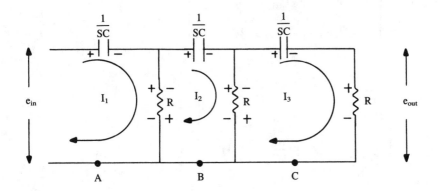

Fig. A.5

At A c.c.w.

$$I_1R - I_2R + I_1\frac{1}{SC} - e_{in} = 0$$

At B c.c.w.

$$I_2R - I_3R + I_2\frac{1}{SC} + I_2R - I_1R = 0$$

At C c.c.w.

$$I_3R + I_3\frac{1}{SC} + I_3R - I_2R = 0$$

$$I_1\left(R + \frac{1}{SC}\right) + I_2(-R) + I_3(0) = e_{in}$$

$$I_1(-R) + I_2\left(2R + \frac{1}{SC}\right) + I_3(-R) = 0$$

$$I_1(0) + I_2(-R) + I_3\left(2R + \frac{1}{SC}\right) = 0$$

$$\Delta = \begin{vmatrix} R + \dfrac{1}{SC} & -R & 0 \\[2mm] -R & 2R + \dfrac{1}{SC} & -R \\[2mm] 0 & -R & 2R + \dfrac{1}{SC} \end{vmatrix}$$

$$I_3 = \frac{\begin{vmatrix} R + \dfrac{1}{SC} & -R & e_{in} \\[2mm] -R & 2R + \dfrac{1}{SC} & 0 \\[2mm] 0 & -R & 0 \end{vmatrix}}{\begin{vmatrix} R + \dfrac{1}{SC} & -R & 0 \\[2mm] -R & 2R + \dfrac{1}{SC} & -R \\[2mm] 0 & -R & 2R + \dfrac{1}{SC} \end{vmatrix}}$$

$$= \frac{e_{in}R^2}{\left(R + \dfrac{1}{SC}\right)\left(2R + \dfrac{1}{SC}\right)^2 - \left[R^2\left(2R + \dfrac{1}{SC}\right) + R^2\left(R + \dfrac{1}{SC}\right)\right]}$$

$$= \frac{e_{in}\,R^2}{R\left(2R + \dfrac{1}{SC}\right)^2 + \dfrac{1}{SC}\left(2R + \dfrac{1}{SC}\right)^2 - R^2\left(2R + \dfrac{1}{SC}\right) - R^2\left(R + \dfrac{1}{SC}\right)}$$

$$= \frac{e_{in}\,R^2}{R\left(2R + \dfrac{1}{SC}\right)\left(2R + \dfrac{1}{SC} - R\right) + \dfrac{1}{SC}\left(2R + \dfrac{1}{SC}\right)^2 - R^2\left(R + \dfrac{1}{SC}\right)}$$

$$I_3 = \frac{e_{in} R^2}{R\left(2R + \dfrac{1}{SC}\right)\left(R + \dfrac{1}{SC}\right) + \dfrac{1}{SC}\left(2R + \dfrac{1}{SC}\right)^2 - R^2\left(R + \dfrac{1}{SC}\right)}$$

$$= \frac{e_{in} R^2}{R\left(R + \dfrac{1}{SC}\right)\left[\left(2R + \dfrac{1}{SC}\right) - R\right] + \dfrac{1}{SC}\left(2R + \dfrac{1}{SC}\right)^2}$$

$$= \frac{e_{in} R^2}{R\left(R + \dfrac{1}{SC}\right)\left[R + \dfrac{1}{SC}\right] + \dfrac{1}{SC}\left(2R + \dfrac{1}{SC}\right)^2}$$

$$= \frac{e_{in} R^2}{R\left(R + \dfrac{1}{SC}\right)^2 + \dfrac{1}{SC}\left(2R + \dfrac{1}{SC}\right)^2}$$

$$\frac{e_{out}}{e_{in}} = \frac{I_3 R}{e_{in}}$$

$$I_3 = \frac{e_{in} R^2}{R\left(R^2 + \dfrac{2R}{SC} + \dfrac{1}{S^2C^2}\right) + \dfrac{1}{SC}\left(4R^2 + \dfrac{4R}{SC} + \dfrac{1}{S^2C^2}\right)}$$

$$= \frac{e_{in} R^2}{R^3 + \dfrac{2R^2}{SC} + \dfrac{R}{S^2C^2} + \dfrac{4R^2}{SC} + \dfrac{4R}{S^2C^2} + \dfrac{1}{S^3C^3}}$$

$$= \frac{e_{in} R^2}{R^3 + \dfrac{5R}{S^2C^2} + \dfrac{6R^2}{SC} + \dfrac{1}{S^3C^3}}$$

Replacing S with $j\omega$,

$$I_3 = \frac{e_{in} R^2}{R^3 + \dfrac{5R}{(j\omega c)^2} + \dfrac{6R^2}{j\omega c} + \dfrac{1}{(j\omega c)^3}}$$

$$= \frac{e_{in} R^2}{R^3 - \dfrac{5R}{\omega^2 C^2} + j\left[\dfrac{1}{\omega^3 C^3} - \dfrac{6R^2}{\omega C}\right]}$$

Imaginary term must be zero for oscillation:

$$\frac{1}{\omega^3 C^3} - \frac{6R^2}{\omega C} = 0$$

Thus,

$$\frac{1}{\omega^3 C^3} = \frac{6R^2}{\omega C}$$

or

$$\frac{1}{\omega^2 C^2} = 6R^2 \qquad \omega^2 = \frac{1}{6R^2 C^2}$$

$$2\pi f = \frac{1}{\sqrt{6}RC}$$

Therefore,

$$f = \frac{1}{2\pi \sqrt{6}RC}$$

Also, the voltage loss in the phase shift network is

$$\frac{e_{out}}{e_{in}} = \frac{I_3 R}{e_{in}} = \frac{R^3}{R^3 - \dfrac{5R}{\omega^2 C^2}} = \frac{R^3}{R^3 - 5R(6R^2)}$$

$$\frac{e_{out}}{e_{in}} = -\frac{1}{29}$$

$$Z_{in} = \frac{e_{in}}{I_1}$$

$$I_1 = \frac{\begin{vmatrix} e_{in} & -R & 0 \\ 0 & 2R + \dfrac{1}{SC} & -R \\ 0 & -R & 2R + \dfrac{1}{SC} \end{vmatrix}}{\Delta}$$

$$I_1 = \frac{e_{in}\left(2R + \dfrac{1}{SC}\right)^2 - e_{in}R^2}{R^3 + \dfrac{5R}{S^2C^2} + \dfrac{6R^2}{SC} + \dfrac{1}{S^3C^3}}$$

$$Z_{in} = \frac{R^3 + \dfrac{5R}{S^2C^2} + \dfrac{6R^2}{SC} + \dfrac{1}{S^3C^3}}{\left(2R + \dfrac{1}{SC}\right)^2 - R^2}$$

$$Z_{in} = \frac{R^3 + \dfrac{5R}{S^2C^2} + \dfrac{6R^2}{SC} + \dfrac{1}{S^3C^3}}{4R^2 + \dfrac{4R}{SC} + \dfrac{1}{S^2C^2} - R^2} = \frac{R^3 + \dfrac{5R}{S^2C^2} + \dfrac{6R^2}{SC} + \dfrac{1}{S^3C^3}}{3R^2 + \dfrac{1}{S^2C^2} + \dfrac{4R}{5C}}$$

However, $SC = j\omega C$,

$$Z_{in} = \frac{R^3 - \dfrac{5R}{\omega^2C^2} + j\left(\dfrac{1}{\omega^3C^3} - \dfrac{6R^2}{\omega C}\right)}{3R^2 - \dfrac{1}{\omega^2C^2} - j\dfrac{4R}{\omega C}}$$

where

$$\frac{1}{\omega C} = \sqrt{6}R$$

$$\frac{1}{\omega^2C^2} = 6R^2$$

$$\frac{1}{\omega^3C^3} = (6R^2)^{3/2} = 6^{3/2}R^3$$

Thus, we derive

$$Z_{in} = \frac{R^3 - 5R(6R^2) + j(6^{3/2}R^3 - 6R^2\sqrt{6}R)}{3R^2 - 6R^2 - j\,4R\sqrt{6}R}$$

$$= R\left[\frac{1 - 30 + j(6^{3/2} - 6\sqrt{6})}{3 - 6 - j\,4\sqrt{6}}\right] = \frac{-29\,R}{-3 - j\,4\sqrt{6}}$$

$$= \frac{29R}{3 + j\,4\sqrt{6}} = \frac{29R}{3 + j\,4\sqrt{6}}\left(\frac{3 - j\,4\sqrt{6}}{3 - j\,4\sqrt{6}}\right) = \frac{29R(3 - j\,4\sqrt{6})}{9 + 16(6)}$$

$$Z_{in} = \frac{29R(3 - j\,4\sqrt{6})}{105} = 2.83R\;\underline{/-72.976134°}$$

APPENDIX A-3

Derivation of Resonant Frequency and Voltage Loss in a Three-Element R-C Network

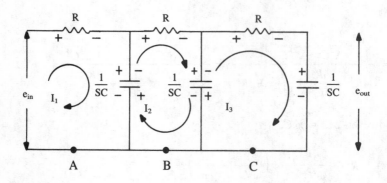

Fig. A.6

At A c.c.w.

$$I_1\frac{1}{SC} + I_2\left(-\frac{1}{SC}\right) + I_1R - e_{in} = 0$$

At B c.c.w.

$$I_2\frac{1}{SC} + I_3\left(-\frac{1}{SC}\right) + I_2R + I_2\frac{1}{SC} + I_1\left(-\frac{1}{SC}\right) = 0$$

At C c.c.w

$$I_3\left(\frac{1}{SC}\right) + I_3R + I_3\frac{1}{SC} + I_2\left(-\frac{1}{SC}\right) = 0$$

Collecting terms:

$$I_1\left(R + \frac{1}{SC}\right) + I_2\left(-\frac{1}{SC}\right) + I_3(0) = e_{in}$$

$$I_1\left(-\frac{1}{SC}\right) + I_2\left(R + \frac{2}{SC}\right) + I_3\left(-\frac{1}{SC}\right) = 0$$

$$I_1(0) + I_2\left(-\frac{1}{SC}\right) + I_3\left(R + \frac{2}{SC}\right) = 0$$

$$\Delta = \begin{vmatrix} R + \dfrac{1}{SC} & -\dfrac{1}{SC} & 0 \\[2mm] -\dfrac{1}{SC} & R + \dfrac{2}{SC} & -\dfrac{1}{SC} \\[2mm] 0 & -\dfrac{1}{SC} & R + \dfrac{2}{SC} \end{vmatrix}$$

$$= \left(R + \frac{1}{SC}\right)\left(R + \frac{2}{SC}\right)^2 - \left[\left(\frac{1}{SC}\right)^2\left(R + \frac{2}{SC}\right)\right.$$
$$\left. + \left(\frac{1}{SC}\right)^2\left(R + \frac{1}{SC}\right)\right]$$

$$= R\left(R + \frac{2}{SC}\right)^2 + \frac{1}{SC}\left(R + \frac{2}{SC}\right)^2 - \left(\frac{1}{SC}\right)^2\left(R + \frac{2}{SC}\right)$$
$$- \left(\frac{1}{SC}\right)^2\left(R + \frac{1}{SC}\right)$$

$$= R\left(R + \frac{2}{SC}\right)^2 + \frac{1}{SC}\left(R + \frac{2}{SC}\right)\left(R + \frac{2}{SC} - \frac{1}{SC}\right)$$
$$- \left(\frac{1}{SC}\right)^2\left(R + \frac{1}{SC}\right)$$

$$= R\left(R + \frac{2}{SC}\right)^2 + \frac{1}{SC}\left(R + \frac{2}{SC}\right)\left(R + \frac{1}{SC}\right) - \left(\frac{1}{SC}\right)^2\left(R + \frac{1}{SC}\right)$$

$$= R\left(R + \frac{2}{SC}\right)^2 + \frac{1}{SC}\left(R + \frac{1}{SC}\right)\left(R + \frac{2}{SC} - \frac{1}{SC}\right)$$

$$= R\left(R + \frac{2}{SC}\right)^2 + \frac{1}{SC}\left(R + \frac{1}{SC}\right)\left(R + \frac{1}{SC}\right)$$

$$= R\left(R + \frac{2}{SC}\right)^2 + \frac{1}{SC}\left(R + \frac{1}{SC}\right)^2$$

$$I_3 = \frac{\begin{vmatrix} R + \dfrac{1}{SC} & -\dfrac{1}{SC} & e_{in} \\[2ex] -\dfrac{1}{SC} & R + \dfrac{2}{SC} & 0 \\[2ex] 0 & -\dfrac{1}{SC} & 0 \end{vmatrix}}{\Delta}$$

$$= \frac{e_{in}\left(\dfrac{1}{SC}\right)^2}{R\left(R + \dfrac{2}{SC}\right)^2 + \dfrac{1}{SC}\left(R + \dfrac{1}{SC}\right)^2}$$

$$= \frac{e_{in}\left(\dfrac{1}{SC}\right)^2}{R\left(R^2 + \dfrac{4R}{SC} + \dfrac{4}{S^2C^2}\right) + \dfrac{1}{SC}\left(R^2 + \dfrac{2R}{SC} + \dfrac{1}{S^2\,C^2}\right)}$$

$$I_3 = \frac{e_{in}\left(\dfrac{1}{SC}\right)^2}{R^3 + \dfrac{4R^2}{SC} + \dfrac{4R}{S^2C^2} + \dfrac{R^2}{SC} + \dfrac{2R}{S^2C^2} + \dfrac{1}{S^3C^3}}$$

$$= \frac{e_{in}\left(\dfrac{1}{SC}\right)^2}{R^3 + \dfrac{6R}{S^2C^2} + \dfrac{1}{S^3C^3} + \dfrac{5R^2}{SC}}$$

$$\frac{e_{out}}{e_{in}} = \frac{I_3\dfrac{1}{SC}}{e_{in}} = \frac{\dfrac{1}{S^3C^3}}{R^3 + \dfrac{6R}{S^2C^2} + \dfrac{1}{S^3C^3} + \dfrac{5R^2}{SC}}$$

$$= \frac{1}{R^3(S^3C^3) + 6RSC + 1 + 5R^2S^2C^2}$$

$$\frac{e_{out}}{e_{in}} = \frac{1}{1 + 5R^2S^2C^2 + 6RSC + R^3S^3C^3}$$

$$= \frac{1}{1 - 5R^2\omega^2C^2 + j(6R\omega C - R^3\omega^3C^3)}$$

The imaginary term must be zero for oscillation. Thus,

$$6R\omega C = R^3\omega^3C^3$$
$$6 = R^2\omega^2C^2$$

Therefore,

$$\omega^2 = \frac{6}{R^2C^2}$$

$$\omega = \frac{\sqrt{6}}{RC}$$

$$2\pi f = \frac{\sqrt{6}}{RC}$$

$$f = \frac{\sqrt{6}}{2\pi RC}$$

$$f = \frac{.39}{RC}$$

$$\frac{e_{out}}{e_{in}} = \frac{1}{1 - 5R^2\omega^2C^2} = \frac{1}{1 - 5R^2\dfrac{6}{R^2}} = -\frac{1}{29}$$

$$Z_{in} = \frac{e_{in}}{I_1}$$

$$
I_1 = \frac{\begin{vmatrix} e_{in} & -\dfrac{1}{SC} & 0 \\[2mm] 0 & R + \dfrac{2}{SC} & -\dfrac{1}{SC} \\[2mm] 0 & -\dfrac{1}{SC} & R + \dfrac{2}{SC} \end{vmatrix}}{R^3 + \dfrac{6R}{S^2C^2} + \dfrac{1}{S^3C^3} + \dfrac{5R^2}{SC}}
$$

$$
I_1 = \frac{e_{in}\left(R + \dfrac{2}{SC}\right)^2 - e_{in}\dfrac{1}{S^2C^2}}{R^3 + \dfrac{6R}{S^2C^2} + \dfrac{1}{S^3C^3} + \dfrac{5R^2}{SC}}
$$

$$
= \frac{e_{in}\left(R^2 + \dfrac{4R}{SC} + \dfrac{4}{S^2C^2}\right) - e_{in}\dfrac{1}{S^2C^2}}{R^3 + \dfrac{6R}{S^2C^2} + \dfrac{1}{S^3C^3} + \dfrac{5R^2}{SC}} = \frac{e_{in}\left(R^2 + \dfrac{4R}{SC} + \dfrac{3}{S^2C^2}\right)}{R^3 + \dfrac{6R}{S^2C^2} + \dfrac{1}{S^3C^3} + \dfrac{5R^2}{SC}}
$$

$$
Z_{in} = \frac{R^3 + \dfrac{6R}{S^2C^2} + \dfrac{1}{S^3C^3} + \dfrac{5R^2}{SC}}{R^2 + \dfrac{4R}{SC} + \dfrac{3}{S^2C^2}}
$$

but $\quad \omega^2C^2 = \dfrac{6}{R^2}$

$$
\frac{1}{\omega^2C^2} = \frac{R^2}{6}
$$

$$
\frac{1}{\omega C} = \frac{R}{\sqrt{6}}
$$

$$
\frac{1}{\omega^3C^3} = \frac{R^3}{6^{3/2}}
$$

Thus,

$$Z_{in} = \frac{R^3 - \dfrac{6R}{\omega^2 C^2} + j\left(\dfrac{1}{\omega^3 C^3} - \dfrac{5R^2}{\omega C}\right)}{R^2 + \dfrac{3}{\omega^2 C^2} - j\,\dfrac{4R}{\omega C}}$$

$$= \frac{R^3 - 6R\left(\dfrac{R^2}{6}\right) + j\left(\dfrac{R^3}{6^{3/2}} - 5R^2\,\dfrac{R}{\sqrt{6}}\right)}{R^2 - 3\,\dfrac{R^2}{6} - j\,4R\left(\dfrac{R}{\sqrt{6}}\right)}$$

$$= \frac{R\{1 - 1 + j\,[6^{-3/2} - 5(6^{-1/2})]\}}{1 - \dfrac{1}{2} - j\,4(6^{-1/2})} = \frac{jR\,[6^{-3/2} - 5(6^{-1/2})]}{\dfrac{1}{2} - j\,4(6^{-1/2})}$$

$$Z_{in} = \frac{-R\,[6^{-3/2} - 5(6^{-1/2})]}{4(6^{-1/2}) + j^{1/2}} = \frac{R\,[5(6^{-1/2}) - 6^{-3/2}]}{4(6^{-1/2} + j^{1/2})}, \text{ where } K_4 = 6^{-3/2}$$

Then, for example,

$$Z_{in} = R\,\frac{2.0412415 - .0680413}{1.6329932 + j.5} = R\,\frac{1.9732001}{1.6329932 + j.5}$$

where $K_2 = 2.0412415$, $K_3 = .0680413$, $K_1 = 1.6329932$. Hence,

$$Z_{in} = R\,\frac{(1.973)}{1.633 + j.5}\left(\frac{1.633 - j.5}{1.633 - j.5}\right) = \frac{R(1.973)(1.633 - j.5)}{2^{2/3} + \dfrac{1}{4}}$$

$$= \frac{R(1.973)(1.633 - j.5)}{8/3 + 1/4} = \frac{R(1.973)(1.633 - j.5)}{35/12}$$

Therefore,

$$Z_{in} = 1.158R\ \underline{/-17.02°}$$

where $K_2 = 1.7078251\ \underline{/17.023813}$

APPENDIX A-4
Derivation of Resonant Frequency and Voltage Loss for a Four-Element Network

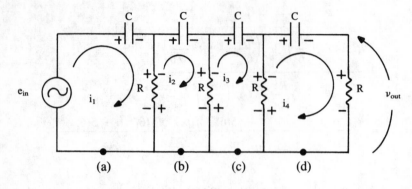

Fig. A.7

From Fig. A.7, we have the following:
At (a) c.c.w.

$$i_1R + i_1\frac{1}{SC} - i_2R - i_3(0) - i_4(0) = e_{in}$$

At (b) c.c.w.

$$-i_1R + i_2\left(2R + \frac{1}{SC}\right) - i_3R + i_4(0) = 0$$

At (c) c.c.w.

$$i_1(0) - i_2R + i_3\left(2R + \frac{1}{SC}\right) - i_4(R) = 0$$

At (d) $i_1(0) + i_2(0) - i_3R + i_4\left(2R + \frac{1}{SC}\right) = 0$

$$\Delta = \begin{vmatrix} R + \dfrac{1}{SC} & -R & 0 & 0 \\[2ex] -R & 2R + \dfrac{1}{SC} & -R & 0 \\[2ex] 0 & -R & 2R + \dfrac{1}{SC} & -R \\[2ex] 0 & 0 & -R & 2R + \dfrac{1}{SC} \end{vmatrix}$$

$$\Delta = \left(R + \dfrac{1}{SC}\right) \begin{vmatrix} 2R + \dfrac{1}{SC} & -R & 0 \\[2ex] -R & 2R + \dfrac{1}{SC} & -R \\[2ex] 0 & -R & 2R + \dfrac{1}{SC} \end{vmatrix}$$

$$+ R \begin{vmatrix} -R & -R & 0 \\[2ex] 0 & 2R + \dfrac{1}{SC} & -R \\[2ex] 0 & -R & 2R + \dfrac{1}{SC} \end{vmatrix}$$

$$\Delta = \left(R + \dfrac{1}{SC}\right) \left\{ \left(2R + \dfrac{1}{SC}\right)^2 - \left[2R^2\left(2R + \dfrac{1}{SC}\right)\right] \right\}$$
$$+ R\left\{ -R\left(2R + \dfrac{1}{SC}\right)^2 + R^3 \right\}$$

$$\Delta = \left(R + \dfrac{1}{SC}\right)\left(2R + \dfrac{1}{SC}\right)\left\{ \left(2R + \dfrac{1}{SC}\right)^2 - 2R^2 \right\}$$
$$+ R^2\left\{ -\left(2R + \dfrac{1}{SC}\right)^2 + R \right\}$$

$$\Delta = \left(2R^2 + \frac{R}{SC} + \frac{2R}{SC} + \frac{1}{S^2C^2}\right)\left[4R^2 + \frac{4R}{SC} + \frac{1}{S^2C^2} - 2R^2\right]$$

$$+ R^2\left\{-\left(4R^2 + \frac{4R}{SC} + \frac{1}{S^2C^2}\right) + R^2\right\}$$

$$= \left(2R^2 + \frac{3R}{SC} + \frac{1}{S^2C^2}\right)\left(2R^2 + \frac{4R}{SC} + \frac{1}{S^2C^2}\right)$$

$$+ R^2\left[-3R^2 - \frac{4R}{SC} - \frac{1}{S^2C^2}\right]$$

$$= 4R^4 + \frac{8R^3}{SC} + \frac{2R^2}{S^2C^2} + \frac{6R^3}{SC} + \frac{12R^2}{S^2C^2} + \frac{3R}{S^3C^3} + \frac{2R^2}{S^2C^2}$$

$$+ \frac{4R}{S^3C^3} + \frac{1}{S^4C^4} - 3R^4 - \frac{4R^3}{SC} - \frac{R^2}{S^2C^2}$$

$$= R^4 + \frac{10R^3}{SC} + \frac{15R^2}{S^2C^2} + \frac{7R}{S^3C^3} + \frac{1}{S^4C^4}$$

$$i_4 = \frac{\begin{vmatrix} R + \dfrac{1}{SC} & -R & 0 & e_{in} \\ -R & 2R + \dfrac{1}{SC} & -R & 0 \\ 0 & -R & 2R + \dfrac{1}{SC} & 0 \\ 0 & 0 & -R & 0 \end{vmatrix}}{\Delta}$$

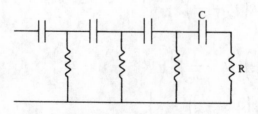

Fig. A.8 4-Section, R to Ground

$$\Delta i_4 = \left(R + \frac{1}{SC}\right) 0 + R \begin{vmatrix} -R & -R & 0 \\ 0 & 2R + \frac{1}{SC} & 0 \\ 0 & -R & 0 \end{vmatrix}$$

$$e_{in} \begin{vmatrix} -R & 2R + \frac{1}{SC} & -R \\ 0 & -R & 2R + \frac{1}{SC} \\ 0 & 0 & -R \end{vmatrix}$$

$$\Delta i_4 = -e_{in}\{-R^3\}; \text{ thus, } i_4 = \frac{e_{in}R^3}{\Delta}, \ v_o = i_4 R, \text{ and } v_o = \frac{e_{in}R^4}{\Delta}$$

$$\frac{v_o}{v_{in}} = \frac{R^4}{\Delta} = \frac{R^4}{R^4 + \frac{10R^3}{SC} + \frac{15R^2}{S^2C^2} + \frac{7R}{S^3C^3} + \frac{1}{S^4C^4}}$$

$$= \frac{R^4}{R^4 + 10R^3 \frac{1}{j\omega C} - \frac{15R^2}{\omega^2C^2} - \frac{7R}{j\omega^3C^3} + \frac{1}{\omega^4C^4}}$$

$$= \frac{R^4}{R^4 - 15\frac{R^2}{\omega^2C^2} + \frac{1}{\omega^4C^4} + j\left[-\frac{10R^3}{\omega C} + \frac{7R}{\omega^3C^3}\right]}$$

The j term is zero when

$$\frac{10R^3}{\omega C} = \frac{7R}{\omega^3C^3} \qquad 10R^2 = \frac{7}{\omega^2C^2}$$

$$\omega^2 = \frac{7}{10R^2C^2} \Rightarrow \omega = \sqrt{\frac{7}{10}}\frac{1}{RC}$$

$$f_0 = \frac{1}{2\pi}\sqrt{\frac{7}{10}}\frac{1}{RC} = (.159154)(.83666)\frac{1}{RC}$$

574

Thus,

$$f_0 = \frac{.1331}{RC}$$

Also,

$$\omega^2 C^2 = \frac{7}{10R^2} \qquad \omega^4 C^4 = \frac{49}{100R^4}$$

A.4-1 For Oscillation

$$\frac{v_0}{v_{in}} = \frac{R^4}{R^4 - 15\dfrac{R^2}{\omega^2 C^2} + \dfrac{1}{\omega^4 C^4}} = \frac{R^4}{R^4 - 15R^2\left(\dfrac{10R^2}{7}\right) + \dfrac{100R^4}{49}}$$

$$= \frac{1}{1 - \dfrac{150}{7} + \dfrac{100}{49}} = \frac{1}{\dfrac{49}{49} - \dfrac{1050}{49} + \dfrac{100}{49}} = \frac{49}{149 - 1050}$$

$$= -\frac{49}{901} = -.054384$$

Voltage gain of amp must be

$$A_V = -\frac{901}{49} = 18.38$$

APPENDIX A-5
Input Impedance of Four-Section Phase Shift Network (R to Ground)

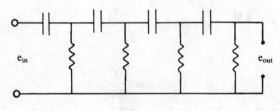

Fig. A.9

$$Z_n = \frac{v_{in}}{i_1}$$

$$i_1 = \frac{\begin{vmatrix} e_{in} & -R & 0 & 0 \\ 0 & 2R + \dfrac{1}{SC} & -R & 0 \\ 0 & -R & 2R + \dfrac{1}{SC} & -R \\ 0 & 0 & -R & 2R + \dfrac{1}{SC} \end{vmatrix}}{\Delta}$$

where

$$\omega^3 C^3 = \frac{1}{R^3}\left(\frac{7}{10}\right)^{3/2}$$

$$\frac{1}{\omega^3 C^3} = R^3\left(\frac{7}{10}\right)^{-3/2}$$

$$\omega C = \frac{1}{R}(7/10)^{1/2}$$

$$\frac{1}{\omega C} = R(7/10)^{-1/2}$$

$$i_1 = \frac{\begin{vmatrix} 2R + \dfrac{1}{SC} & -R & 0 \\ e_{in} & -R & 2R + \dfrac{1}{SC} & -R \\ & 0 & -R & 2R + \dfrac{1}{SC} \end{vmatrix}}{\Delta}$$

$$= \frac{e_{in}\left\{\left[\left(2R + \frac{1}{SC}\right)^3\right] - \left[2R^2\left(2R + \frac{1}{SC}\right)\right]\right\}}{R^4 - \frac{15R^2}{\omega^2 C^2} + \frac{1}{\omega^4 C^4}}$$

where, $\omega C = \frac{1}{R}\sqrt{\frac{7}{10}}, \frac{1}{\omega C} = \frac{R}{\sqrt{.7}}$

$\omega^2 C^2 = (7/10)R^2, \omega^4 C^4 = (49/100)R^4$

$$\times \left(2R + \frac{1}{j\omega C}\right)^2$$

$$\times 4R^2 + \frac{4R}{j\omega C} - \frac{1}{\omega^2 C^2}$$

$$\times \frac{2R + j\omega C}{\dfrac{4R^2}{j\omega C} - \dfrac{4R}{\omega^2 C^2} - \dfrac{1}{j\omega^3 C^3}}$$

$$\times \frac{8R^3 + \dfrac{8R^2}{j\omega C} - \dfrac{2R}{\omega^2 C^2}}{8R^3 + \dfrac{12R^2}{j\omega C} - \dfrac{6R}{\omega^2 C^2} - \dfrac{1}{j\omega^3 C^3}}$$

Therefore,

$$i_1 = \frac{e_{in}\left\{\left[2R + \dfrac{1}{j\omega C}\right]^3 - 2R^2\left(2R + \dfrac{1}{j\omega C}\right)\right\}}{R^4 - 15\dfrac{R^2}{\omega^2 C^2} + \dfrac{1}{\omega^4 C^4}}$$

$$\frac{i_1}{e_{in}} = \frac{8R^3 + \dfrac{12R^2}{j\omega C} - \dfrac{6R}{\omega^2 C^2} - \dfrac{1}{j\omega^3 C^3} - 4R^3 - \dfrac{2R^2}{j\omega C}}{R^4 - 15\dfrac{R^2}{\omega^2 C^2} + \dfrac{1}{\omega^4 C^4}}$$

$$\frac{i_1}{e_{in}} = \frac{4R^3 + \dfrac{10R^2}{j\omega C} - \dfrac{6R}{\omega^2 C^2} - \dfrac{1}{j\omega^3 C^3}}{R^4 - 15\dfrac{R^2}{\omega^2 C^2} + \dfrac{1}{\omega^4 C^4}}$$

A-5.1 Input Impedance for 4-Section

$$\frac{i_1}{e_{in}} = \frac{\left(2R + \dfrac{1}{SC}\right)^2 - 2R^2\left(2R + \dfrac{1}{SC}\right)}{\Delta}$$

$$= \frac{\left(2R + \dfrac{1}{SC}\right)\left[\left(2R + \dfrac{1}{SC}\right)^2 - 2R^2\right]}{\Delta}$$

$$= \frac{\left(2R + \dfrac{1}{SC}\right)\left[4R^2 + \dfrac{4R}{SC} + \dfrac{1}{S^2 C^2} - 2R^2\right]}{\Delta}$$

$$= \frac{\left(2R + \dfrac{1}{SC}\right)\left(2R^2 + \dfrac{4R}{SC} + \dfrac{1}{S^2 C^2}\right)}{\Delta}$$

$$= \frac{4R^3 + \dfrac{8R^2}{SC} + \dfrac{2R}{S^2 C^2} + \dfrac{2R^2}{SC} + \dfrac{4R}{S^2 C^2} + \dfrac{1}{S^3 C^3}}{\Delta}$$

$$= \frac{4R^3 + 10\dfrac{R^2}{SC} + 6\dfrac{R}{S^2 C^2} + \dfrac{1}{S^3 C^3}}{\Delta}$$

$$= \frac{4R^3 + 10R^2\left(\dfrac{1}{j\omega C}\right) + 6R\left(-\dfrac{1}{\omega^2 C^2}\right) - \dfrac{1}{j\omega^3 C^3}}{R^4 - 15\dfrac{R^2}{\omega^2 C^2} + \dfrac{1}{\omega^4 C^4}}$$

$$\frac{e_{in}}{i_{in}} = \frac{R^4 - 15\dfrac{R^2}{\omega^2 C^2} + \dfrac{1}{\omega^4 C^4}}{4R^3 - j\,10R^2\left(\dfrac{1}{\omega C}\right) - 6R\left(\dfrac{1}{\omega^2 C^2}\right) + j\left(\dfrac{1}{\omega^3 C^3}\right)}$$

$$\frac{e_{in}}{i_{in}} = \frac{R^4 - 15R^2\left(\dfrac{10R^2}{7}\right) + \dfrac{100R^4}{49}}{4R^3 - j\,10R^2\,\sqrt{10R^2/7} - 6R\left(\dfrac{10R^2}{7}\right) + j\left(\dfrac{10R^2}{7}\right)^{3/2}}$$

$$= R\left[\frac{1 - \dfrac{150}{7} + \dfrac{100}{49}}{4 - j\,10\,\sqrt{10/7} - \dfrac{60}{7} + j\left(\dfrac{10}{7}\right)^{3/2}}\right]$$

$$= \frac{R\left[\dfrac{49}{49} - \dfrac{1050}{49} + \dfrac{100}{49}\right]}{4 - 60/7 + j\left[\left(\dfrac{10}{7}\right)^{3/2} - \dfrac{10^{3/2}}{\sqrt{7}}\right]}$$

$$= \frac{R\left(-\dfrac{901}{49}\right)}{4\left(\dfrac{49}{49}\right) - \dfrac{60(7)}{49} + j\,10^{3/2}\left(\dfrac{1}{7^{3/2}} - \dfrac{1}{7^{1/2}}\right)}$$

$$= \frac{R\left(-\dfrac{901}{49}\right)}{4\left(\dfrac{49}{49}\right) - \dfrac{60(7)}{49} + j\,10^{3/2}\left[\dfrac{\sqrt{7}}{49} - \dfrac{7^{3/2}}{49}\right]}$$

$$= \frac{-R901}{4 - 420 + j\,10^{3/2}\,(7^{1/2} - 7^{3/2})} = \frac{-901R}{-416 - j\,10^{3/2}\,(7^{3/2} - 7^{1/2})}$$

Therefore, where $K_1 = 501.99601$, we have

$$Z_{in} = \frac{901R}{416 + j\,501.99601} = \frac{901R}{651.96318\ \underline{/50.35^\circ}} = 1.38R\ \underline{/-50.35^\circ}$$

A-5.2 Four-Section R-C Network (C to Ground)

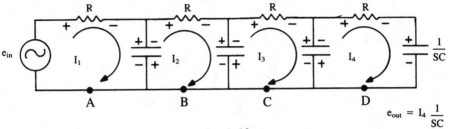

Fig. A.10

Given $e_o = I_4 \dfrac{1}{SC}$

At A c.c.w.

$$I_1 \frac{1}{SC} + I_2 \left(-\frac{1}{SC}\right) + I_1 R - e_{in} = 0$$

At B c.c.w.

$$I_2 \frac{1}{SC} + I_3 \left(-\frac{1}{SC}\right) + I_2 R + I_2 \frac{1}{SC} + I_1 \left(-\frac{1}{SC}\right) = 0$$

At C c.c.w.

$$I_3 \frac{1}{SC} + I_4 \left(-\frac{1}{SC}\right) + I_3 R + I_3 \left(\frac{1}{SC}\right) + I_2 \left(-\frac{1}{SC}\right) = 0$$

At D c.c.w.

$$I_4 \left(\frac{1}{SC}\right) + I_4 R + I_4 \frac{1}{SC} + I_3 \left(-\frac{1}{SC}\right) = 0$$

Collecting terms:

$$I_1 \left(R + \frac{1}{SC}\right) + I_2 \left(-\frac{1}{SC}\right) + I_3(0) + I_4(0) = e_{in}$$

$$I_1\left(-\frac{1}{SC}\right) + I_2\left(R + \frac{2}{SC}\right) + I_3\left(-\frac{1}{SC}\right) + I_4(0) = 0$$

$$I_1(0) + I_2\left(-\frac{1}{SC}\right) + I_3\left(R + \frac{2}{SC}\right) + I_4\left(-\frac{1}{SC}\right) = 0$$

$$I_1(0) + I_2(0) + I_3\left(-\frac{1}{SC}\right) + I_4\left(R + \frac{2}{SC}\right) = 0$$

$$\Delta = \begin{vmatrix} R + \dfrac{1}{SC} & -\dfrac{1}{SC} & 0 & 0 \\[2mm] -\dfrac{1}{SC} & R + \dfrac{2}{SC} & -\dfrac{1}{SC} & 0 \\[2mm] 0 & -\dfrac{1}{SC} & R + \dfrac{2}{SC} & -\dfrac{1}{SC} \\[2mm] 0 & 0 & -\dfrac{1}{SC} & R + \dfrac{2}{SC} \end{vmatrix}$$

$$\Delta = \left(R + \frac{1}{SC}\right) \begin{vmatrix} R + \dfrac{2}{SC} & -\dfrac{1}{SC} & 0 \\[2mm] -\dfrac{1}{SC} & R + \dfrac{2}{SC} & -\dfrac{1}{SC} \\[2mm] 0 & -\dfrac{1}{SC} & R + \dfrac{2}{SC} \end{vmatrix}$$

$$+ \frac{1}{SC} \begin{vmatrix} -\dfrac{1}{SC} & -\dfrac{1}{SC} & 0 \\[2mm] 0 & R + \dfrac{2}{SC} & -\dfrac{1}{SC} \\[2mm] 0 & -\dfrac{1}{SC} & R + \dfrac{2}{SC} \end{vmatrix}$$

$$= \left(R + \frac{1}{SC}\right) \left\{ \left(R + \frac{2}{SC}\right)^3 - \left[\frac{2}{S^2C^2}\left(R + \frac{2}{SC}\right)\right]\right\}$$

$$+ \frac{1}{SC}\left[-\frac{1}{SC}\left(R + \frac{2}{SC}\right)^2 + \frac{1}{S^3C^3}\right]$$

$$\Delta = \left(R + \frac{1}{SC}\right)\left(R + \frac{2}{SC}\right)\left[\left(R + \frac{2}{SC}\right)^2 - \frac{2}{S^2C^2}\right]$$
$$+ \frac{1}{S^2C^2}\left[\frac{1}{S^2C^2} - \left(R + \frac{2}{SC}\right)^2\right]$$

$$= \left(R^2 + \frac{2R}{SC} + \frac{R}{SC} + \frac{2}{S^2C^2}\right)\left(R^2 + \frac{4R}{SC} + \frac{4}{S^2C^2} - \frac{2}{S^2C^2}\right)$$
$$+ \frac{1}{S^2C^2}\left[\frac{1}{S^2C^2} - R^2 - \frac{4R}{SC} - \frac{4}{S^2C^2}\right]$$

$$= \left(R^2 + \frac{3R}{SC} + \frac{2}{S^2C^2}\right)\left(R^2 + \frac{4R}{SC} + \frac{2}{S^2C^2}\right) - \frac{R^2}{S^2C^2} - \frac{4R}{S^3C^3} - \frac{3}{S^4C^4}$$

$$= R^4 + \frac{4R^3}{SC} + \frac{2R^2}{S^2C^2} + \frac{3R^3}{SC} + \frac{12R^2}{S^2C^2} + \frac{6R}{S^3C^3} + \frac{2R^2}{S^2C^2} + \frac{8R}{S^3C^3}$$
$$+ \frac{4}{S^4C^4} - \frac{R^2}{S^2C^2} - \frac{4R}{S^3C^3} - \frac{3}{S^4C^4}$$

$$= R^4 + 7\frac{R^3}{SC} + 15\frac{R^2}{S^2C^2} + 10\frac{R}{S^3C^3} + \frac{1}{S^4C^4}$$

$$\frac{e_o}{e_{in}} = \frac{I_4\dfrac{1}{SC}}{e_{in}}$$

$$I_4 = \frac{\begin{vmatrix} R + \dfrac{1}{SC} & -\dfrac{1}{SC} & 0 & e_{in} \\[2mm] -\dfrac{1}{SC} & R + \dfrac{2}{SC} & -\dfrac{1}{SC} & 0 \\[2mm] 0 & -\dfrac{1}{SC} & R + \dfrac{2}{SC} & 0 \\[2mm] 0 & 0 & -\dfrac{1}{SC} & 0 \end{vmatrix}}{\Delta}$$

$$\frac{e_o}{e_{in}} = \frac{-e_{in} \begin{vmatrix} -\dfrac{1}{SC} & R + \dfrac{2}{SC} & -\dfrac{1}{SC} \\ 0 & -\dfrac{1}{SC} & R + \dfrac{2}{SC} \\ 0 & 0 & -\dfrac{1}{SC} \end{vmatrix}}{\Delta} = \frac{-e_{in}\left(-\dfrac{1}{S^3C^3}\right)}{\Delta} = \frac{e_{in}\dfrac{1}{S^3C^3}}{\Delta}$$

$$= +\frac{\dfrac{1}{S^4C^4}}{\Delta} = \frac{\dfrac{1}{S^4C^4}}{R^4 + 7\dfrac{R^3}{SC} + \dfrac{15R^2}{S^2C^2} + \dfrac{10R}{S^3C^3} + \dfrac{1}{S^4C^4}}$$

$$= \frac{1}{+1 + 10RSC + 15R^2S^2C^2 + 7R^3S^3C^3 + R^4S^4C^4}$$

$$= \frac{1}{+1 + 15R^2S^2C^2 + R^4S^4C^4 + 10RSC + 7R^3S^3C^3}$$

Substituting $j\omega = S$

$$\frac{e_o}{e_{in}} = \frac{1}{+1 - 15R^2\omega^2C^2 + R^4\omega^4C^4 + j(10R\omega C - 7R^3\omega^3C^3)}$$

A-5.3 For Oscillation

$$10R\omega C = 7R^3\omega^3C^3$$

$$\frac{10}{7} = R^2\omega^2C^2$$

$$\frac{1}{\omega^2C^2} = \frac{7R^2}{10}$$

$$\frac{1}{\omega^3C^3} = R^3\left(\frac{7}{10}\right)^{3/2}$$

$$\omega^2C^2 = \frac{10}{7R^2}, \ \omega^4C^4 = \frac{100}{49R^4},$$

$$\omega = \frac{\sqrt{10}}{\sqrt{7}\,RC}, f = \frac{\sqrt{10}}{2\pi\,\sqrt{7}\,RC}, f = \frac{.19}{RC}, \text{ where } \frac{1}{\omega C} = R\left(\frac{7}{10}\right)^{1/2}$$

Thus,

$$\frac{e_o}{e_{in}} = \frac{1}{+1 - 15R^2\omega^2C^2 + R^4\omega^4C^4}$$

$$= \frac{1}{+1 - 15R^2\dfrac{10}{7R^2} + R^4\dfrac{100}{49R^4}} = \frac{1}{+1 - \dfrac{150}{7} + \dfrac{100}{49}}$$

$$= \frac{1}{+\left(\dfrac{49}{49}\right) - 150\left(\dfrac{7}{49}\right) + \dfrac{100}{49}} = \frac{49}{(49) - 150(7) + 100}$$

$$\frac{e_o}{e_{in}} = \frac{49}{49 - 1050 + 100} = \frac{49}{-901} = -.054384$$

$$Z_{in} = \frac{e_{in}}{I_1}$$

$$I_1 = \frac{\begin{vmatrix} e_{in} & -\dfrac{1}{SC} & 0 & 0 \\[2mm] 0 & R + \dfrac{2}{SC} & -\dfrac{1}{SC} & 0 \\[2mm] 0 & -\dfrac{1}{SC} & R + \dfrac{2}{SC} & -\dfrac{1}{SC} \\[2mm] 0 & 0 & -\dfrac{1}{SC} & R + \dfrac{2}{SC} \end{vmatrix}}{\Delta}$$

$$= \frac{e_{in}\begin{vmatrix} R + \dfrac{2}{SC} & -\dfrac{1}{SC} & 0 \\[2mm] -\dfrac{1}{SC} & R + \dfrac{2}{SC} & -\dfrac{1}{SC} \\[2mm] 0 & -\dfrac{1}{SC} & R + \dfrac{2}{SC} \end{vmatrix}}{\Delta}$$

A-5.4 At Resonance Only

$$I_1 = \frac{e_{in}\left[\left(R + \dfrac{2}{SC}\right)^3 - \dfrac{2}{S^2C^2}\left(R + \dfrac{2}{SC}\right)\right]}{R^4 + 15\dfrac{R^2}{S^2C^2} + \dfrac{1}{S^4C^4}}$$

$$Z_{in} = \frac{R^4 + 15\dfrac{R^2}{S^2C^2} + \dfrac{1}{S^4C^4}}{\left(R + \dfrac{2}{SC}\right)\left[\left(R + \dfrac{2}{SC}\right)^2 - \dfrac{2}{S^2C^2}\right]}$$

$$= \frac{R^4 + 15\dfrac{R^2}{S^2C^2} + \dfrac{1}{S^4C^4}}{\left(R + \dfrac{2}{SC}\right)\left[R^2 + \dfrac{4R}{SC} + \dfrac{4}{S^2C^2} - \dfrac{2}{S^2C^2}\right]}$$

$$= \frac{R^4 + 15\dfrac{R^2}{S^2C^2} + \dfrac{1}{S^4C^4}}{\left(R + \dfrac{2}{SC}\right)\left[R^2 + \dfrac{4R}{SC} + \dfrac{2}{S^2C^2}\right]}$$

$$= \frac{R^4 + 15\dfrac{R^2}{S^2C^2} + \dfrac{1}{S^4C^4}}{R^3 + \dfrac{4R^2}{SC} + \dfrac{2R}{S^2C^2} + \dfrac{2R^2}{SC} + \dfrac{8R}{S^2C^2} + \dfrac{4}{S^3C^3}}$$

$$= \frac{R^4 + 15\dfrac{R^2}{S^2C^2} + \dfrac{1}{S^4C^4}}{R^3 + \dfrac{6R^2}{SC} + 10\dfrac{R}{S^2C^2} + \dfrac{4}{S^3C^3}}$$

$$= \frac{R^4 - 15\dfrac{R^2}{\omega^2C^2} + \dfrac{1}{\omega^4C^4}}{R^3 - 10\dfrac{R}{\omega^2C^2} + j\left(\dfrac{4}{\omega^3C^3} - \dfrac{6R^2}{\omega C}\right)}$$

$$= \frac{R^4 - 15R^2\left(\dfrac{7R^2}{10}\right) + \dfrac{49R^4}{100}}{R^3 - 10R\left(\dfrac{7R^2}{10}\right) + j\,[4R^3\,(7/10)^{3/2} - 6R^2(R)\,(7/10)^{1/2}]}$$

$$Z_{in} = \frac{R\left[1 - \dfrac{105}{10} + \dfrac{49}{100}\right]}{1 - 7 + j\left[4(7/10)^{3/2} - 6\dfrac{(7)^{1/2}}{(10)^{1/2}}\right]} = \frac{R\left[\dfrac{100}{100} - \dfrac{1050}{100} + \dfrac{49}{100}\right]}{-\dfrac{600}{100} - j\dfrac{267.7312}{100}}$$

where $K_2 = 267.7312$. Thus,

$$Z_{in} = \frac{-901}{-600 - j267.7} = \frac{+901R}{657\ \underline{/+24°}} = +1.37R\ \underline{/-24°}$$

APPENDIX A-6

Common Emitter Analysis (with Unbypassed Emitter Resistor)

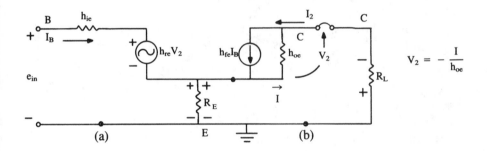

Fig. A.11

Find Z_1:

Input Loop: $I_B R_E + I_2 R_E + h_{re} V_2 + I_B h_{ie} - e_{in} = 0$ (1)

At (a) c.c.w. $Z_1 = \dfrac{e_{in}}{I_b}$

At (b) c.w. $I_b R_E + I_2 R_\epsilon + V_2 + I_2 R_L = 0$ (2)

$$I_b R_E + I_2 R_E + \frac{I_2}{h_{oe}} - \frac{h_{fe} I_B}{h_{oe}} + I_2 R_2 = 0 \qquad (2)$$

$$\Rightarrow I_b \left(R_E - h_{fe}/h_{oe} \right) + I_2 \left(R_E + \frac{1}{h_{oe}} + R_L \right) = 0 \qquad (2)$$

$$I_B R_E + I_2 R_E - \frac{h_{re} I_2}{h_{oe}} + I_B h_{ie} = e_{in} \qquad (1)$$

$$\Rightarrow I_B \left(h_{ie} - \frac{h_{re} h_{fe}}{h_{oe}} + R_E \right) + I_2 \left(\frac{h_{re}}{h_{oe}} + R_E \right) = e_{in} \qquad (1)$$

$$I_B = \frac{\begin{vmatrix} e_{in} & -\dfrac{h_{re}}{h_{oe}} + R_E \\[2ex] 0 & R_E + \dfrac{1}{h_{oe}} + R_L \end{vmatrix}}{\begin{vmatrix} h_{ie} - \dfrac{h_{re} h_{fe}}{h_{oe}} + R_E & +\dfrac{h_{re}}{h_{oe}} + R_E \\[2ex] R_E - \dfrac{h_{fe}}{h_{oe}} & R_E + \dfrac{1}{h_{oe}} + R_L \end{vmatrix}}$$

At node E:
$$h_{fe} I_B = I_2 + I$$
$$h_{fe} I_B = I_2 - V_2 h_{oe}$$
Thus,
$$V_2 h_{oe} = -h_{fe} I_B + I_2$$

$$V_2 = -\frac{h_{fe}}{h_{oe}} I_B + \frac{I_2}{h_{oe}}$$

$$V_2 = -\frac{I}{h_{oe}} - \frac{h_{fe} I_2}{h_{oe}} + \frac{I_2}{h_{oe}}$$

$$V_2 = \frac{I_2}{h_{oe}} - \frac{h_{fe} I_2}{h_{oe}}$$

$$\frac{e_{in}}{I_B} = \frac{\left(h_{ie} + \dfrac{h_{re} h_{fe}}{h_{oe}} + R_E \right) \left(R_e + \dfrac{1}{h_{oe}} + R_L \right) - \left(-\dfrac{h_{fe}}{h_{oe}} + R_E \right) \left(+\dfrac{h_{re}}{h_{oe}} + R_E \right)}{R_E + 1/h_{oe} + R_L}$$

$$\frac{e_{in}}{I_B} = h_{ie}R_E + \frac{h_{ie}}{h_{oe}} + h_{ie}R_L - \frac{h_{re}h_{fe}}{h_{oe}}R_E - \frac{h_{re}h_{fe}}{h_{oe}^2} - \frac{h_{re}h_{fe}}{h_{oe}}R_L + R_E^2$$

$$+ \frac{R_E}{h_{oe}} + R_E R_L - \left[-\frac{h_{fe}h_{re}}{h_{oe}^2} - \frac{h_{fe}R_E}{h_{oe}} + \frac{h_{re}}{h_{oe}}R_e + R_E^2 \right]$$

D

$$= \frac{h_{ie}\left(R_E + \dfrac{1}{h_{oe}} + R_L\right) + [\ldots]}{R_\epsilon + 1/h_{oe} + R_L}$$

$$= h_{ie} + \frac{\left[-\dfrac{h_{re}h_{fe}R_E}{h_{oe}} - h_{re}h_{fe}R_L + \dfrac{R_E}{h_{oe}} + \dfrac{R_E R_L h_{oe}}{h_{oe}} + \dfrac{h_{fe}R_E}{h_{oe}} - \dfrac{h_{re}R_E}{h_{oe}} \right]}{\dfrac{h_{oe}R_E}{h_{oe}} + \dfrac{1}{h_{oe}} + \dfrac{R_L h_{oe}}{h_{oe}}}$$

$$= h_{ie} + \frac{[-h_{re}h_{fe}R_E - h_{re}h_{fe}R_L + R_E + R_E R_L h_{oe} + h_{fe}R_E - h_{re}R_E]}{h_{oe}R_E + 1 + R_L h_{oe}}$$

$$= h_{ie} + \frac{[-h_{re}h_{fe}R_L + R_E(-h_{fe}h_{re} + 1 + h_{oe}R_L + h_{fe} - h_{re})]}{h_{oe}R_E + 1 + R_L h_{oe}}$$

$$Z_1 = h_{ie} + \left[\frac{-h_{re}h_{fe}R_L + R_E(+1 + h_{oe}R_L + h_{fe} - h_{re} - h_{fe}h_{re})}{+1 + h_{oe}(R_E + R_L)} \right]$$

$$= h_{ie} + \left\{ \frac{-h_{fe}h_{re}R_L + R_E[h_{oe}R_L + (h_{fe} + 1) - h_{re}(h_{fe} + 1)]}{h_{oe}(R_L + R_e) + 1} \right\}$$

$$= h_{ie} + \frac{-h_{fe}h_{re}R_L + R_E[h_{oe}R_L + (h_{fe} + 1)(1 - h_{re})]}{h_{oe}(R_L + R_E) + 1}$$

Hence, if $R_E = 0$,

$$Z_1 = h_{ie} - \frac{h_{fe}h_{re}Z_L}{1 + h_{oe}Z_L}$$

$$A_I = \frac{I_2}{I_B}$$

$$I_2 = \cfrac{\begin{vmatrix} h_{ie} + \dfrac{h_{re}h_{fe}}{h_{oe}} + R_E & e_{in} \\[2ex] R_E - \dfrac{h_{fe}}{h_{oe}} & 0 \end{vmatrix}}{\begin{vmatrix} h_{ie} - \dfrac{h_{re}h_{fe}}{h_{oe}} + R_E & \times \dfrac{h_{re}}{h_{oe}} + R_E \\[2ex] R_E - \dfrac{h_{fe}}{h_{oe}} & R_E + \dfrac{1}{h_{oe}} + R_L \end{vmatrix}}$$

$$A_I = \cfrac{\begin{vmatrix} h_{ie} - \dfrac{h_{fe}h_{re}}{h_{oe}} + R_E & e_{in} \\[2ex] R_E - \dfrac{h_{fe}}{h_{oe}} & 0 \end{vmatrix}}{\begin{vmatrix} e_{in} & \times \dfrac{h_{re}}{h_{oe}} + R_E \\[2ex] 0 & R_E + 1/h_{oe} + R_L \end{vmatrix}} = -\cfrac{e_{in}\left(R_E - \dfrac{h_{fe}}{h_{oe}}\right)}{e_{in}(R_E + 1/h_{oe} + R_L)}$$

$$A_I = \frac{h_{fe}/h_{oe} - R_E}{R_E + 1/h_{oe} + R_L} = \frac{h_{fe} - h_{oe}R_E}{1 + h_{oe}(R_L + R_E)}$$

Hence, if $R_e = 0$

$$A_I = \frac{h_{fe}}{1 + h_{oe} Z_L}$$

$$A_V = \frac{V_0}{V_1} = \left(-\frac{I_2}{I_B}\right)\frac{R_L}{Z_1} = A_I \frac{R_L}{Z_1} = \frac{h_{fe} - h_{oe} R_E}{1 + h_{oe}(R_L + R_E)} \frac{R_L}{Z_1}$$

$$= -\frac{(h_{fe} - h_{oe}R_E)(R_L)}{[1 + h_{oe}(R_L + R_E)]}$$

$$\frac{[1 + h_{oe}(R_L + R_E)]}{[1 + h_{oe}(R_L + R_E)]\, h_{ie} - h_{fe}\, h_{re}\, R_L}$$

$$c + R_E\,[h_{oe}R_L + (1 - h_{re})(h_{fe} + 1)]$$

$$= -\frac{(h_{fe} - h_{oe}\,R_E)\,R_L}{h_{ie} + h_{ie}\,h_{oe}\,R_L + h_{ie}h_{oe}\,R_E - h_{fe}h_{re}\,R_L}$$

$$x\ \frac{1}{R_E\,h_{oe}R_L + R_e\,h_{fe} - R_e\,h_{re} - R_Eh_{re}h_{fe} + 1}$$

$$= -\frac{(h_{fe} - h_{oe}\,R_E)\,R_L}{h_{ie} + (h_{ie}h_{oe} - h_{fe}h_{re})R_L + R_E}$$

$$\times\ \frac{1}{(h_{ie}h_{oe} + h_{oe}\,R_L + h_{fe} - h_{re} - h_{fe}h_{re} + 1)}$$

$$= -\frac{(h_{fe} - h_{oe}R_E)R_L}{h_{ie} + (h_{ie}h_{oe} - h_{fe}h_{re})R_L + R_E}$$

$$\times\ \frac{1}{(h_{ie}h_{oe} - h_{fe}h_{re}) + R_E\,[h_{oe}R_L + (h_{fe} + 1) - h_{re}]}$$

$$A_V = -\frac{(h_{fe} - h_{oe}R_E)R_L}{h_{ie} + (h_{ie}h_{oe} - h_{fe}h_{re})(R_L + R_E) + R_E[h_{oe}R_L + (h_{fe} + 1) - h_{re}]}$$

$$A_V = -\frac{(h_{fe} - h_{oe}R_E)R_L}{h_{ie} + (h_{ie}h_{oe} - h_{fe}h_{re})(R_L + R_E) + R_e(1 + h_{fe} + h_{oe}R_L - h_{re})}$$

APPENDIX A-7

A_V *(with Emitter Resistor): Approximation*

$$A_V = -\frac{(h_{fe} - h_{oe}R_E)R_L}{h_{ie} + (h_{ie}h_{oe} - h_{fe}h_{re})(R_L + R_E) + R_f(1 + h_{fe} + h_{oe}R_L - h_{re})}$$

If $h_{oe} \to 0$, $h_{re} \to 0$, $h_{fe} >> 1$:

$$A_V =$$

$$-\frac{(h_{fe} - h_{oe}R_E)R_L}{h_{ie} + (h_{ie}h_{oe} - h_{fe}h_{re})(R_L + R_E) + R_E(1 - h_{fe} + h_{oe}R_L - h_{re})}$$

$$A_V = \frac{-h_{fe}R_L}{h_{ie} + R_Eh_{fe}} = \frac{-R_L}{\dfrac{h_{ie}}{h_{fe}} + R_E} = -\frac{R_L}{R_E}$$

$$Z_1 = \frac{V_2}{I_2}$$

$$Z_2 = \frac{V_2}{I_2}$$

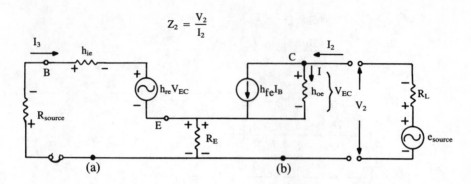

Fig. A.12

At (a) c.c.w. $\Sigma V = 0$

$$I_B R_E + I_2 R_E + h_{re} V_{ec} + I_B h_{ie} + I_B R_S = 0$$

At (c) $\Sigma I = 0$

$$I_2 = I + h_{fe} I_B \Rightarrow I_2 = h_{oe} V_{eC} + h_{fe} I_B \Rightarrow V_{eC} = \frac{I_2 - h_{fe} I_B}{h_{oe}}$$

$$I_B R_E + I_2 R_E + \frac{h_{re} I_2}{h_{oe}} - \frac{h_{re} h_{fe} I_B}{h_{oe}} + I_B h_{ie} + I_B R_S = 0$$

At (b) c.c.w.

$$I_B \left(R_E - \frac{h_{fe} h_{re}}{h_{oe}} + h_{ie} + R_S \right) + I_2 \left(R_E + \frac{h_{re}}{h_{oe}} \right) = 0 \qquad (1)$$

$$I_B R_E + I_2 R_E + V_{eC} - V_2 = 0$$
$$I_B R_E + I_2 R_E + \frac{I_2 - h_{fe} I_B}{h_{oe}} - V_2 = 0$$

$$I_B R_E + I_2 R_E + \frac{I_2}{h_{oe}} - \frac{h_{fe} I_B}{h_{oe}} - V_2 = 0$$

$$I_B (R_E - h_{fe}/h_{oe}) + I_2 (R_E + 1/h_{oe}) = V_2$$

$$I_2 = \cfrac{\begin{vmatrix} R_E - \dfrac{h_{fe}h_{re}}{h_{oe}} + h_{ie} + R_S & 0 \\[2mm] R_E - h_{fe}/h_{oe} & V_2 \end{vmatrix}}{\begin{vmatrix} R_E - \dfrac{h_{fe}h_{re}}{h_{oe}} + h_{ie} + R_S & R_E + h_{re}/h_{oe} \\[2mm] R_E - \dfrac{h_{fe}}{h_{oe}} & R_E + 1/h_{oe} \end{vmatrix}}$$

$$I_2 = \cfrac{V_2\left(R_E - \dfrac{h_{fe}h_{re}}{h_{oe}} + h_{ie} + R_S\right)}{\left(R_E - \dfrac{h_{fe}h_{re}}{h_{oe}} + h_{ie} + R_S\right)(R_E + 1/h_{oe}) - \left(R_E + \dfrac{h_{re}}{h_{oe}}\right)\left(R_E - \dfrac{h_{fe}}{h_{oe}}\right)}$$

$$\frac{V_2}{I_2} = \cfrac{\left(R_E - \dfrac{h_{fe}h_{re}}{h_{oe}} + h_{ie} + R_S\right)(R_E + 1/h_{oe}) - \left(R_E + \dfrac{h_{re}}{h_{oe}}\right)\left(R_E - \dfrac{h_{fe}}{h_{oe}}\right)}{R_e - \dfrac{h_{fe}h_{re}}{h_{oe}} + h_{ie} + R_S}$$

$$\frac{V_2}{I_2} = \cfrac{R_E^2 + \dfrac{R_E}{h_{oe}} - \dfrac{h_{fe}h_{re}}{h_{oe}}R_E - \dfrac{h_{fe}h_{re}}{h_{oe}^2} + h_{ie}R_E + \dfrac{h_{ie}}{h_{oe}} + R_E R_S + \dfrac{R_S}{h_{oe}}}{R_E - \dfrac{h_{fe}h_{re}}{h_{oe}} + h_{ie} + R_S}$$

$$- \cfrac{R_E^2 + \dfrac{h_{fe}}{h_{oe}}R_E - \dfrac{h_{re}}{h_{oe}}R_E + \dfrac{h_{re}h_{fe}}{h_{oe}^2}}{D}$$

$$= \frac{R_E(1 - h_{fe}h_{re} + h_{ie}h_{oe} + R_S h_{oe} + h_{fe} - h_{re}) + h_{ie} + R_S}{R_E h_{oe} - h_{fe}h_{re} + h_{oe}h_{ie} + h_{oe}R_S}$$

$$Z_2 = \frac{h_{ie} + R_S + R_E(1 + \Delta^{he} + h_{oe}R_S + h_{fe} - h_{re})}{\Delta^{he} + h_{oe}(R_S + R_E)}$$

where $\Delta^{he} = h_{ie}h_{oe} - h_{fe}h_{re}$

APPENDIX A-8

Common Emitter with Collector Feedback

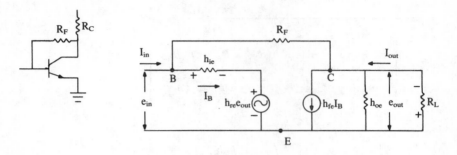

Fig. A.13

Find $Z_1 = \dfrac{e_{in}}{I_{in}}$:

At B Nodal Analysis ("+" out):

$$-I_{in} + \left(\frac{e_{in} - e_o}{R_F}\right) + I_B = 0 \tag{1}$$

$$h_{re}e_o + I_B h_{ie} - e_{in} = 0$$

$$I_B = \frac{e_{in} - h_{re}e_o}{h_{ie}}$$

At C:

$$h_{fe}I_B + \left(\frac{e_o - e_{in}}{R_F}\right) + e_o h_{oe} + \frac{e_o}{R_L} = 0 \tag{2}$$

$$+ \frac{e_{in}}{R_F} - \frac{e_o}{R_F} + \left[\frac{e_{in} - h_{re}e_o}{h_{ie}}\right] = I_{in} \tag{1}$$

$$- \frac{e_{in}}{R_F} + e_o\left(\frac{1}{R_F} + h_{oe} + \frac{1}{R_L}\right) + h_{fe}\left[\frac{e_{in} - h_{re}e_o}{h_{ie}}\right] = 0 \tag{2}$$

$$e_{in}\left[\frac{1}{R_F} + \frac{1}{h_{ie}}\right] + e_o\left[-\frac{1}{R_F} - \frac{h_{re}}{h_{ie}}\right] = I_{in}$$

$$e_{in}\left[-\frac{1}{R_F} + \frac{h_{fe}}{h_{ie}}\right] + e_o\left[\frac{1}{R_F} + h_{oe} + \frac{1}{R_L} - \frac{h_{fe}h_{re}}{h_{ie}}\right] = 0$$

$$e_{in} = \frac{\begin{vmatrix} I_{in} & -1/R_F - h_{re}/h_{ie} \\ 0 & \dfrac{1}{R_F} + \dfrac{1}{R_L} + h_{oe} - \dfrac{h_{fe}h_{re}}{h_{ie}} \end{vmatrix}}{\begin{vmatrix} \dfrac{1}{R_F} + \dfrac{1}{h_{ie}} & -\dfrac{1}{R_F} - \dfrac{h_{re}}{h_{ie}} \\ -\dfrac{1}{R_F} + \dfrac{h_{fe}}{h_{ie}} & \dfrac{1}{R_F} + \dfrac{1}{R_L} + h_{oe} - \dfrac{h_{fe}h_{re}}{h_{ie}} \end{vmatrix}}$$

$$e_{in} = \frac{I_{in}\left[\dfrac{1}{R_F} + \dfrac{1}{R_L} + h_{oe} - \dfrac{h_{fe}h_{re}}{h_{ie}}\right]}{\left[\dfrac{1}{R_F} + \dfrac{1}{h_{ie}}\right]\left[\dfrac{1}{R_F} + \dfrac{1}{R_L} + h_{oe} - \dfrac{h_{fe}h_{re}}{h_{ie}}\right] + \left[\dfrac{1}{R_F} + \dfrac{h_{re}}{h_{ie}}\right]\left[-\dfrac{1}{R_F} + \dfrac{h_{fe}}{h_{ie}}\right]}$$

$$\frac{e_{in}}{I_{in}} = \frac{\dfrac{1}{R_F} + \dfrac{1}{R_L} + h_{oe} - \dfrac{h_{fe}h_{re}}{h_{ie}}}{\dfrac{1}{R_F^2} + \dfrac{1}{R_F R_L} + \dfrac{h_{oe}}{R_F} - \dfrac{h_{fe}h_{re}}{R_F h_{ie}} + \dfrac{1}{h_{ie}R_F} + \dfrac{1}{h_{ie}R_L}}$$

$$+ \frac{\dfrac{1}{R_F} + \dfrac{1}{R_L} + h_{oe} - \dfrac{h_{fe}h_{re}}{h_{ie}}}{\dfrac{h_{oe}}{h_{ie}} - \dfrac{h_{fe}h_{re}}{h_{ie}^2} - \dfrac{1}{R_F^2} + \dfrac{h_{fe}}{R_F h_{ie}} - \dfrac{h_{re}}{h_{ie}R_F} + \dfrac{h_{re}h_{fe}}{h_{ie}^2}}$$

$$\frac{e_{in}}{I_{in}} = \frac{\dfrac{h_{ie}}{R_F} + \dfrac{h_{ie}}{R_L} + h_{ie}h_{oe} - h_{fe}h_{re}}{\dfrac{h_{ie}}{R_F R_L} + \dfrac{h_{ie}h_{oe}}{R_F} - \dfrac{h_{fe}h_{re}}{R_F} + \dfrac{1}{R_F} + \dfrac{1}{R_L} + h_{oe} + \dfrac{h_{fe}}{R_F} - \dfrac{h_{re}}{R_F}}$$

$$= \frac{h_{ie}\left(1 + \dfrac{R_F}{R_L}\right) + \Delta^{he}R_F}{h_{ie}\left(\dfrac{1}{R_L}\right) + \Delta^{he} + 1 + \dfrac{R_F}{R_L} + h_{oe}R_F + h_{fe} - h_{re}}$$

$$= \frac{h_{ie}(R_L + R_F) + \Delta^{he} R_F R_L}{h_{ie} + \Delta^{he} R_L + R_L + R_F + h_{oe}R_F R_L + (h_{fe} - h_{re})R_L}$$

$$= \frac{h_{ie}(R_F + R_L) + \Delta^{he} R_F R_L}{h_{ie} + \Delta^{he}R_L + R_L(h_{fe} + 1 - h_{re}) + R_F(1 + h_{oe}R_L)}$$

If $h_{oe} = 0$, $h_{ie} = 0$, then

$$\frac{e_{in}}{I_{in}} = \frac{h_{ie}\left(1 + \dfrac{R_F}{R_L}\right)}{\dfrac{h_{ie}}{R_L} + 1 + \dfrac{R_F}{R_L} + h_{fe}}$$

$$Z_1 = \frac{h_{ie}(R_L + R_F)}{h_{ie} + R_L + R_F + h_{fe}R_L} = \frac{h_{ie}(R_L + R_F)}{h_{ie} + R_L(h_{fe} + 1) + R_F}$$

$$\frac{e_o}{e_{in}} = \frac{\begin{vmatrix} \dfrac{1}{R_F} + \dfrac{1}{h_{ie}} & I_{in} \\[2ex] -\dfrac{1}{R_F} + \dfrac{h_{fe}}{h_{ie}} & 0 \end{vmatrix}}{\begin{vmatrix} I_{in} & -\dfrac{1}{R_F} - \dfrac{h_{re}}{h_{ie}} \\[2ex] 0 & \dfrac{1}{R_F} + \dfrac{1}{R_L} + h_{oe} - \dfrac{h_{fe}h_{re}}{h_{ie}} \end{vmatrix}}$$

$$= \frac{-I_{in}\left(-\dfrac{1}{R_F} + \dfrac{h_{fe}}{h_{ie}}\right)}{I_{in}\left[\dfrac{1}{R_F} + \dfrac{1}{R_L} + h_{oe} - \dfrac{h_{fe}h_{re}}{h_{ie}}\right]}$$

$$= \frac{-\left(-\dfrac{h_{ie}}{R_F} + h_{fe}\right)}{\dfrac{h_{ie}}{R_F} + \dfrac{h_{ie}}{R_L} + h_{ie}h_{oe} - h_{fe}h_{re}}$$

$$\frac{e_o}{e_{in}} = A_V = \frac{-\left(h_{fe} - \dfrac{h_{ie}}{R_F}\right)}{h_{ie}\left(\dfrac{1}{R_F} + \dfrac{1}{R_L}\right) + \Delta^{he}}$$

If $h_{re} = 0$, $h_{oe} = 0$, then

$$A_V = \frac{-\left(h_{fe} - \dfrac{h_{ie}}{R_F}\right)}{h_{ie}\left(\dfrac{1}{R_F} + \dfrac{1}{R_L}\right)}$$

If $R_F \gg R_L$ and $R_F \gg h_{ie}$, then

$$A_V = \frac{-\left(h_{fe} - \dfrac{h_{ie}}{R_F}\right)R_L}{h_{ie}\left(\dfrac{R_L}{R_F} + 1\right)} \doteq \frac{-h_{fe}R_L}{h_{ie}}$$

$$A_I = \frac{I_o}{I_{in}} = \frac{-e_o/R_L}{I_{in}} = \frac{-e_o}{I_{in}R_L}$$

$$e_o = \frac{\begin{vmatrix} \dfrac{1}{R_F} + \dfrac{1}{h_{ie}} & I_{in} \\[2ex] -\dfrac{1}{R_F} + \dfrac{h_{fe}}{h_{ie}} & 0 \end{vmatrix}}{\begin{vmatrix} \dfrac{1}{R_F} + \dfrac{1}{h_{ie}} & -\dfrac{1}{R_F} - \dfrac{h_{re}}{h_{ie}} \\[2ex] -\dfrac{1}{R_F} + \dfrac{h_{fe}}{h_{ie}} & \dfrac{1}{R_F} + \dfrac{1}{R_L} + h_{oe} - \dfrac{h_{fe}h_{re}}{h_{ie}} \end{vmatrix}}$$

$$e_o = \frac{-I_{in}\left(-\dfrac{1}{R_F} + \dfrac{h_{fe}}{h_{ie}}\right)}{\left(\dfrac{1}{R_F} + \dfrac{1}{h_{ie}}\right)\left(\dfrac{1}{R_F} + \dfrac{1}{R_L} + h_{oe} - \dfrac{h_{fe}h_{re}}{h_{ie}}\right) + \left(\dfrac{1}{R_F} + \dfrac{h_{re}}{h_{ie}}\right)\left(\dfrac{h_{fe}}{h_{ie}} - \dfrac{1}{R_F}\right)}$$

$$= \frac{I_{in}\left(1/R_F - \dfrac{h_{fe}}{h_{ie}}\right)}{\dfrac{1}{R_F{}^2} + \dfrac{1}{R_F R_L} + \dfrac{h_{oe}}{R_F} - \dfrac{h_{fe}h_{re}}{h_{ie}R_F} + \dfrac{1}{h_{ie}R_F} + \dfrac{1}{h_{ie}R_L} + \dfrac{h_{oe}}{h_{ie}} - \dfrac{h_{fe}h_{re}}{h_{ie}{}^2}}$$

$$+ \frac{-I_{in}\left(1/R_F - \dfrac{h_{fe}}{h_{ie}}\right)}{\dfrac{h_{fe}}{h_{ie}R_F}} - \dfrac{1}{R_F{}^2} + \dfrac{h_{re}h_{fe}}{h_{ie}{}^2} - \dfrac{h_{re}}{h_{ie}R_F}$$

$$e_o = \frac{I_{in}(h_{ie} - h_{fe}R_F)}{\dfrac{h_{ie}}{R_L} + (h_{ie}h_{oe} - h_{fe}h_{re}) + 1 + \dfrac{R_F}{R_L} + h_{oe}R_F + h_{fe} - h_{re}}$$

$$A_I = \frac{-e_o}{I_{in}R_L} = \frac{-h_{ie} + h_{fe}R_F}{h_{ie} + \Delta^{h^e}R_L + R_L + R_F + h_{oe}R_FR_L + h_{fe}R_L - h_{re}R_L}$$

$$A_I = \frac{-h_{ie} + h_{fe}R_F}{h_{ie} + \Delta^{h^e}R_L + R_L(1 + h_{fe} - h_{re}) + R_F(1 + h_{oe}R_L)}$$

$$A_I = \frac{-h_{ie} + h_{fe}R_F}{h_{ie} + \Delta^{h^e}R_L + R_L(h_{fe} + 1 - h_{re}) + R_F(1 + h_{oe}R_L)}$$

If $h_{oe} = 0$, $h_{re} = 0$, then

$$A_I = \frac{-h_{ie} + h_{fe}R_F}{h_{ie} + (h_{fe} + 1)R_L + R_F}$$

$$A_I = \frac{h_{fe} - \dfrac{h_{ie}}{R_F}}{\dfrac{h_{ie}}{R_F} + \Delta^{h^e}\dfrac{R_L}{R_F} + \dfrac{R_L}{R_F}(h_{fe} + 1 - h_{re}) + 1 + h_{oe}Z_L}$$

If $R_F \rightarrow \infty$:

$$A_I = \frac{+h_{fe}}{1 + h_{oe}Z_L}$$

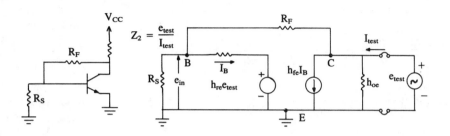

Fig. A.14

At node B:

$$\frac{e_{in}}{R_S} + I_B + \frac{e_{in} - e_{Test}}{R_F} = 0 \qquad (a)$$

At node C:

$$\frac{e_{Test} - e_{in}}{R_F} + h_{fe}I_B + e_{Test}h_{oe} - I_{Test} = 0 \qquad (b)$$

ΣV around input loop:
Thus,

$$h_{re}e_{Test} + I_Bh_{ie} - e_{in} = 0$$

$$I_Bh_{ie} = e_{in} - h_{re}e_{Test}$$

$$I_B = \frac{e_{in} - h_{re}e_{Test}}{h_{ie}}$$

Substituting into (a):

$$\frac{e_{in}}{R_S} + e_{in} - \frac{h_{re}e_{test}}{h_{ie}} + \frac{e_{in} - e_{test}}{R_F} = 0 \qquad (c)$$

Substituting into (b):

$$\frac{e_{Test} - e_{in}}{R_F} + h_{fe}\left(\frac{e_{in} - h_{re}e_{Test}}{h_{ie}}\right) + e_{Test}h_{oe} - I_{Test} = 0 \qquad \text{(d)}$$

Expanding (c):

$$\frac{e_{in}}{R_S} + \frac{e_{in}}{h_{ie}} - \frac{h_{re}e_{Test}}{h_{ie}} + \frac{e_{in}}{R_F} - \frac{e_{Test}}{R_F} = 0$$

$$e_{in}\left(\frac{1}{R_S} + \frac{1}{h_{ie}} + \frac{1}{R_F}\right) + e_{Test}\left(-\frac{h_{re}}{h_{ie}} - \frac{1}{R_F}\right) = 0 \qquad \text{(e)}$$

Expanding (d):

$$\frac{e_{Test}}{R_F} - \frac{e_{in}}{R_F} + e_{in}\frac{h_{fe}}{h_{ie}} - e_{Test}\frac{h_{fe}h_{re}}{h_{ie}} + e_{Test}\,h_{oe} = I_{Test}$$

$$e_{in}\left(\frac{h_{fe}}{h_{ie}} - \frac{1}{R_F}\right) + e_{Test}\left(\frac{1}{R_F} - \frac{h_{fe}h_{re}}{h_{ie}} + h_{oe}\right) = I_{Test}$$

From (e) and (f):

$$e_{Test} = \frac{\begin{vmatrix} \dfrac{1}{R_S} + \dfrac{1}{h_{ie}} + \dfrac{1}{R_F} & 0 \\[2ex] \dfrac{h_{fe}}{h_{ie}} - \dfrac{1}{R_F} & I_{Test} \end{vmatrix}}{\begin{vmatrix} \dfrac{1}{R_S} + \dfrac{1}{h_{ie}} + \dfrac{1}{R_F} & -\dfrac{h_{re}}{h_{ie}} - \dfrac{1}{R_F} \\[2ex] \dfrac{h_{fe}}{h_{ie}} - \dfrac{1}{R_F} & \dfrac{1}{R_F} - \dfrac{h_{fe}h_{re}}{h_{ie}} + h_{oe} \end{vmatrix}}$$

$$\frac{e_{Test}}{I_{Test}} = \frac{\dfrac{1}{R_S} + \dfrac{1}{h_{ie}} + \dfrac{1}{R_F}}{\left(\dfrac{1}{R_S} + \dfrac{1}{h_{ie}} + \dfrac{1}{R_F}\right)\left(\dfrac{1}{R_F} - \dfrac{h_{fe}h_{re}}{h_{ie}} + h_{oe}\right) + \left(\dfrac{h_{fe}}{h_{ie}} - \dfrac{1}{R_F}\right)\left(\dfrac{h_{re}}{h_{ie}} + \dfrac{1}{R_F}\right)}$$

$$\frac{e_{Test}}{I_{Test}} = \frac{\dfrac{1}{R_S} + \dfrac{1}{h_{ie}} + \dfrac{1}{R_F}}{\dfrac{1}{R_S R_F} - \dfrac{h_{fe}h_{re}}{R_S h_{ie}} + \dfrac{h_{oe}}{R_S} + \dfrac{1}{h_{ie}R_F} - \dfrac{h_{fe}h_{re}}{h_{ie}{}^2} + \dfrac{1}{R_F{}^2} - \dfrac{h_{fe}h_{re}}{R_F h_{ie}} + \dfrac{h_{oe}}{R_F}}$$

$$+ \frac{\dfrac{1}{R_S} + \dfrac{1}{h_{ie}} + \dfrac{1}{R_F}}{\dfrac{h_{fe}h_{re}}{h_{ie}{}^2} - \dfrac{h_{fe}}{h_{ie}R_F} - \dfrac{h_{re}}{R_F h_{ie}} - \dfrac{1}{R_F{}^2}}$$

$$= \frac{\dfrac{1}{R_S} + \dfrac{1}{h_{ie}} + \dfrac{1}{R_F}}{\dfrac{1}{R_S R_F} - \dfrac{h_{fe}h_{re}}{R_S h_{ie}} + \dfrac{h_{oe}}{R_S} + \dfrac{1}{h_{ie}R_F} + \dfrac{h_{oe}}{h_{ie}}}$$

$$- \frac{\dfrac{1}{R_S} + \dfrac{1}{h_{ie}} + \dfrac{1}{R_F}}{\dfrac{h_{fe}h_{re}}{R_F h_{ie}} + \dfrac{h_{oe}}{R_F} + \dfrac{h_{fe}}{h_{ie}R_F} - \dfrac{h_{re}}{R_F h_{ie}}}$$

$$= \frac{\dfrac{h_{ie}R_F + R_S R_F + R_S h_{ie}}{R_S h_{ie}R_F}}{\dfrac{h_{ie}}{R_S R_F h_{ie}} - \dfrac{h_{fe}h_{re}R_F}{R_S R_F h_{ie}} + \dfrac{h_{oe}R_F h_{ie}}{R_S R_F h_{ie}} + \dfrac{R_S}{R_S R_F h_{ie}} + \dfrac{h_{oe}R_S R_F}{R_S R_F h_{ie}}}$$

$$- \frac{\dfrac{h_{ie}R_F + R_S R_F + R_S h_{ie}}{R_S h_{ie}R_F}}{\dfrac{h_{fe}h_{re}R_S}{R_S R_F h_{ie}} + \dfrac{R_S h_{fe}}{R_S R_F h_{ie}} - \dfrac{R_S h_{re}}{R_S R_F h_{ie}}}$$

$$= \frac{h_{ie}R_F + R_S R_F + R_S h_{ie}}{h_{ie} - h_{fe}h_{re}R_F + h_{oe}h_{ie}R_F + R_S + h_{oe}R_S R_F}$$

$$- \frac{h_{ie}R_F + R_S R_F + R_S h_{ie}}{h_{fe}h_{re}R_S + R_S h_{fe} - R_S h_{re}}$$

$$= \frac{h_{ie}\,R_F + R_S R_F + R_S h_{ie}}{h_{ie} + R_F\,\Delta^{h^e} + R_S\,(1 + h_{oe}R_F - h_{fe}h_{re} + h_{fe} - h_{re})}$$

$$Z_2 = \frac{h_{ie}R_F + R_S R_F + R_S h_{ie}}{h_{ie} + R_F\,\Delta^{h^e} + R_S\,[1 + h_{fe} + h_{oe}R_F - h_{re}\,(1 + h_{fe})]}$$

Then, dividing by R_F,

$$Z_2 = \frac{-h_{ie} + R_S + \dfrac{R_S h_{ie}}{R_F}}{\dfrac{h_{ie}}{R_F} + \Delta^{h^e} + \dfrac{R_S}{R_F}[(1 + h_{fe})(1 - h_{re}) + h_{oe}R_F]}$$

For $R_F \to \infty$, we have

$$= \frac{h_{ie} + R_S}{\Delta^{h^e} + h_{oe}R_S} = \frac{h_{ie} + R_S}{h_{oe}h_{ie} - h_{fe}h_{re} + h_{oe}R_S}$$

$$= \frac{h_{ie} + R_S}{h_{oe}(h_{ie} + R_S) - h_{fe}h_{re}} = \frac{1}{h_{oe} - \dfrac{h_{fe}h_{re}}{h_{ie} + R_S}}$$

Note that this result is identical to the equation for Z_2 for a simple common-emitter stage. Therefore, allowing $R_F \to \infty$ is a partial check on our results.

Index

DATE DUE

N